AF800442

EcoProduction

Environmental Issues in Logistics and Manufacturing

Series Editor

Paulina Golinska-Dawson, Poznań, Poland

The EcoProduction Series is a forum for presenting emerging environmental issues in Logistics and Manufacturing. Its main objective is a multidisciplinary approach to link the scientific activities in various manufacturing and logistics fields with the sustainability research. It encompasses topical monographs and selected conference proceedings, authored or edited by leading experts as well as by promising young scientists. The Series aims to provide the impulse for new ideas by reporting on the state-of-the-art and motivating for the future development of sustainable manufacturing systems, environmentally conscious operations management and reverse or closed loop logistics.

It aims to bring together academic, industry and government personnel from various countries to present and discuss the challenges for implementation of sustainable policy in the field of production and logistics.

More information about this series at http://www.springer.com/series/10152

Paulina Golinska-Dawson · Kune-Muh Tsai ·
Monika Kosacka-Olejnik
Editors

Smart and Sustainable Supply Chain and Logistics – Trends, Challenges, Methods and Best Practices

Volume 1

Springer

Editors
Paulina Golinska-Dawson
Faculty of Engineering Management
Poznań University of Technology
Poznań, Poland

Kune-Muh Tsai
Department of Logistics Management
National Kaohsiung University of Science and Technology
Kaohsiung, Taiwan

Monika Kosacka-Olejnik
Faculty of Engineering Management
Poznań University of Technology
Poznań, Poland

ISSN 2193-4614 ISSN 2193-4622 (electronic)
EcoProduction
ISBN 978-3-030-61946-6 ISBN 978-3-030-61947-3 (eBook)
https://doi.org/10.1007/978-3-030-61947-3

© The Editor(s) (if applicable) and The Author(s), under exclusive license to Springer Nature Switzerland AG 2020
This work is subject to copyright. All rights are solely and exclusively licensed by the Publisher, whether the whole or part of the material is concerned, specifically the rights of translation, reprinting, reuse of illustrations, recitation, broadcasting, reproduction on microfilms or in any other physical way, and transmission or information storage and retrieval, electronic adaptation, computer software, or by similar or dissimilar methodology now known or hereafter developed.
The use of general descriptive names, registered names, trademarks, service marks, etc. in this publication does not imply, even in the absence of a specific statement, that such names are exempt from the relevant protective laws and regulations and therefore free for general use.
The publisher, the authors and the editors are safe to assume that the advice and information in this book are believed to be true and accurate at the date of publication. Neither the publisher nor the authors or the editors give a warranty, expressed or implied, with respect to the material contained herein or for any errors or omissions that may have been made. The publisher remains neutral with regard to jurisdictional claims in published maps and institutional affiliations.

This Springer imprint is published by the registered company Springer Nature Switzerland AG
The registered company address is: Gewerbestrasse 11, 6330 Cham, Switzerland

Preface

The integration of the physical infrastructure and information systems is challenging but necessary in order to achieve more environmentally friendly, efficient, safe management of logistics and supply chain operations. In the last decade, the development of trends, such as Internet of Things, Industry 4.0 and advanced data analytics, opens new possibilities in order to achieve more timely and reliable deliveries of goods. However, expectations of the customers with regards to frequent and door-to-door deliveries (e-commerce) must be combined with the policy guidelines on CO_2 reductions, especially in urban areas.

Pressure on the reduction of greenhouse gases emissions encourages development of new methods, tools and new technologies, which are cleaner and more efficient.

This book entitled *Smart and Sustainable Supply Chain and Logistics – Trends, Challenges, Methods and Best Practices* presents a multidisciplinary approach. It contains selected theoretical and empirical studies. The Authors in the individual chapters discuss the original methods and tools, as well as practical case studies on topics, as follows:

- Smart Supply Chain Management,
- Sustainable supply chain and logistics,
- Human factor in logistics and supply chain management,
- Modelling and optimization of the supply chain and logistics operations.

This monograph includes the selected papers of the Authors who have submitted their work to the 15th International Congress on Logistics and SCM Systems (ICLS 2020). Although not all of the received chapters appear in this book, the efforts spent and the work done for this book are very much appreciated.

The ICLS 2020 is organized by the Faculty of Engineering Management, Poznań University of Technology and the International Federation of Logistics and SCM Systems (IFLS). We would like to express our gratitude to the Board of the International Federation of Logistics and SCM Systems (IFLS) for the invaluable contribution to the volume:

- Honorary Chairman: Prof. Karasawa, Yutaka, Kanagawa University, Japan.
- Advisors: Prof. Kachitvichyanukul, Voratas, Asian Institute of Technology, Thailand; Prof. Katayama, Hiroshi, Waseda University, Japan.
- Chairman: Prof. Tsai, Kune-Muh, National Kaohsiung University of Science and Technology, Taiwan.
- Vice Chairmen: Prof. Lai, Kin Keung, City University of Hong Kong; Prof. Liu, Xiaohong, Central University of Finance and Economics, China; Prof. Rim, Suk-Chul, Ajou University, Korea; Sethanan, Kanchana, Khon Kaen Univesity, Thailand; Prof. Wakabayashi, Keizo, Nihon University, Japan and Wu, Yenchun Jim, National Taiwan Normal University, Taiwan.
- The Board Members.

This scientific monograph has been double-blind reviewed. We would like to thank all reviewers whose names are not listed in the volume due to the confidentiality of the process. Their voluntary service and comments helped the authors to improve the quality of the manuscripts.

Paulina Golinska-Dawson
Kune-Muh Tsai
Monika Kosacka-Olejnik

Contents

Sustainable Supply Chain and Logistics

Human Factor in Logistics and Supply Chain Management

Smart Supply Chain Management and Logistics

Smart Contracts in the Context of Procure-to-Pay

Lorenz Trautmann and Rainer Lasch

Abstract Smart contract technology is beginning to alter the way supply chain processes are designed and executed. Blockchains enable peer-to-peer networks where contractual terms are automatically executed without the need for unbiased intermediaries. Due to the immutability of blockchains, smart contracts reduce risks and provide traceable audit trails of all transactions. Moreover, smart contracts can be automatically triggered, resulting in decreasing administration and service costs while improving process efficiencies. These attributes make smart contract technology a promising invention for the poorly automated procure-to-pay process. That is, if the existing technological limitation, like poor scalability, high energy consumption, and limited transaction throughput, can be solved. The primary purpose of this chapter is to investigate how smart contracts can be implemented into existing procure-to-pay processes, what advantages can be gained from the application, and which technological restrictions are yet to overcome. To this end, a systematic literature review was conducted. Academic and practitioner literature were analyzed based on a qualitative and deductive methodology. Additionally, this chapter presents a use case for smart contracts in procure-to-pay in the context of international trade. That way, it lays a foundation for future research, provides valuable insight for procurement managers, and offers approaches to overcome the most pressing challenges.

Keywords Smart contracts · Blockchain · Procurement · Procure-to-Pay · Distributed-ledger

L. Trautmann (✉) · R. Lasch
Chair of Business Management, esp. Logistics, Technische Universität Dresden, Dresden, Germany
e-mail: lorenz.trautmann@tu-dresden.de

© The Author(s), under exclusive license to Springer Nature Switzerland AG 2020

P. Golinska-Dawson et al. (eds.), *Smart and Sustainable Supply Chain and Logistics – Trends, Challenges, Methods and Best Practices*, EcoProduction,
https://doi.org/10.1007/978-3-030-61947-3_1

1 Introduction

Since the inception of cryptocurrencies, blockchain-based technologies, such as smart contracts, have gained significant attention for having possible applications in other fields. One of the most promising areas that may gain from adoption is the supply chain. Due to their safe and autonomous execution, smart contracts are suitable for the automation of various intercompany processes throughout the supply chain. Especially the integration of smart contracts into the procure-to-pay process holds the promise of automating procurement procedures and creating an inter-company single-source-of-truth supplier network. The procure-to-pay process is composed of sequential steps necessary for a company to undertake during the procurement of goods or services. Because these steps consist of many standardizeable processes that involve a variety of parties, the use of smart contract technology offers great potential for optimization. This paper explores the improvements that smart contract technology can bring to generic procure-to-pay processes, which technological limitations prevent the technology from being used in practice, and which approaches exist for solving these issues. Additionally, this paper demonstrates the findings by developing a use case in the context of international trade.

The term "smart contracts" was coined in 1994 by Nick Szabo, a computer scientist, who defined a smart contract as a "computerized transaction protocol that executes the terms of a contract" (Szabo 1994). However, up until recently, smart contracts lacked the opportunity of a real-world application. With the invention of *Bitcoin* in 2008 (Nakamoto 2008), the underlying blockchain network emerged as a disruptive technology for many industries. The combination of a distributed-ledger technology with consensus mechanisms, and cryptographic algorithms form a perfect platform for the implementation of smart contracts, and a breeding pool for further development.

Generally speaking, smart contracts are self-executing "if-then-statements", that are written into code and are deployed on and secured by a blockchain. Although the name suggests differently, smart contracts are not contracts in a legal sense. They are software that can enforce conditions stemming from legal documents made by two or more parties. Compared to conventional contracts, smart contracts offer distinct advantages. Once implemented into a blockchain, smart contracts cannot be altered or manipulated. This fact mitigates the risk of fraud while simultaneously providing an audit trail of all transactions. Furthermore, smart contracts are executed automatically, resulting in decreasing administration costs, elimination of the need for unbiased third parties, more efficient process execution and reduced turnaround times (Brody 2017; Capgemini 2019; DHL & Accenture 2018; Prinz and Schulte 2017). While the potential upside of smart contracts is undisputed, several challenges and limitations are yet to be solved. These challenges include legal issues (general civil law, data protection law, and questions of legal provability), functional requirements (human-readable code, block randomness), economic considerations (cost-benefit-analysis), and technological restrictions. This paper focuses on the technological issues of smart contracts and the IT system landscape (e.g. scalability, energy consumption, and performance restrictions).

The paper is structured as follows. Section 2 describes the methodology used in this study. Section 3 provides the technological background of blockchains and smart contracts and presents popular types and platforms. Section 4 then summarizes the findings, including the positive impact smart contract technology can have on procure-to-pay processes, which technological challenges currently limit smart contract technology, as well as possible solutions to these challenges. Section 5 demonstrates a use case for smart contracts in procure-to-pay. Section 6 concludes the paper.

2 Methodology

This section illustrates the procedure adopted for examining smart contracts in the context of the procure-to-pay process. To achieve a holistic, transdisciplinary understandinng of the topic and to guarantee relevant results, a systematic literature review, as proposed by Tranfield et al. (2003), complemented by several authors, (Kitchenham and Charters 2007; Brocke et al. 2015), was conducted. To this end, a qualitative and deductive approach was chosen. Following Tranfield et al. (2003), the steps shown in Fig. 1 were carried out.

Stage 1—Planning the Review

The main goal of this paper is to investigate the way smart contract technology can impact the procure-to-pay process. The lack of a classification of the immense potential and the severe existing limitations, as well as an insufficient amount of applicable use cases, justify the need for a systematic review and this study. To reduce possible researcher bias, a search protocol was created. The following search string was chosen after conducting pilot searches: ["blockchain" OR "distributed ledger" OR "smart contracts"] AND ["procurement" Or "purchase" OR "supply chain" OR "logistics"]. The term "bitcoin" was not included since it led to papers with a focus on the economic topic of cryptocurrencies. The chosen databases were *Ebsco Host (Academic Search Complete, Business Source Complete, Econ Lit), Elsevier Science Direct, Emerald Insight, Wiso Wirtschaftswissenschaften, Springer Link*, and *Scopus*.

Stage 2—Conducting the Review

After applying the search protocol in the scientific databases, 691 papers were identified. After removing doubles, the literature was screened based on titles, leaving 90 papers. The following screening of the abstracts yielded 38 papers that suggested proximity to the research questions. These papers were analyzed based on the full text. Finally, 17 scientific papers were deemed relevant for answering the research questions. However, to review the current state of new technologies, a wide range of sources must be included in the review (Webster and Watson 2002). Since the

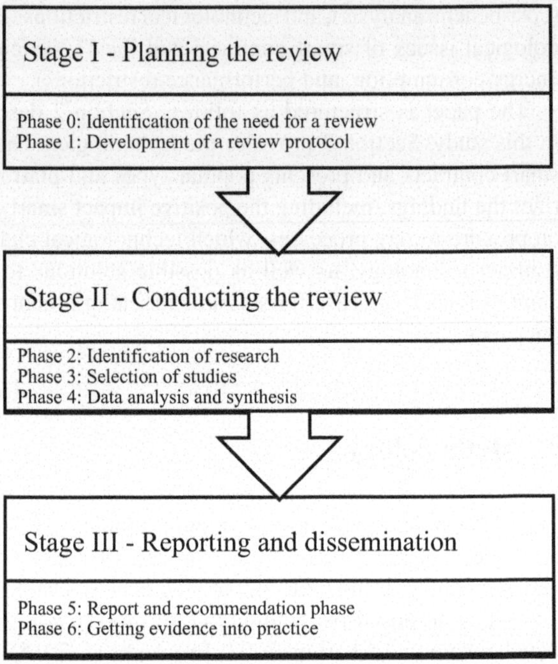

Fig. 1 Stages for conducting the systematic literature review (own depiction based on Tranfield et al. 2003)

academic assessment of technological innovation tends to lag behind practitioner literature and to emphasize the exploratory character of this paper, as well as the practical relevance of smart contract technologies, several white papers by business consultancies and research institutions were added to the relevant literature. Subsequently, the identified literature was analyzed based on a qualitative and deductive approach. The extant information was analyzed and synthesized according to the analysis-based literature review proposed by Seuring and Gold (2012). Codes were created by reading the texts and extracting the strengths and weaknesses of smart contracts in procure-to-pay processes. To create the codes, a qualitative data analysis software (*MaxQDA* 2018, Ver. 18.2.0) was used.

Stage 3—Reporting and Dissemination

The last stage consists of the presentation of the findings. To lay the foundation, the technological basics for smart contracts and blockchain in general are outlined, as well as different types of blockchains and smart contract platforms. Based on that, improvements that smart contract technology can bring to procure-to-pay processes, technological limitations that prevent the technology from being used in practice, and approaches for solving these issues are presented. Building on those findings, a use case, according to Cockburn (2000), for smart contracts in an international trading process is discussed. According to Cockburn, a use case is a "description of the possible sequences of interactions between the system under discussion and its external actors, related to a particular goal" (Cockburn 2000). Finally, conclusions about the results, limitations, and possible future research fields will be drawn.

3 Technological Background

Current smart contract technology relies on blockchains for implementation and safe execution (DeRose 2016). For that reason, the following section introduces blockchain technology and types of blockchains, followed by an overview of smart contracts and development platforms.

3.1 Blockchain Technology

The concept of blockchain was first described in a paper by Satoshi Nakamoto (Nakamoto 2008). Due to versatile further development in various disciplines, no consistent definition of the term can be found in the literature. Nevertheless, several organizations are working on developing a uniform definition for blockchains. In 2017 the technical committee *ISO/TC 307: Blockchain and distributed ledger technologies* was founded by the International Organization for Standardization (ISO) with the goal of defining the concepts of *identity, interoperability, governance, security* and *privacy*, *use cases* and *smart contracts* in the context of blockchain technology (Bundesverband and e.V. 2017; Deutsche Institut für Normung 2017; International Organization for Standardization 2017).

The blockchain is a kind of database that consists of a chain of chronologically arranged data blocks, stored on users' computers in the same form and with identical content. It is a special form of a distributed ledger that uses cryptographic technologies, consensus mechanisms and peer-to-peer connections, to ensure the integrity and authenticity of the data without central control (Drescher and Lenz 2017; Meier and Stormer 2018; Mitschele 2018; Pearson et al. 2019). Transactions that occur in the network are combined with other transaction and implemented into a block. This block is then irreversibly attached to the previous block using a specific communication and verification protocol. In this way, a chain of blocks is created, thus naming the technology "blockchain". A change in a previous block would affect every following block, making it virtually impossible to change or delete previous transactions (Mik 2017; Wang et al. 2019).

The blockchain is not governed by one single party, nor does a central administration for complying with the rules exist. Rather, the blockchain is stored within the distributed system as an identical copy on every node. Every participant of the distributed systems has equal rights and contributes to the adherence of the protocol (Bosch and Penthin 2018; DHL & Accenture 2018; Mullender 1993). A peer-to-peer connection directly connects the nodes for them to communicate, execute consensus protocols, exchange data and even calculation resources among each other. Because of the data being distributed in the network, no single-point-of- failure exists. Consequentially, the data is protected against system failure and unwanted interventions (Drescher and Lenz 2017).

To ensure data integrity, secure transmission and user authenticity, blockchains use two cryptographic elements as essential parts, digital signatures and hash functions. Digital signatures aim to verify the integrity of messages that are exchanged via the blockchain and ensure the authenticity of participants. To achieve this goal, digital signatures are created using two keys via a two-stage, asynchronous encryption method. The private key generates the signature and is used to sign the message. It is only known to the sender of the message. The public key, on the other hand, is accessible to everyone and serves to verify the authenticity of the generated signature (Badev and Chen 2014; DHL & Accenture 2018; Meier and Stormer 2018). So-called hash functions tackle another challenge blockchain-technology faces. Huge amounts of data circulate within the network and algorithms are needed to make the data identifiable and comparable in a simple way. Hash functions can generate a unique fingerprint of a character string of any length. Small changes in the text trigger a large change in the hash value. Therefore, by comparing the original hash value with a newly created one, users can validate the integrity of messages and determine possible interferences. These fingerprints can easily be calculated based on the content of the message, but it is almost impossible to deduce the content of the message knowing only the hash value. Because of this property, hash functions are often referred to as "one-way-functions" (Casino et al. 2019; Drescher and Lenz 2017; Hald and Kinra 2019).

For a large network of users with equal rights and without central management, mechanisms for reaching agreement on the current network status must be installed. A democratic process, including all participants, would be costly and time-consuming. Therefore, Blockchains use consensus mechanisms to define the status of the network. Since the inception of blockchain technology, many different consensus mechanisms were invented. They mainly differ in the way in which the node that defines the new status of the network is selected. All mechanisms have in common that the creator of the new block (called "miner") must prove himself worthy of the task, for example, by solving a difficult mathematical task (Proof-of-Work) or by blocking a large amount of the owned cryptocurrency (Proof-of-Stake) (Casino et al. 2019; Cole et al. 2019).

3.2 Types of Blockchains

Blockchains can be divided into three categories according to access and usage rights. The *public blockchain* can be used by anyone without access restrictions. Each node can participate in consensus building and transaction verification. For this purpose, the program code is available as open-source. Each transaction within the network is publicly accessible. An example of a public blockchain is *Bitcoin* (O'Leary 2017).

For certain use cases, however, it can be useful to restrict access to certain users. These blockchains are referred to as *private blockchains*. They are used, for example, within a company or across companies, where it is necessary to limit the users to the participating contractual partners. Transactions can be presented transparently to all participants and stored in a forgery-proof manner without being disclosed to the public. Furthermore, the reduced size of private blockchains require less computing capacity which results in increased performance (DHL & Accenture 2018; O'Leary 2017).

A mixture of the two is the so-called *consortium blockchain*. This hybrid blockchain tries to combine the advantages of private and public networks. In terms of access restriction, this hybrid behaves like private blockchains. However, the management of the network is not the responsibility of one participant, but a group. They decide on access regulations, rights for participants and make decisions for the entire network. The consortium blockchain has the efficiency and data protection of private blockchains but removes the autonomy of the leading user of the private blockchain (O'Leary 2017). The exact design of the blockchain should be based on the objective pursued, the size of the network and data protection aspects.

3.3 Smart Contract Basics

The paper by Nick Szabo defined the term "smart contract" as "a set of promises, specified in digital form, including protocols within which the parties perform on these promises" (Szabo 1994). In other words, a smart contract is a piece of written code recording contracting parties' mutual promises and enforces those promises on a suitable platform. The blockchain offers for the first time a platform on which smart contracts can be stored and executed (DeRose 2016). Just like other transactions, smart contracts are integrated into blocks and remain within the blockchain after completion of the transaction. The miner initiates the program code in the validation process of the corresponding block (Bundesamt für Sicherheit in der Informationstechnik 2019; Wang et al. 2019).

Although the name suggests a legally binding contract, smart contracts are computer programs that can merely realize legal contracts by performing actions based on the agreement between multiple parties as soon as "if-then" conditions have been met (Allam 2019; Debono 2019). The rigidity of these "if-then-statements" and the fact that smart contracts must be understood by machines directly influences the way how legal contracts, which are to be used as the basis, must be structured. In business transactions, circumstances and business relationships are constantly changing. Therefore, in some cases, it is necessary to deviate from contractual clauses, by mutual agreement, for the benefit of the business relationship. However, the immutability of the blockchain and the automatic execution of smart contracts by means of "if-then-conditions" leave no room for non-contractual arrangements. Therefore, smart contracts will have to be programmed in a way that allows for a certain degree of leeway for the fulfillment of contractual terms. This in turn means, that legal contracts must be formulated more precisely and possible deviations from the regular fulfillment of the contract must already be decided and agreed upon when the legal contract is concluded. The result will be much more detailed legal contracts covering a wider range of different business conditions and resulting actions (Fertig and Schütz 2019).

3.4 Smart Contract Platforms

Smart contracts cannot be implemented on any arbitrary blockchain but need a specific type of platform. These blockchain platforms differ from pure cryptocurrency platforms in that they offer a broader application spectrum than the pure transfer of digital money. They are able to execute generic code, necessary for executing smart contracts (Bocek et al. 2017). The following section presents three smart contract platforms that were chosen because of their technological maturity: *Ethereum, Hyperledger Fabric* and *R3 Corda* (Bocek and Stiller 2017).

Ethereum is a blockchain platform that executes general decentralized applications. It uses Turing-complete programming languages, such as *Solidity, Serpent* and *Mutan,* allowing for the implementation of smart contracts and decentralized applications. Similar to *Bitcoin, Ethereum* uses the Proof-of-Work algorithm for generating consensus and an account-based data model for identifying users. The miners that perform the algorithms necessary for reaching consensus are rewarded with the cryptocurrency *Ether*. Users pay for their transaction to be included in the blockchain (Luu et al. 2016; Zheng et al. 2020). Laurence deems *Ethereum* as one of the "most developed and accessible blockchains in the ecosystem and an industry leader in innovation and blockchain applications" (Laurence 2017).

Hyperledger Fabric is an open-source protocol for private and restricted decentralized databases, which was launched in December 2015 by the Linux Foundation (Cachin 2016). Unlike *Ethereum, Hyperledger* utilizes conventional programming languages like *Java* and *Go,* which also are Turing-complete. For executing smart contracts, *Hyperledger* uses *Docker container.* The modular structure of *containers* simplifies the designing process, allows for the precise allocation of access rights for individual users and therefore, helps with the scalability of the blockchain. However, the modular principle only allows for a limited degree of individualization. The platform is designed for networks where members are at least partially trusted, and consensus can be reached easily. *Hyperledger* uses the *Practical Byzantine Fault Tolerance* (pBFT) consensus mechanism. The drawback of this algorithm is a prolonged consensus process (Zheng et al. 2020).

The *R3 Corda* protocol is specialized for the application in the financial industry and was developed by various banks and financial providers in 2016. The platform utilizes high-level programming languages like *Java* and *Kotlin*, which are carried out by the *Java Virtual Machine* (Brown 2018; Zheng et al. 2020). This platform is usually used for private blockchains, where trust between participants is high and consensus can be reached more easily. To this end, *Corda* uses the *Raft* consensus algorithm

(Howard et al. 2015). The advantage of this mechanism is the high data security since exclusively parties involved in the transaction have access to the data. The validity and uniqueness of the transaction are ensured by central network participants, so-called "Notary Nodes". The transaction parties themselves guarantee the correctness of the transaction. Another significant advantage of this platform is the possibility to integrate contracts in written form as metadata into the blockchain. This way, a legal legitimation of the transaction is directly included in the blockchain. However, the drawbacks for the use in procurement include recentralization through the "notary nodes" and the focus on the financial market (Brown 2018; Howard et al. 2015).

4 Smart Contracts for Procure-to-Pay

The procure-to-pay process is part of the operational procurement activities of businesses. It describes the necessary steps for a company to undertake when ordering goods or services. The process includes various parties, such as buyers, suppliers, banks, and transportation providers, resulting in a complex process with a wide range of different application and communication systems, considerable interface effort as well as lengthy and error-prone manual activities (Arnolds et al. 2010). The following section illustrates how current procure-to-pay solutions and communications protocols can be improved by blockchain and smart contracts and in which way these technologies can help to tackle typical issues and improve existing procure-to-pay processes. Afterwards, technological limitations and possible solutions are discussed.

4.1 Current Procure-to-Pay Solutions

Business activities in most companies nowadays are managed by Enterprise Resource Planning (ERP) systems. These systems are used for a wide range of functions such as accounting, human resource and also procurement activities, like the procure-to-pay process. Because of the uniqueness of business procedures ERP systems must be individually adapted to the requirements of each company. The ERP system plays a vital role in transforming an organization´s internal processes. However, these ERP systems do not connect the company to other organizations. Supply chains are ecosystems with a multitude of participating companies, which contribute to the value chain in various types and with different processes. As each company uses individual ERP systems and these systems are often not connected, it is difficult to get an overview of the entire supply chain with the current ERP technology (Banerjee 2018; Linke and Strahringer 2018). This can change fundamentally through the use of blockchain and smart contracts. These technologies can function as the missing links between the various ERP systems. By connecting the existing ERP systems to a blockchain network, a secure, reliable, and immutable single-source-of-truth, for all supply chain participants to attain, emerges. With the inclusion of smart contracts, transactions between companies can now be automated across the entire supply chain including various IT systems (Banerjee 2018; Bosch and Penthin 2018; Brody 2017; DHL & Accenture 2018; Linke and Strahringer 2018).

The development and introduction of ERP systems are cost-intensive and usually take several years. It is quite difficult to make changes to ERP systems afterwards or to add new functions. That is why, when integrating blockchains into existing ERP systems, it must be ensured that minimal modifications to the existing ERP systems and minimal disruption to regular business processes are made. That way, companies do not have to change their internally used systems and still can reap the benefits from the transparency of a blockchain-based supply chain (Linke and Strahringer 2018). As soon as the ERP systems are connected to the blockchain and smart contracts are embedded in the blockchain, this enables the automation of entire process chains along the supply chain, while at the same time increasing trust between participants and transparency of supply chain processes.

4.2 Electronic Data Interchange (EDI) and Blockchain

Communication between participants of procure-to-pay processes is oftentimes based on EDI. EDI is a universal language that refers of the exchange of data between application systems of different companies using electronic transfer procedures (Banerjee 2018; Boschi et al. 2018; Fiaidhi et al. 2018). In procure-to-pay, EDI is used in a variety of processes, such as orders, shipping and involves. Business documents based on the international EDI standards (e.g. UN/EDIFACT and ANSI X12.4) can be transferred electronically between participants without the need to re-enter information (Boschi et al. 2018; Fiaidhi et al. 2018).

However, with the advent of the fourth industrial revolution and new technologies, like the Internet of Things, many business processes have changed fundamentally. Supply chains are more complex and dynamic than ever before. Supply networks consist of international companies, large numbers of goods and products, various means of transportation, as well as an abundance of payment and information flows. Current EDI systems cannot adequately support these complex supply chains for several reasons. One problem with EDI systems is that once the EDI message is transmitted, the submitting company loses all visibility of the information and the current system status. Only when the EDI message comes back as an invoice, for example, conclusions can be drawn (Boschi et al. 2018). Furthermore, conventional EDI systems are based on one-way, point-to-point communication which means that the messages can only be read by the two parties directly involved. Third parties, for whom the content of the messages could also have consequences are left out. Additionally, if a participant makes an error in an EDI message, for example, if data is entered incorrectly, fields are swapped, or master data is loaded into another participant's system, other partners are not informed, and the error travels through the system. Therefore, additional methods are currently required to maintain data integrity. These include manual maintenance, communication via e-mails and third-party services that keep the data synchronized. In today's dynamic business world, customers and suppliers need shared, real-time information about the actual situation of supply chains, as well as automated early detection systems for incidents. For EDI to cope with the new complex processes of supply chains, they need a secure, and shared data source. Without that, EDI messages will have great difficulty in accurately mapping the communication between all the various supply chain participants (Boschi et al. 2018; Fiaidhi et al. 2018).

The blockchain can complement EDI systems by providing increased transparency and reliable data through distributed-ledger technology. By presenting information in a secure, auditable, and transparent way, the blockchain can overcome the weaknesses of one-way, point-to-point, and batch approaches of existing communication protocols (Banerjee 2018). This will enable companies to jointly identify and solve problems and overall reduce costly bottlenecks (Boschi et al. 2018; Fiaidhi et al. 2018).

Whether blockchain technology will replace or supplement existing messaging systems is questionable. Some argue that it would be more successful to supplement existing communication systems rather than fully replace them with blockchains while others suppose that blockchain technology will completely replace EDI systems and will institute a completely new way of B2B communication (Banerjee 2018; Fiaidhi et al. 2018).

4.3 Process Automation

Nowadays, the flow of materials through the supply chain is highly automated, thanks to intelligent planning concepts using the latest technology (e.g. ERP systems). The flow of cash and information, on the other hand, is completely detached from the high degree of automation of the material flow (Schütte et al. 2017). To illustrate this point, the average receivable of a U.S. Fortune 100 company is 60 days old, even though the average payment period is only 30 days long (Brody 2017; Capgemini 2019). This delay in payment results in a drastic increase in necessary operational assets and leads to delays in corporate decision making. Responsible are time-consuming and error-prone steps of the procure-to-pay process. The individual activities are often cumbersome, as they are largely initiated manually and involve coordination with many parties. For instance, invoice processing for 60% of corporate transactions is conventionally done by manually writing an invoice on a sheet of paper by the supplier, and manually initiating payment by the buyer (Schütte et al. 2017). Through the use of smart contracts, the entire invoice processing can be automated and synchronized with the material flow, eradicating relating inefficiencies. Beyond that, unnecessary process steps and intermediaries can be removed, thereby reducing cost, time, data sources, interfaces and IT systems (Capgemini 2019; Guyonnet and

Mohammed 2016; Prinz and Schulte 2017). In that way, smart contracts can help to close the "analogous gap" and contribute to a synchronized flow of materials, information and money (Apptus 2018; Eickemeyer et al. 2018; Stahlbock et al. 2018).

4.4 Transparency and Traceability

During the course of globalization, the number of goods and commodities transported worldwide has increased exponentially. There are more than 500,000 shipping companies in the USA alone (DHL & Accenture 2018). The results are complex supply chain processes with large numbers of parties, non-standardized processes, various IT systems, databases and a low level of transparency. Companies face the challenge of obtaining reliable information about the origin and condition of materials and the status of payment as a basis for planning and decision-making. Despite enormous investments in digital infrastructures, continuous insight into the entire supply chain cannot be provided adequately with conventional technologies (Apptus 2018; Bosch and Penthin 2018; Brody 2017).

Smart contract technology, in combination with blockchains and IoT sensors, can monitor the manufacturing and transportation conditions as well as whereabouts of goods in real-time while simultaneously serving as a consistent source of information for all participants (Apptus 2018; Capgemini 2019). The blockchain makes information visible, for instance, the identity of the manufacturer as well as production and transportation conditions of the product. Given this information, smart contracts can verify whether safety and environmental regulations were complied with. In this way, the growing demand of customers for transparent and sustainable supply chains can be met, without the high costs for certification by auditors (Apptus 2018; Eickemeyer et al. 2018; Guyonnet and Mohammed 2016; Kouhizadeh et al. 2018; Stahlbock et al. 2018).

Several examples of projects using the blockchain for increasing supply chain transparency are found in the literature. The project "Modum" aspires to track and ensure transportation quality of pharmaceuticals. IoT sensors measure data such as temperature and humidity during the entire transportation process. This information is submitted automatically to the blockchain, where smart contracts compare it to the product's requirements (Bocek et al. 2017; DHL & Accenture 2018). Another company, called "Everledger" offers systems for tracking and protecting valuable goods based on their characteristics. Ownership can be verified by comparing the unique properties of the luxury items with data in the blockchain, similar to a fingerprint (Prinz and Schulte 2017; Straube et al. 2018).

Besides, the blockchain not only holds promises for materialistic goods. The transfer of ownership rights of digital goods, such as copyrights and legal claims, can be processed and stored by smart contracts. This becomes relevant wherever licenses and other rights are acquired. Application examples are found in the context of warranty and guarantee claims, spare parts business or the paperless processing of dangerous goods transports (Bosch and Penthin 2018; Camelot 2017; DHL & Accenture 2018; Eickemeyer et al. 2018; Guyonnet and Mohammed 2016).

The traceability of the individual process steps can also have effects on the provability in legal disputes. The most common cases in procurement include delays in deliveries, the disappearance of goods or the failure to make payments. With the help of blockchains, responsible parties can be identified (Eickemeyer et al. 2018). Besides, the transparent presentation of relevant data serves to prevent corruption, bribery, money laundering, fraud and other legal violations, such as child labor. This increases the trustworthiness towards end customers and helps to identify sources of complaints and to check for compliant processes (Bocek et al. 2017; Straube et al. 2018).

Additionally, the transparent audit trail left in the blockchain has major implications for future procurement reporting. The high number of transactions within purchasing departments makes a complete overview difficult. Various sub-departments, employees and in special cases, even other companies carry out the purchase orders. Large companies face the challenge of collecting all necessary data for comprehensive reporting. In some cases, up to 60 employees are on the payroll for obtaining information regarding procurement activities (Brody 2017). The blockchain offers a database on which all transactions are stored and displayed, regardless of who executed them. Negotiations and contract processing between companies and their suppliers leave behind chronological and unchangeable audit trails within the blockchain that all participants can access and relay on Apptus (2018), Bosch and Penthin (2018), Capgemini (2019), DHL & Accenture (2018), Guyonnet and Mohammed (2016), Schütte et al. (2017).

Not only large companies would benefit from blockchain technology. Access to origin and manufacturing data allows for transparent value creation processes and reduces the incentive for companies to withdraw excessive profits from the individual stages of the supply chain, resulting in better pricing and balancing of information asymmetries (Saberi et al. 2019; Yoo and Won 2018). A more transparent and market-conform pricing based on the actual market in procurement would be the result. Particularly small companies are affected by the asymmetries that currently prevent them from competing with their larger rivals (Apptus 2018).

4.5 Data Security and Integrity

The role of cybersecurity is growing immensely. Companies must find ways of ensuring data accuracy, validity and integrity, while simultaneously being hyper-connected to all necessary participants along the supply chain. In the financial sector, where blockchain applications are showing promising results, the modern methods of data encryption and cryptography have helped the blockchain being nicknamed "The Trust Machine" by the Economist. Blockchains can secure databases against human-caused errors, intentional manipulation and system failure (Bosch and Penthin 2018;

Camelot 2017; DHL & Accenture 2018; Eickemeyer et al. 2018; Guyonnet and Mohammed 2016; Prinz and Schulte 2017).

By automating procurement processes, smart contracts reduce manual intervention, and thereby, the susceptibility to human-caused errors. Furthermore, a typical security gap of central databases called the "single point of failure", can be closed. This weak spot leads, in the event of its own malfunctioning, to the failure of the entire system. The redundant distribution of the database on all nodes of the network and the immutability of integrated blocks protects from the risk of total system failure, limits the impact of hacker attacks on the entire system, and secures against intentional manipulation (Capgemini 2019; Kouhizadeh et al. 2018; Stahlbock et al. 2018).

4.6 Technological Restrictions

The technological restrictions of smart contracts are closely linked to the development of the blockchain. Particularly with large numbers of participants, blockchains currently come up against their technical boundaries. With each new block, the data volume, and therefore, calculation difficulty increases. Consequently, large networks require great storage space, robust internet connection, high computing power and immense energy. These fundamental limitations reduce the scalability of blockchain networks nowadays and prevent smart contract technology currently for being used for extensive networks with many participants and transactions (Bosch and Penthin 2018; DHL and Accenture 2018).

Several solutions to tackle these issues are promoted in the literature. One approach to reducing the amount of memory needed is using external data storage. The blockchain would only contain a reference to the original files. This reference is stored as a hash value, which can access external information while ensuring its integrity (Schütte et al. 2017).

To reduce the required computing power and therefore, energy consumption, different consensus mechanisms are being proposed. With the so-called Proof-of-Stake, a functioning alternative to the Proof-of-Work exists. This approach reduces the impact of large numbers of users on scalability. Further approaches are the deletion of expired data (Unspent Transaction Output) and the partitioning of the blockchain (Prinz and Schulte 2017).

Another point of criticism is the reduced performance of blockchain systems. The *Bitcoin* blockchain balances itself out at approximately seven transactions per second, compared to the transaction of the financial services corporation Visa with more than 2,500 transactions per second (Straube et al. 2018). The reason for the limited amount of transactions are computing processes needed for guaranteeing the safety and integrity of the data. These cryptographic means are necessary to ensure a peer-to-peer network without the need for intermediaries. A possible solution is the partial centralization of the calculation to increase processing speed. While this approach does increase computing velocity, it is accompanied by a reduction of the decentralization of the network and the associated advantages (Schütte et al. 2017).

The fundamental dilemma of prioritizing between decentralization, security and scalability was named by the founder of the *Ethereum* blockchain "the scalability trilemma". It states that at most, two of the three criteria can be fully met (Perboli et al. 2018).

5 Use Case

In this section, an exemplary use case for smart contracts implemented into the procure-to-pay process is introduced. Using the example of international trade, it is demonstrated how smart contracts can simplify, automate, and shorten complex procure-to-pay processes and remove intermediaries. Firstly, the conventional process, as shown in Fig. 2, is demonstrated. Afterwards, smart contracts are implemented, and the proposed process, as shown in Fig. 3, is explained. The technological features of the implementation are addressed, as well.

Importer "I" buys goods from a foreign supplier "S", that are delivered by a transportation service provider "T". After determining the requirements and receiving the product catalog from "S", "I" manually selects the vendor and places the purchase order. The delivery terms and business agreements are negotiated and agreed on, via traditional communication channels, such as e-mail and telephone. Afterwards "I" requests a letter of credit from its bank "BI", which is then issued to "BS", the bank of "S". After confirming the reception to "S", the goods are handed over to "T" in exchange for a proof of shipment. Bank "BS" then pays "S" the agreed amount. After receiving the goods, "I" manually processes the invoice and orders "BI" to pay the agreed amount to "BS". Finally, "I" must collect all necessary data for comprehensive reporting.

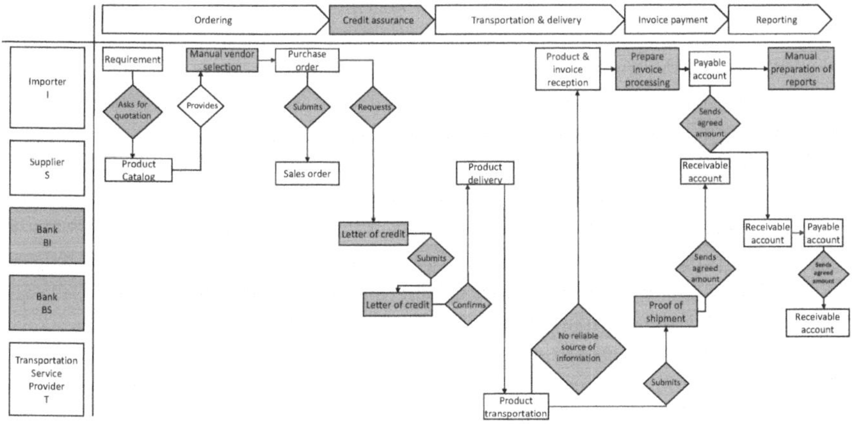

Fig. 2 Conventional Procure-to-Pay process

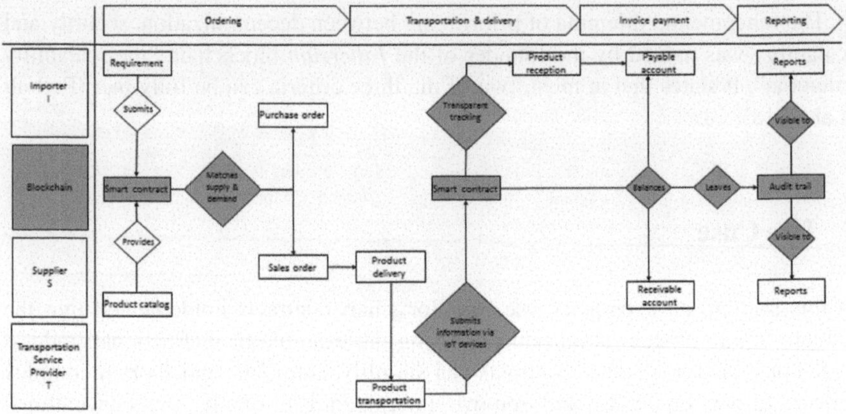

Fig. 3 Proposed Procure-to-Pay process with Smart Contract Technology

In addition to a large number of process steps, the parties in this example use a variety of different IT systems to handle internal processes and external communication. For monitoring of resources and planning of requirements, most companies use ERP systems, like the "SAP S/4 HANA". Additionally, larger companies use specifically developed systems for data transfer and communication with external stakeholders. At Daimler, for example, this system is called the "Daimler Supplier Portal" (Linke and Strahringer 2018). All goods receipts, payments and delivery information are, additionally handled by means of conventional communication, such as telephone and e-mail. The various IT Systems paired with a large number of steps, including many participants, make the procure-to-pay process heterogenic and complicated. The goal of the implementation of smart contracts is to reduce the number of necessary participants, process steps, and actions for executing the procure-to-pay process. Figure 2 depicts the aspects of the process that may be replaced or automated with the use of blockchain-based smart contracts in grey. The banks "BI" and "BS" may no longer be vital, as well as the entire process of "credit assurance". Furthermore, several process steps may be automated, including "manual vendor selection", "invoice processing", several of the payment steps as well as the "manual preparation of reports".

The following section proposed the implementation of smart contract technology into the conventional procure-to-pay process. Figure 3 shows the new aspects of the proposed procure-to-pay process with the use of smart contract technology in grey. The banks are replaced by a blockchain platform on which smart contracts are implemented. The blockchain acts as a distributed ledger for all parties to access and several smart contracts are implemented into that blockchain, executing "supply and demand matching", "transparent tracking of shipped goods", "automatic invoice processing and payment", as well as "leaving a traceable audit trail".

However, for the technology to be implemented in the procure-to-pay process, the various technological limitations of smart contracts and further restrictions must be addressed. For external communication and processes, blockchain-technology offers excellent potential. However, for internal tasks, where trust between the participants is established, blockchain applications provide no significant advantage over traditional technology. Therefore, these internal tasks should not be replaced by blockchain technology. However, this, in turn, has an impact on how blockchain applications must be designed. Most of the internal tasks are commonly performed by ERP systems, which link all company departments and function as common databases. For blockchains to be able to communicate with ERP systems and access off-chain data, appropriate software-connectors must be implemented. The advantage of using software connectors is that neither the existing ERP system nor the blockchain needs to be adapted. Established ERP systems, like the "SAP S/4 HANA", usually offer integrated connectors.

In the case of a supplier-buyer network, closed consortium blockchains should be used. These blockchains are jointly managed by all participants, which increases the willingness to cooperate between companies. The participants decide jointly which new members are admitted and which rules are to apply in the network. In order to achieve complete process automation, including invoice payments, the use of cryptocurrencies is advisable. A possible alternative would be to detach the invoice payment from the blockchain process and carry it out the conventional way.

To increase data throughput and to keep data volume as small as possible, not all data should be stored on the blockchain. Instead, the data should be saved locally on the companies' servers, with only the hash values being stored on the blockchain. With these references, the companies can access the corresponding data and check their integrity.

After the successful implementation of the smart contract, the proposed process is carried out as follows. To begin with, "I" submits its requirement and "S" his product catalog to the blockchain, where a smart contract matches the two. Simultaneously, "I" can verify the authenticity and evaluate the supplier using the audit trail of past transactions. The contractual agreements are defined in a smart contract, which ensures the fulfillment of contractual clauses. After handing off the goods to "T", IoT devices will continuously report the whereabouts and transportation conditions of the goods to the blockchain where the information is easily accessible. "I" has reliable information about the order. After receiving the goods, the smart contract compares the order to the delivery and invoice data. The confirmation of the correctness of the transaction automatically triggers the smart contract to send the agreed amount to the account of the supplier while leaving a transparent and accessible audit trail for all involved parties to use for reporting.

The example shows that smart contracts can help streamline processes. First and foremost, the banks "BI" and "BS" are no longer needed. The blockchain provides the necessary security and trust between the two contractual partners "I" and "S". Furthermore, through the implementation of self-enforcing smart contracts, unnecessary process steps are eliminated (e.g. request for quotation and letter of credit), and several remaining processes can be automated (e.g. vendor selection, invoice processing, and payment).

6 Conclusions

This paper examined the extent to which blockchain-based smart contracts are suitable for the application in the procure-to-pay process. To this end, the basic and operating principles of blockchain and smart contract technology were explained. The paper presented the potential benefits of the new technology for the procure-to-pay process as well as technical limitations and recent advances in solving these challenges. Based on a use case, it was shown that procure-to-pay processes could be optimized through smart contracts. With the use of smart contracts, it was possible to streamline the process, eliminate unnecessary intermediaries and process steps, and automate various tasks. The connection of the digital with the real world in combination with the security mechanisms and the distributed storage of data leads to process automation, supply chain transparency, data security, and increased trust among all participants of the supply chain. In that way, smart contract technology goes beyond the sole automation of internal process steps as current procure-to-pay solutions (e.g. ERP systems).

However, it was determined that for smart contracts to be market-ready for the logistics industry, several challenges are yet to overcome. Technological limitations include limited scalability, poor transaction throughput, and high energy consumption. Approaches for overcoming these issues include external storage of data, alternative consensus protocols, and partial centralization of computing power. Besides that, future research should focus on closing the gap between industry professionals and blockchain developers. Promising solutions are the usage of artificial intelligence, natural language processing, and human-readable code. The question of how blockchain technologies will replace or supplement existing systems, such as ERP systems and EDI communication, also needs to be clarified. Future studies should focus on integrating legal, functional, and economic aspects of smart contracts to achieve a transdisciplinary understanding.

References

Allam Z (2019) On smart contracts and organisational performance: a review of smart contracts through the blockchain technology. Rev Econ Bus Stud 11(2):137–156. https://doi.org/10.1515/rebs-2018-0079

Apptus (2018) Procurement: why blockchain may be the key to a bright future. https://apttus.com/blog/procurement-why-blockchain-may-be-the-key-to-a-bright-future/

Arnolds H, Heege F, Röh C, Tussing W (2010) Materialwirtschaft und einkauf—grundlagen—spezialthemen—übungen. Materialwirtschaft Und Einkauf, 464 https://doi.org/10.1524/9783486592535

Badev A, Chen M (2014) Bitcoin: technical background and data analysis. FEDS Working Paper No. 2014–104, 7 October 2014. https://dx.doi.org/10.2139/ssrn.2544331

Banerjee A (2018) Blockchain technology: supply chain insights from ERP. In: Advances in computers, vol 111. Academic Press Inc., pp 69–98. https://doi.org/10.1016/bs.adcom.2018.03.007

Blockchain Bundesverband e.V (2017) Blockchain—chancen und herausforderungen einer neuen digitalen Infrastruktur für Deutschland. https://bundesblock.de/wp-content/uploads/2019/01/bundesblock_positionspapier_v1.1.pdf

Bocek T, Stiller B (2017) Smart contracts—blockchains in the wings. Digital marketplaces unleashed, pp 169–184. https://doi.org/10.1007/978-3-662-49275-8_19

Bocek T, Rodrigues BB, Straser T, Stiller B (2017) Blockchains everywhere—a use-case of blockchains in the pharma supply-chain. In: Proceedings of the im 2017–2017 IFIP/IEEE international symposium on integrated network management, Lisbon, Portugal, 8–12 May 2017

Bosch R, Penthin S (2018) Blockchain als treiber im modernen supply chain management 4.0. In: Bearingpoint.com. https://www.bearingpoint.com/files/Blockchain_im_SCM.pdf?download=0&itemId=552008

Boschi AA, Borin R, Raimundo JC, Batocchio A (2018) An exploration of blockchain technology in supply chain management. In: 22nd Cambridge international manufacturing symposium, 27–28

Brody P (2017) How blockchain is revolutionizing supply chain management. Ernst & young. https://www.ey.com/Publication/vwLUAssets/ey-blockchain-and-the-supply-chain-three/$FILE/ey-blockchain-and-the-supply-chain-three.pdf

Brown RG (2018) The corda platform: An introduction. Corda platform white paper, 1–21. https://www.corda.net/wp-content/uploads/2018/05/corda-platform-whitepaper.pdf

Bundesamt für Sicherheit in der Informationstechnik. (2019).Bundesamt für sicherheit in der informationstechnik (2019) Blockchain sicher gestalten. bundesamt für sicherheit in der informationstechnik, 98. https://www.bsi.bund.de/SharedDocs/Downloads/DE/BSI/Krypto/Blockchain_Analyse.pdf?__blob=publicationFile&v=5

Cachin C (2016) Architecture of the hyperledger blockchain fabric. In: Workshop in distributed cryptocurrencies and consensus ledgers, vol 70 2016. https://doi.org/10.4230/LIPIcs.OPODIS.2016.24

Camelot (2017) Blockchain and smart contracts: what´s in it for procurement organizations? Camelot. https://www.camelot-mc.com/de/study/blockchain-and-smart-contracts-whats-in-for-procurement-organizations/

Capgemini (2019) Blockchain als erfolgsfaktor-auswirkungen blockchain als erfolgsfaktor-auswirkungen der technologie auf den Beschaffungsprozess der Technologie auf den Beschaffungsprozess. https://www.capgemini.com/de-de/2019/02/blockchain-als-erfolgsfaktor/

Casino F, Dasaklis TK, Patsakis C (2019) A systematic literature review of blockchain-based applications: Current status, classification and open issues. In: Telematics and informatics, vol 36. Elsevier Ltd, pp 55–81. https://doi.org/10.1016/j.tele.2018.11.006

Cockburn A (2000) Writing effective use cases. 3, 204. https://www.infor.uva.es/~mlaguna/is1/materiales/BookDraft1.pdf

Cole R, Stevenson M, Aitken J (2019) Blockchain technology: implications for operations and supply chain management. Supply Chain Manag 24(4):469–483. https://doi.org/10.1108/SCM-09-2018-0309

Debono P (2019) Transforming public procurement contracts into smart contracts. Int J Inf Technol Proj Manag 10(2):16–28. https://doi.org/10.4018/IJITPM.2019040103

DeRose C (2016) Bankthink "smart contracts" are the future of blockchain. https://www.americanbanker.com/opinion/smart-contracts-are-the-future-of-blockchain

Deutsche Institut für Normung (2017) ISO/TC 307 Blockchain and distributed ledger technologies. https://www.iso.org/committee/6266604.html

DHL & Accenture (2018) Blockchain in logistics: perspectives on the upcoming impact of blockchain technology and use cases for the logistics industry. In: DHL Customer Solutions & Innovation. https://www.logistics.dhl/content/dam/dhl/global/core/documents/pdf/glo-core-blockchain-trend-report.pdf

Drescher D, Lenz G (2017) Blockchain Grundlagen: Eine Einführung in die elementaren Konzepte in 25 Schritten. mitp Verlag

Eickemeyer SC, Halaszovich T, Lattemann C (2018) Blockchain technologies for ensuring material, information and monetary flows in logistics—success factors for the Chinese "Belt road initiative." HMD Praxis Der Wirtschaftsinformatik 55(6):1260–1273. https://doi.org/10.1365/s40702-018-00465-7

Fertig T, Schütz A (2019) Blockchain für entwickler—grundlagen, programmierung, Anwendung. Rheinwerk Verlag

Fiaidhi J, Mohammed S, Mohammed S (2018) EDI with blockchain as enabler for extreme automation. IT Professional.

Guyonnet P (Accenture) Mohammed H (Accenture). (2016) How blockchain can bring greater value to procure-to-pay processes. In: Accenture. https://www.accenture.com/t20170103T200504Z__w__/us-en/_acnmedia/PDF-37/Accenture-How-Blockchain-Can-Bring-Greater-Value-Procure-to-Pay.pdf

Hald KS, Kinra A (2019) How the blockchain enables and constrains supply chain performance. Int J Phys Distrib Logistics Manag 49(4):376–397. https://doi.org/10.1108/IJPDLM-02-2019-0063

Howard H, Schwarzkopf M, Madhavapeddy A, Crowcroft J (2015) Raft refloated: do we have consensus? ACM SIGOPS Operating Sys Rev 49(1):12–21. https://doi.org/10.1145/2723872.2723876

International Organization for Standardization (2017) ISO/TC 307 Blockchain and distributed ledger technologies. https://www.iso.org/committee/6266604.html

Kitchenham B, Charters S (2007) Source: "Guidelines for performing systematic literature reviews in SE", kitchenham et al guidelines for performing systematic literature reviews in software engineering.

Kouhizadeh M, Saberi S, Sarkis J, Foisie RA (2018) Blockchain technology and sustainable supply chains. northeast decision science institute 2018 Annual conference, providence, Rhode Island, USA

Laurence T (2017) Blockchain for dummies. Wiley-VCH Verlag GmbH & CO, KGaA

Linke D, Strahringer S (2018) Blockchain integration Within an ERP system for the procure-to-pay process: implementation of a prototype with SAP S/4HANA and hyperledger fabric in the daimler AG. HMD Praxis Der Wirtschaftsinformatik 55(6):1341–1359. https://doi.org/10.1365/s40702-018-00472-8

Luu L, Chu DH, Olickel H, Saxena P, Hobor A (2016) Making smart contracts smarter. In: Proceedings of the ACM conference on computer and communications security, 24–28, 254–269 October 2016. https://doi.org/10.1145/2976749.2978309

Meier A, Stormer H (2018) Blockchain = distributed ledger + consensusblockchain = distributed ledger + consensus. HMD Praxis Der Wirtschaftsinformatik 55(6):1139–1154. https://doi.org/10.1365/s40702-018-00457-7

Mik E (2017) Smart contracts: terminology, technical limitations and real world complexity. Law Innovation Technol 9(2):269–300. https://doi.org/10.1080/17579961.2017.1378468

Mitschele A (2018) Blockchain- definition

Mullender SJ (1993) Distributed Systems, 2nd ed. Addison-Wesley

Nakamoto S (2008) Bitcoin: a peer-to-peer electronic cash system. SSRN Electron J 1–9. https://doi.org/10.2139/ssrn.3440802

O'Leary DE (2017) Configuring blockchain architectures for transaction information in blockchain consortiums: the case of accounting and supply chain systems. Intell Sys Account Finance Manag 24(4):138–147. https://doi.org/10.1002/isaf.1417

Pearson S, May D, Leontidis G, Swainson M, Brewer S, Bidaut L, Frey JG, Parr G, Maull R, Zisman A (2019) Are distributed ledger technologies the panacea for food traceability? In Global food security, vol 20. Elsevier B.V, pp 145–149. https://doi.org/10.1016/j.gfs.2019.02.002

Perboli G, Musso S, Rosano M (2018) Blockchain in logistics and supply chain: a lean approach for designing real-world use cases. IEEE Access 6:62018–62028. https://doi.org/10.1109/ACCESS.2018.2875782

Prinz W, Schulte AT (2017) Blockchain und smart contracts—technologien, forschungsfragen und Anwendungen. Fraunhofer-Gesellschaft, 50

Saberi S, Kouhizadeh M, Sarkis J, Shen L (2019) Blockchain technology and its relationships to sustainable supply chain management. Int J Prod Res 57(7):2117–2135. https://doi.org/10.1080/00207543.2018.1533261

Schütte J, Fridgen G, Prinz W, Rose T, Urbach N, Hoeren T, Guggenberger N, Welzel C, Holly S, Schulte A, Sprenger P, Schwede C, Weimert B, Otto B, Dalheimer M, Harz M, Kreutzer M (2017) Blockchain- technologien, forschungsfragen und anwendungen. In Fraunhofer blockchain positionspapier (Issue Iml). https://www.aisec.fraunhofer.de/content/dam/aisec/Dokumente/Publikationen/Studien_TechReports/deutsch/FhG-Positionspapier-Blockchain.pdf

Seuring S, Gold S (2012) Conducting content-analysis based literature reviews in supply chain management. Supply Chain Manag 17(5):544–555. https://doi.org/10.1108/13598541211258609

Stahlbock R, Heilig L, Voß S (2018) Blockchain technology in maritime logistics. HMD Praxis Der Wirtschaftsinformatik 55(6):1185–1203. https://doi.org/10.1365/s40702-018-00464-8

Straube F, Junge AL, Grunow O (2018) Blockchain in der logistik—status quo und anwendungsbereiche. www.logistik.tu-berlin.de

Szabo N (1994) Smart contracts, vol 40(4). https://doi.org/10.1007/s00287-017-1045-2

Tranfield D, Denyer D, Smart P (2003) Towards a methodology for developing evidence-informed management knowledge by means of systematic review. British J Manag

vom Brocke J, Simons A, Riemer K, Niehaves B, Plattfaut R, Cleven A (2015) Standing on the shoulders of giants: challenges and recommendations of literature search in information systems research. Commun Assoc Inf Sys 37:205–224. https://doi.org/10.17705/1cais.03709

Wang S, Ouyang L, Yuan Y, Ni X, Han X, Wang FY (2019) Blockchain-enabled smart contracts: architecture, applications, and future trends. IEEE Trans Sys Man Cybern Sys 49(11):2266–2277. https://doi.org/10.1109/TSMC.2019.2895123

Wang Y, Han JH, Beynon-Davies P (2019) Understanding blockchain technology for future supply chains: a systematic literature review and research agenda. Supply Chain Manag 24(1):62–84. Emerald Group Publishing Ltd. https://doi.org/10.1108/SCM-03-2018-0148

Webster J, Watson RT (2002) Analyzing the past to prepare for the future: writing a literature review. MIS Q 26(2). https://www.misq.org/misreview/announce.html

Yoo M, Won Y (2018) A study on the transparent price tracing system in supply chain management based on blockchain. Sustain 10(11). (Switzerland). https://doi.org/10.3390/su10114037

Zheng Z, Xie S, Dai HN, Chen W, Chen X, Weng J, Imran M (2020) An overview on smart contracts: challenges, advances and platforms. Future Gener Comput Sys 105:475 491 https://doi.org/10.1016/j.future.2019.12.019

Saberi S, Kouhizadeh M, Sarkis J, Shen L (2019) Blockchain technology and its relationships to sustainable supply chain management. Int J Prod Res 57(7):2117–2135. https://doi.org/10.1080/00207543.2018.1533261

[illegible]

Out-of-Home Delivery as a Solution of the Last Mile Problem in E-commerce

Arkadiusz Kawa

Abstract According to the report Urban Logistics Opportunities-Last-Mile Innovation, prepared by Frost & Sullivan, expenditure on logistics in the world is expected to reach USD 10.6 trillion in 2020. 70% of these expenses will be generated by transport and as much as 40% by last mile deliveries. In addition, there are many social costs associated with urban supplies—traffic jams, noise, pollution, etc. As a result, more and more logistics service companies and e-tailers are making changes to eliminate the effects of problems arising from deliveries to individual customers. These include both relatively easy organizational improvements and more complex solutions that require investment. One of them is out-of-home delivery. The subject of out-of-home delivery in e-commerce is a relatively new research area. There are very few studies conducted on last mile so far have focused on alternative delivery methods, especially its impact on satisfaction and loyalty of e-customers. That is why the goals of this chapter are to identify the components of out-of-home delivery, and to present their influence on satisfaction and loyalty in e-commerce. The studies are empirical and are based on primary data. CATI (computer-assisted telephone interview) was selected as a technique of information collection, which had been preceded by FGIs (focus group interviews).

Keywords Last mile · E-commerce · Out-of-home delivery · PUDO · Parcel lockers

1 Introduction

The Internet and its accompanying services have had a major impact not only on the society but also on the economy since the very beginning of their development. Recently, this impact has been increasingly noticeable. Nowadays, more and more managers are indicating that the digital transformation is taking place in their companies. This applies to communication, sales and distribution channels. Digital

A. Kawa (✉)
Poznan University of Economics and Business, al. Niepodległości 10, 61-875 Poznań, Poland
e-mail: arkadiusz.kawa@gmail.com

© The Author(s), under exclusive license to Springer Nature Switzerland AG 2020
P. Golinska-Dawson et al. (eds.), *Smart and Sustainable Supply Chain and Logistics – Trends, Challenges, Methods and Best Practices*, EcoProduction,
https://doi.org/10.1007/978-3-030-61947-3_2

and physical processes are increasingly intertwined, creating omni-channel systems. This is particularly evident in commerce, which is gradually moving to the Internet.

The value of the world trade was USD 25.04 trillion in 2019. USD 3.54 trillion was generated by e-commerce, representing 14.1% of the total retail sales. E-commerce is projected to grow to $6.54 trillion and will account for 22% of the total retail sales in 2023 (eMarketer Global Ecommerce 2019).

Currently, almost all products that can be bought in stationary stores are sold via the Internet, such as: apparel, footwear, consumer electronics, books, movies, music & games, personal care products, furniture & homeware, household appliances, food & beverages, toys & baby products.

Due to the lack of geographical limitations, the possibilities of buying and selling are even greater, as almost everyone can be a customer. The only restriction is logistics, especially product delivery to the customer. It is not always possible to deliver a product to every place, because not all sellers offer such an option.

E-commerce logistics is associated with many challenges. Until the time of the dynamic development of e-commerce, CEP (courier, express and parcel) operators served mainly companies, i.e. the senders and recipients were businesses (B2B). Currently, a large proportion of shipments is sent to individual customers (B2C). The main difference between the two segments lies in the different time windows (time slots) (Faugere and Montreuil 2016) in deliveries to individuals and companies. Very often, at the time of delivery, recipients are outside the indicated address, which may cause inconvenience to the courier company and generate higher costs. In addition, CEP operators frequently compete with each other with low prices. B2C also causes gradual cannibalisation of B2B—end customers skip middlemen. Another challenge for e-commerce logistics is the relatively high variability of demand (seasonality of demand, Christmas peak, accumulation of orders during the day, week, etc.) and fragmentation of orders, which means frequent shipments of small batches of products. It is troublesome to send non-standard goods and difficult to deal with shipments incorrectly labelled by shippers, e.g. recipient's wrong address, incorrect postal code, no phone number of the recipient.

Online shopping has accustomed customers to convenience. They can look for products and order them anytime and anywhere. They also have similar expectations in relation to the way they have their parcels delivered—they want to have influence on where and when they receive their e-shopping. In addition, this overlaps with the extended working time and an increasingly active lifestyle. Therefore, e-commerce and its operation have an impact on the logistics services. E-customers do not only want to receive the ordered product quickly and get free shipment, but also to pick it up at any place and time.

Most of the goods ordered via the Internet are delivered by courier to the indicated location. Door-to-door deliveries are not always convenient because either the customer has to wait for the courier at home or s/he is away when the courier tries to deliver. Therefore, alternative services have been created to solve this problem. These include late hour deliveries and pick-up and drop-off points, automatic parcel terminals (Kawa and Różycki 2018).

Out-of-home delivery services are certainly more convenient for CEP operators because they reduce the last mile problem. They are therefore increasingly popular and promoted by e-commerce entities. However, a question arises whether customers are willing to use such services and whether they are satisfied with them.

The subject of out-of-home delivery in e-commerce is a relatively new research area. There have been very few studies conducted on the last mile so far, and they have focused on alternative delivery methods, especially their impact on satisfaction and loyalty of e-customers. That is why the goals of this chapter are to identify the components of out-of-home delivery, and to present their influence on satisfaction and loyalty in e-commerce.

The studies are empirical and are based on primary data. CATI (computer-assisted telephone interview) was selected as the technique of information collection, which had been preceded by FGIs (focus group interviews).

The chapter consists of 9 sections. The structure of this chapter is as follows. The last mile issue is described in Sect. 2. In Sect. 3 the door-to-door delivery method is presented. Section 4 introduces out-of-home delivery method, especially parcel lockers and PUDO points. The relationship between out-of-home delivery and customer satisfaction is outlined in Sect. 5. In Sect. 6 methodology of research (research stages, data gathering, and measures) are described. The analysis and results of research are discussed in Sect. 7. In Sect. 8 the development directions are given. Section 9 concludes the chapter and presents managerial implication, research limitations, and further research steps.

2 Last Mile

At the beginning, the Internet was mainly used for cooperation between companies. As already mentioned, B2C transactions are now beginning to dominate e-commerce, where the product is ordered by individual customers and sent by businesses. In both B2B and B2C, the so-called last mile service emerges. It is the last stage in the transport of a parcel by courier to the place designated by the customer. Most often, the parcel goes to the customer's place of residence or work.

The last mile is also called "the last leg of the delivery process from a regional depot to the recipient" (Orenstein et al. 2019) and "the final leg of the journey where a product lands in a consumer's hands" (Capgemini 2019).

The last mile is one of the most important phases of the online order process and is a critical touchpoint in the e-customer journey (Vakulenko et al. 2019) because that is when the only direct contact between the courier company and the recipient takes place. The courier becomes a representative of the seller.

This is often the weakest link in the supply chain—the service quality at the last stage is decided by the person who delivers the shipment on time and in the right condition, while logistics processes at other stages are automated and optimized. It is worth to add, however, that the last mile phenomenon is not only a feature of e-commerce. It also includes supply from municipal warehouses or distribution centres and transfers between branches. According to KennisDC Logistiek (2017), 80% of urban transports are currently performed by companies on their own, not by professional logistics service providers. This leads to lower efficiency in urban transport networks because of the lack of load consolidation from many senders. It creates many social costs associated with urban supplies–traffic jams, noise, pollution, etc. (Kawa 2019a).

Handing the last mile is the most difficult process in e-commerce logistics. Each parcel is first picked up by a courier from the warehouse and then delivered directly to the customer. As a result, about 100 parcels require almost the same number of repeated actions. What is important is that the courier still has to correctly identify the recipient's address and set the delivery route accordingly, so that it is optimal. In addition, the customer must be at home at that time to be able to pick it up.

That is why the last mile is related to a significant part of the costs in the CEP industry (McKinsey 2016; Orenstein et al. 2019). According to the report Urban Logistics Opportunities—Last-Mile Innovation, prepared by Frost Sullivan (2017), expenditure on logistics in the world is expected to reach USD 10.6 trillion in 2020. 70 percent of these expenses will be generated by transport, and as much as 40 percent by last mile deliveries. That is why more and more logistics service companies and e-tailers are making changes to eliminate the effects of problems arising from deliveries to individual customers. These include both relatively easy organizational improvements and more complex solutions that require investment (Kawa 2019a).

Just like companies, local authorities are trying to reduce the negative effects of vehicle traffic, particularly in city centres (Sitek and Wikarek 2019). They regulate the principles of organizing the delivery of goods and courier shipments in selected areas of the city and indicate the types of vehicles and their drives. Restrictions apply to the times of the day in which deliveries can be made, and the time of loading and unloading goods (Kawa 2019a).

The key to success in the last mile is effective logistics solutions and good communication with the recipient. For example, according to the World Economic Forum, "advanced analytics and Internet of Things-based solutions such as load-pooling and dynamic re-routing could contribute to an overall scenario that reduces emission by 10%, unit cost by 30%, and congestion by 30%" (Vakulenko et al. 2019).

3 Door-to-Door Delivery

Until a few years ago, products ordered via the Internet were delivered mainly by courier companies and postal operators. The dynamic development of e-commerce has made CEP services available to the public (mainly through courier brokers). The biggest advantages of courier services are the door-to-door system and short delivery time. Neither the sender nor the recipient of the shipment have to go out to use this service (for this reason the service is more and more frequently treated as a premium). Delivery is usually made by the next working day. The disadvantages of this solution are quite high a price of the service, lack of information about the exact delivery date and difficulties in determining the correct address, or the addressee's absence. The last problem is the biggest one. Some customers feel the stress of not being at home when a parcel is being delivered (Faugere and Montreuil 2016). That is why avoiding the need to synchronize with the courier is desirable for many recipients (Orenstein et al. 2019). This problem is solved by the customer indicating the workplace as the place of delivery. However, some employers do not allow parcels to be picked up at work due to the fact that these are private matters for employees. The greatest problem, however, is that customers have to adapt their schedule to the wide window of delivery time and wait for the parcel at home (Faugere and Montreuil 2016). In the case of postal delivery, in turn, the biggest advantage is the price of the service. A disadvantage is longer time than in the case of courier deliveries and a lower level of customer service. In addition, parcels often must be collected from the post office, and these places are often crowded by other customers.

CEP operators undertake various actions to eliminate the effects of problems arising from deliveries to individual customers (Vakulenko et al. 2019). These include both relatively easy organizational improvements and more complex solutions that require investment. One solution is to send information to the recipient about an incoming shipment (Sitek and Wikarek 2019). This allows the recipient to better prepare for the collection of the shipment, which increases the effectiveness of deliveries at the first attempt. Another solution that improves the final delivery stage is the possibility to pay for COD orders by card. An even better solution is to adjust the delivery time of shipments to the time when the recipient is present in the place of residence. Shipments to individual customers are therefore delivered in the afternoon and evening hours. On the one hand, this reduces the number of undelivered shipments, but does not eliminate them completely, as some of the addressees are not present at the indicated place of delivery at that time, either.

CEP operators also offer the possibility to manage the place and time of delivery. It is possible thanks to special IT tools designed to support interactive delivery management. The customer receives information about the incoming shipment well in advance and can either redirect it to another location or change the delivery date, or select a delivery period to a specific address. Due to the fact that couriers need to service a large number of clients (mainly business) during the day, it is an expensive solution because it requires additional staff (Kawa and Różycki 2018). Another way is to enable the customer to dynamically change the place and time of pickup. The customer receives information about the upcoming shipment in advance and redirects it to another place or delays its delivery. This increases the number of effective deliveries and reduces the number of empty runs. On the other hand, the option of changing the delivery time or adding a time window requires reorganization of the couriers' work, which increases the operating costs of the CEP operator (Kawa 2019a).

4 Out-of-Home Delivery

The solution to the delivery problems related to the door-to-door system is out-of-home (OOH) delivery. It consists in delivering a shipment to a point or machine that is in a convenient place for the customer. Such a parcel can be picked up on the customer's commuting path at convenient time, within a time frame of some days (Faugere and Montreuil 2016). It allows to consolidate last mile shipments (Orenstein et al. 2019; Sitek and Wikarek 2019), which increases delivery efficiency and can reduce delivery costs. Resources (cars, couriers) and processes (dropping off more parcels at the same location decreases the number of stops required and eliminates unsuccessful deliveries due to an absent recipient) (Faugere and Montreuil 2016) are reduced. Moreover, both the delivery costs and the average delivery time are significantly decreased (Orenstein et al. 2019).

However, out-of-home delivery involves the customer in the last mile process. Customers must do some of the work that the courier normally does, i.e. they must go to the OOH point and pick up their parcels by themselves (Orenstein et al. 2019).

OOH is most commonly found in two forms: parcel locker and PUDO (pick up drop off) point. Currently, they are some of the key trends in the last mile logistics. They are also the fastest growing service and the most frequently chosen form of parcel delivery in China, Germany, Great Britain, France, Belgium and, also, in Poland. In the further part of the paper they are described in detail.

Parcel Lockers

An automated parcel machine (APM), or just a parcel locker, is a machine that is used to drop off and pick up parcels. Most often, it consists of several boxes. Therefore, they are also referred to as secure locker banks which group reception box units (Businesswire 2019; Iwan et al. 2016). Lockers are interconnected with the hub and spokes system through the Internet of Things (Faugere and Montreuil 2016). Such devices are most often open 24/7 and are fully self-serviced, so they are also called unattended delivery (McKinsey 2016). They can be accessed via digital pickup codes, QR codes or a mobile application (Orenstein et al. 2019).

Such machines are usually located in easily accessible places and highly frequented areas, such as public premises, public transit stations, public locations in living neighbourhoods (Faugere and Montreuil 2016), e.g. at train or bus stations, gas stations (Orenstein et al. 2019). An advantage of this solution is that the customer can indicate a convenient location and pick up the shipment at any time, so it is a good solution for customers who are more mobile, flexible and want to have freedom. For this reason, they are called "born out of the frustration of failed deliveries to shoppers who aren't home" (Orenstein et al. 2019). A disadvantage is that the customer has to bother to go to such a point, and so "to make the final leg of the journey" (Iwan et al. 2016). There are also limitations in terms of size, shape and weight of shipments. In addition, there are no or relatively few facilities of this kind in rural or less populated areas. This can be an obstacle for people who are unable to collect the parcel, e.g.

because of their disabilities. There is also a problem with larger goods, such as refrigerators, washing machines, garden equipment and tires. Parcel machines restrict the weight and dimensions of a shipment. Large products are still delivered directly to the address indicated (Kawa 2019b).

Delivery services to parcel lockers are provided by both CEP operators (e.g. DHL PackStation in Germany, InPost in Poland), marketplaces (Amazon, JD.com) and independent companies (Hive Box, SwipBox). In this case, the last mile service is much easier and cheaper for the CEP operator. The courier transports a lot of parcels at one time and delivers them to one specific location (Kawa 2019b).

More and more often, parcel lockers are equipped with new functions. For example, they can be used for cash on delivery, to withdraw money (ATM function), charge a car or scooter (e.g. InPost), pick up groceries (automated food locker by Cleveron). For this reason, they are also called smart lockers terminals (Faugere and Montreuil 2016).

Parcel lockers usually have an on-line connection to the system of the network operator and update information when the status of a parcel changes (delivered, waiting for pickup, collected). Thanks to that, it is possible to track which lockers are empty or full (Businesswire 2019).

Apart from parcel lockers, there are other solutions that have a similar function. These are: a reception box—a locker which is permanently fixed to the wall outside the customer's house; a delivery box—a box of goods temporarily attached to the wall of the customer's house, which is accessible by means of a lock; controlled access systems—a solution that allows the courier to access a closed area in order to leave the goods inside, e.g. in a garage, car trunk (Iwan et al. 2016).

In addition to convenience, the benefits for companies and the environment are also important. According to the World Economic Forum, they can "reduce delivery costs by 2–12% and, at the same time, ease congestion by 5–18%" (World Economic Forum 2020).

PUDO Points

PUDO are special points where one can drop off or pick up parcels. These points are located in places that are relatively easy to access and that are regularly visited by customers. They often have long opening hours (Iwan et al. 2016) and they are near the recipient's home or office address, or in the recipient's favourite shopping mall (Orenstein et al. 2019). Special points are located in places such as grocery stores, newsagent's, traffic kiosks, shopping malls and gas stations (Kawa 2019a).

This service originated from the click & collect concept, which was initially developed by companies trading mainly in fixed locations (e.g. retail chains), but gradually moving part of their sales to the Internet. Thanks to it, customers did not have to wait for the courier and pay for the delivery (Kawa and Różycki 2018).

The points where one can pick up and send a parcel are not a new solution. Such places existed much earlier, and they were post offices. It was only a few years ago that CEP companies, which had problems with undelivered parcels to individual customers, noticed them.

PUDO eliminates problems related to determining the correct address or the recipient's absence from home. Using this type of delivery is really simple. When shopping on the Internet, it is enough to select the appropriate point in the area and pick up the shipment there later. When choosing a point, the customer pays attention to the distance from the place of residence or work, so the density and distribution of the facilities are important, too (Kawa and Różycki 2018).

Currently, there is a very strong trend towards the development of the PUDO service, which is why CEP operators continue to expand their network of points independently and through external partners. They promote it as convenient for the customers—they can pick up or send a shipment at a chosen place and time. Due to greater consolidation of shipments, the cost of delivery to the points of sending or collecting a parcel is lower than the cost of door-to-door courier services. However, this does not always mean a lower price for the customer. Some shipments are redirected after the courier cannot effectively deliver the package to the addressee, and some go directly to the PUDO points. This creates the cost of the delivery to the points and the cost of handling shipments at those points. Yet, because of the growing costs of the last mile (mainly due to rising labour costs), the differences between the prices of direct delivery and that to the points will keep increasing.

A drawback of PUDO is the limitation of the service availability through the opening hours of the points, which has become an impulse to automate the service on the basis of the previously described self-service terminals offering 24-h access. PUDO is characterized by, similarly to parcel lockers, limitations on size, weight, etc. of the shipments to be accepted.

Most points are available through only one operator. This means that the customer cannot pick up all products in one place. For this reason, agnostic networks that include multi-brand services are needed. In such points customers can pick up and drop off parcels from different CEP operators. However, this requires cooperation between competitors (World Economic Forum 2020).

The pick up and drop off points are also a very effective solution for the CEP operators and online retailers, as they involve fewer questions from recipients who automatically receive information about the status of their shipment. They also reduce the number of undelivered shipments due to the recipient's absence. PUDO points can be successfully used for returns. Recipients can easily and cheaply, and sometimes even free of charge (depending on the seller's offer), send back the product (Kawa and Różycki 2018). As the OOH network develops, CEP operators expand their services by opening special points. For example, the DPD Pickup's and GLS's offer includes many additional services, including shipment insurance, international shipping and cash on delivery. Customers who buy clothes or shoes online can try their order on in the dressing room and return it immediately, if necessary.

5 Out-of-Home Delivery and Customer Satisfaction

A network of PUDO points and parcel lockers is currently an important element in the development of CEP operators and e-commerce companies. This is related to the growing challenges for city logistics and the requirements of city residents. On the one hand, this solution reduces the volume of courier traffic in cities, and on the other hand, it gives the customer the freedom to decide how and when to collect his/her order sent from the online store. The operators' priority is to give their customers the greatest convenience and flexibility possible in using courier services and the possibility to choose between deliveries: to their own hands, to their home, workplace, parcel locker or PUDO point. This makes the customer shopping online more satisfied (Faugere and Montreuil 2016), and causes a feeling that the e-tailer understands their needs, which later leads to recommendation to buy at a specific shop.

On the basis of the above considerations, the following hypothesis has been formulated:

H1: The out-of-home delivery has a positive impact on customer satisfaction in e-commerce.

Numerous studies show that customer satisfaction leads to increased customer loyalty (Cyr 2008). If a customer is satisfied with their online shopping, it is very likely that they will buy this product again or choose the same seller the next time. This observation leads to another research hypothesis:

H2: Customer satisfaction has a positive impact on loyalty in e-commerce.

The out-of-home delivery can directly affect the loyalty variable. Therefore, a research hypothesis has been put forward which is as follows:

H3: The out-of-home delivery has a positive impact on customer loyalty in e-commerce.

6 Methodology of Research

Research Stages

In this study, a quantitative method was used to investigate the dependencies between out-of-home deliveries and customer satisfaction and loyalty. It was preceded by an in-depth literature analysis, presented in the previous parts of this paper, which was used to prepare a qualitative method for the research. Using this method was aimed at preliminary analysis of the problem of the last mile in e-commerce and providing information necessary for proper organization of the research with the quantitative method, including first of all designing the measuring instrument (questionnaire). A focus group interview (FGI) was chosen as the research technique. The selection of respondents for the study was purposeful. The results of these interviews (Kawa et al. 2019) have become the basis for the preparation of the next step of research—study of online retailers with the use of quantitative research. A tool in the form of a questionnaire was created.

Data Gathering

In the third stage of the research, both computer-assisted web interviews (CAWI) and computer-assisted telephone interviews (CATI) were applied. The investigation was carried out between November 2017 and May 2018 by a research agency. The database of polish online retailers was the sample of the study. Approx. 6 thousand representatives of online retailers were invited to take part in the investigation with usage of non-random purposeful sampling. It was 20% of the total population of e-tailers in Poland. A restrictive condition, as in the FGI, was that each participant in the interview had to conduct selling activity on the Internet for at least one year as a necessary condition. A return ratio was 10%. We received 592 correctly filled questionnaires—392 interviews from CAWI and 200 from CATI. This sample is sufficient to generalise our results for the entire population of the polish e-tailers. An acceptable margin of error is 3.99% with assumption that the confidence level is 95% and the response distribution is 50%.

Measures

Three constructs (latent variables) were distinguished in this research: out-of-home delivery, customer satisfaction and loyalty. Based on an in-depth literature analysis and results of focus group interview, the items of these variables have been developed. They have been included in the questionnaire in the form of statements. The respondents evaluated these statements with the use of a five-point Likert scale. In the case of the loyalty measure, the respondents were asked to compare their parameters with those of other online retailers.

Out-of-home delivery was measured by the possibilities of picking up goods ordered via the Internet (excluding door-to-door deliveries). Customers can buy products from online sellers who offer deliveries to PUDO points, self-service terminals (e.g. parcel lockers), as well as the click & collect solution. Satisfaction was related to the customer's happiness with their purchases, their feeling that the seller understood their needs and that they would recommend purchasing from the same seller to their family or friends. Loyalty, in turn, referred to buying again from the same seller in the near future, even if the conditions changed, i.e. the products, their delivery and payments offered by other vendors would be more competitive (Cyr 2008).

With the results of the interviews, the exploratory factor analysis was used in order to find the indicators with the highest loading values. The validity and reliability analysis by the Cronbach's α method has been used for this purpose. All Cronbach's α were above 0.7, indicating satisfactory internal consistency of variables (see Table 1).

Table 1 Constructs, items and scales of the out-of-home delivery, customer satisfaction and loyalty variables

Out-of-home (OOH) delivery Cronbach's alpha =0.74 • Customers buy from online sellers who offer deliveries to PUDO (pick up drop off) points (e.g. a traffic kiosk. gas station) • Customers buy from online sellers who offer deliveries to self-service terminals (e.g. parcel locker) • Customers buy from online sellers who offer click & collect solution
Customer satisfaction Cronbach's alpha =0.78 • Customers are satisfied with their purchases Customers will buy again at my shop in the near future • Customers feel that we understand their needs • Customers will recommend buying at my shop to their nearest and dearest
Customer loyalty Cronbach's alpha =0.81 • Customers will continue to buy with them. even if the products offered by other online retailers are more competitive • Customers will continue to buy with them. even if the delivery of products offered by other online retailers will be more competitive • Customers will continue to buy with them. even if payments for products offered by other online retailers are more competitive

Source own elaboration

7 Analysis and Results

We analysed the data using one of the statistical measures—Pearson's correlation coefficient, which is particularly appropriate for testing hypotheses. In this way, it was possible to confirm that the correlations were not accidental and allowed to generalise the results of the conducted tests with a sufficiently high degree of probability (Kawa 2019b). The empirical models include out-of-home delivery (OOH), satisfaction, and loyalty variables, and the interaction between them. Table 2 reports means, standard deviation, and Pearson's correlation coefficients for all variables used in this study.

Our empirical testing has shown that all hypotheses have been supported (see Fig. 1). The out-of-home delivery has a positive influence on the satisfaction and the loyalty variables. These correlations have been statistically significant but not too

Table 2 Descriptive statistics

Variable	Mean	S.D	1	2	3
1. OOH delivery	3.62	1.11		0.17*	0.134*
2. Satisfaction	4.01	0.92			0.444*
3. Loyalty	3.27	1.13			

**$p < 0.01$; $N = 592$
Source own elaboration

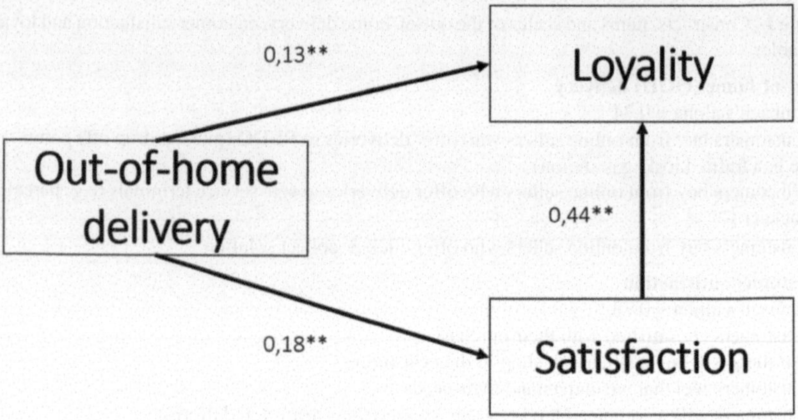

Fig. 1 Confirmed correlations between the out-of-home delivery, customer satisfaction and loyalty variables **p < .01. *Source* own elaboration

strong. The strongest statistically significant relationship has been found between the satisfaction and the loyalty variables. The research conducted has confirmed that delivering the right value to the customers increases their satisfaction, which in turn translates into customer loyalty and, further, into repurchasing (Chiou and pan 2009; Chiu et al. 2009; Vakulenko et al. 2019). So, deliveries in e-commerce are a very important component of value for the customer. It means that the more attention a customer pays to out-of-home delivery, "the more satisfaction they get from shopping and the more loyal they are, and thus the more they spend on shopping" (Kawa 2019b). Consequently, the sellers of goods who take alternative deliveries seriously can expect better performance than their competitors.

It is important that the out-of-home solutions are created by entities supporting e-commerce and providing complementary services, such as logistics, or parcel service. This means that all e-commerce stakeholders should pay special attention to the issue of alternative ways of deliveries, because creating value from them for the customer will bring them benefits.

The results and conclusions are very important from a scientific perspective, because the impact of the out-of-home delivery on the customer satisfaction and loyalty, in particular in e-commerce, is rarely mentioned in the literature (Kawa 2019b).

Out-of-home deliveries are not only PUDOs in the form of traditional points and parcel lockers, but include many other solutions that various e-commerce stakeholders are working on. An interesting idea is personal parcel boxes, which can be located at the place of residence. They fulfil a function similar to traditional mailboxes, but have greater capacity. Thanks to them, the courier or postman can leave a parcel in the box, even if the customer is not at home. The customer can also send a parcel. The box combines the advantages of courier deliveries (delivery to the door) with those of automatic parcel lockers (possibility to pick up the package at any time). The parcel box can also handle returns. For the time being, such solutions are mainly in the phase of testing or initial development, but in a few years' time they may become widespread, especially in the case of detached and terraced houses.

Another way to deal with the last mile problem is to use cars as mobile PUDOs. When ordering products, the customer indicates his/her vehicle as the place of delivery. The courier uses a disposable digital key to open the trunk and place the package in it. Then, information about the delivery is sent to the customer. Such solutions are tested or applied by retailers (Amazon), car manufacturers (Volvo, BMW, Audi) and CEP operators (DHL). Similarly, a parcel can be left in a specific part of the house (e.g. garage).

8 Development Directions

The question about the direction in which alternative means of delivering products ordered over the Internet will develop is very difficult to answer unequivocally. Until a few years ago, PUDO was used marginally and parcel lockers were at an early stage of development. Today, it is difficult to imagine trading over the Internet without them. However, new disruptive solutions may appear which will replace the previous ones.

The role of independent integrators is likely to increase in the near future. They will consolidate the services of different entities in one place according to the one stop shopping concept. In addition to the CEP operators, whose domain will be transporting parcels from point A to point B, entities that will offer competence in sending and receiving parcels will be very important. This is not only about the physical points, but also about technology companies that will provide effective solutions for the integration of many partners and the flow of information between them. Co-opetitive activities, i.e. cooperation between competitors, may also develop in order to reduce costs and compete together for e-customers.

A number of solutions based on automation, autonomy and robotisation (drones, robots, autonomous vehicle deliveries with parcel lockers, etc.) (McKinsey 2016) are currently being tested which are very important in the context of the last mile challenges, but they mainly concern door-to-door deliveries. Undoubtedly, such solutions automate the process of parcel delivery to a customer, but it is difficult to imagine that they will be used on a massive scale. Every day, millions of parcels are delivered and all cannot be transported by drones or robots. So they are used for urgent deliveries, in hard-to-reach places and where there is a low density of recipients. In addition, in many countries there are legal restrictions, e.g. flight at a certain height under human supervision (Kawa 2019a). These solutions are also widely described in the literature, e.g. (Vakulenko et al. 2019), just as using the concept of sharing economics, in particular crowdsourcing. The society and its resources are involved in the delivery of consignment services (Kawa 2019a). However, these are solutions to the problems of door-to-door deliveries, which are not covered by this article.

9 Conclusions

The Internet has not only revolutionised trade but has also had a major impact on logistics. It is now one of the key factors in the development of logistics companies, in particular CEP operators. More and more people buy things online, and that translates into the number of shipments that have to be delivered to the final recipient. The issue of the last mile, which appears at the last stage of the logistics process, is becoming particularly important. Most often, shipments are delivered by door-to-door couriers. As it was shown in the article, this type of delivery is not effective. Hence, out-of-home deliveries are very helpful.

Deliveries to PUDOs and parcel lockers are characterized by flexibility of the delivery place and date. When e-shopping, this is undoubtedly an additional argument for customers who are more mobile and want to be free to choose the place and time of delivery. Unfortunately, this can be an obstacle for those who need to go a long way to such a point, e.g. living in rural, less populated areas or not being able to pick up the parcel. For some customers, the door-to-door courier service will still remain most convenient and reliable. There is also a problem with larger goods such as refrigerators, washing machines, garden equipment, and tyres. PUDOs and parcel lockers obviously limit the weight and dimensions of a potential shipment. Large products are still delivered directly to the indicated address.

For the purpose of this study, interviews were conducted with 592 online retailers. The hypothesis that the out-of-home delivery has a positive impact on customer satisfaction in e-commerce has been confirmed. The customer satisfaction has a positive impact on loyalty in e-commerce. There is also a direct relationship between the out-of-home delivery and loyalty. It leads to the managerial implication that greater attention to out-of-home delivery in e-commerce resulted in greater customer satisfaction, and, in turn, in influence on customer loyalty. Managers representing online sellers should offer alternative forms of delivery options (not only courier services). It leads to a greater sense of control by customers choosing the best delivery option for them and to good experiences in the whole process of online shopping (World Economic Forum 2020).

The presented research has both methodical and substantive limitations. The approach developed for the needs of the paper aimed to identify universal dependencies between delivery methods, customer satisfaction and loyalty. However, such an intention may result in disregarding other aspects of the impact on customer satisfaction and loyalty, such as delivery monitoring, time of delivery. Inclusion thereof may be a future direction of research, leading to an increase in the substantive value of the approach. The limitation of the methodological nature of this study is the essence of the approach itself, which simplifies the economic reality and thus reduces the complexity of the actual state.

The development of the PUDO points and parcel lockers is undoubtedly one of the most important trends in the CEP industry. It can be assumed that their popularity will increase with the development of e-commerce; new solutions will emerge, and they will slowly displace courier services, which will become premium services over time. Surcharges for delivery of parcels to the customer's door are more and more often discussed. This trend is worth observing. Therefore, future work can embrace a rerun of this study with the addition of new forms of out-of-home deliveries.

Acknowledgements This paper has been written with financial support of the National Centre of Science [Narodowe Centrum Nauki]—grant number DEC-2015/19/B/HS4/02287.

References

Businesswire (2019). Global parcel locker systems market insight Report 2018. https://www.businesswire.com/news/home/20190718005526/en/Global-Parcel-Locker-Systems-Market-Insight-Report

Capgemini (2019). The last-mile delivery challenge. https://www.capgemini.com/wp-content/uploads/2019/01/Report-Digital-%E2%80%93-Last-Mile-Delivery-Challenge1.pdf

Chiou J-S, Pan L-Y (2009) Antecedents of internet retailing loyalty: differences between heavy versus light shoppers. J Bus Psychol 24:327–339

Chiu C-M, Lin H-Y, Sun S-Y, Hsu M-H (2009) Understanding customers' loyalty intentions towards online shopping: an integration of technology acceptance model and fairness theory. Behav Inf Technol 28(4):347–360

Cyr D (2008) Modeling web site design across cultures: relationships to trust, satisfaction, and e-loyalty. J Manag Inf Syst 24(4):47–72

eMarketer, Global Ecommerce (2019) Ecommerce continues strong gains amid global economic uncertainty. https://www.emarketer.com/content/global-ecommerce-2019

Faugere L, Montreuil B (2016) Hyperconnected city logistics: smart lockers terminals & last mile delivery networks. In: Proceedings of the 3rd international physical internet conference, Atlanta, GA, vol 29. USA

Frost, Sullivan (2017) Urban logistics opportunities—last-mile innovation. https://store.frost.com/urban-logistics-opportunities-last-mile-innovation.html

Iwan S, Kijewska K, Lemke J (2016) Analysis of parcel lockers' efficiency as the last mile delivery solution–the results of the research in Poland. Transp Res Procedia 12:644–655

Kawa A (2019a) Last mile logistics. in technological revolution. directions in the development of the transport-forwarding-logistics sector. https://pitd.org.pl/wp-content/uploads/2020/01/PITD-Report-technological-revolution-directions-in-the-development-of-the-transport-forwarding-logistics-sector.pdf, pp 76–79

Kawa A (2019b) Returns in e-commerce as a value for customers from different perspectives. Bus. Logistics Mod. Manag. **19** 43–58

Kawa A, Różycki M (2018) PUDO, czyli jak ułatwić klientom nadawanie i odbieranie przesyłek? Magazyn e-Commerce, 2

Kawa A, Pierański B, Zdrenka W (2019) Wartość dla klienta z perspektywy sprzedawców internetowych–wyniki badań z wykorzystaniem FGI. Gospodarka Materiałowa I Logistyka 3:33–40

KennisDC Logistiek (2017) The future of last mile delivery: 10 most important trends https://www.kennisdclogistiek.nl/nieuws/the-future-of-last-mile-delivery-10-most-important-trends

McKinsey (2016) Parcel delivery. The future of last mile. https://www.mckinsey.com/~/media/mckinsey/industries/travel%20transport%20and%20logistics/our%20insights/how%20customer%20demands%20are%20reshaping%20last%20mile%20delivery/parcel_delivery_the_future_of_last_mile.ashx.

Orenstein I, Raviv T, Sadan E (2019) Flexible parcel delivery to automated parcel lockers: models, solution methods and analysis. EURO J Transp Logistics 8(5):683–711

Sitek P, Wikarek J (2019) Capacitated vehicle routing problem with pick-up and alternative delivery (CVRPPAD): model and implementation using hybrid approach. Ann Oper Res 273(1–2):257–277

Vakulenko Y, Shams P, Hellström D, Hjort K (2019) Service innovation in e-commerce last mile delivery: mapping the e-customer journey. J Bus Res 101:461–468

World Economic Forum (2020), The future of the last-mile ecosystem. https://www3.weforum.org/docs/WEF_Future_of_the_last_mile_ecosystem.pdf

Evaluation and Control of a Collaborative Automated Picking System

Mathias Rieder and Richard Verbeet

Abstract Picking is a core process of logistics. The challenge of acquiring personnel for operations and handling steadily changing product ranges can be tackled by partwise automated picking systems to create a cooperative working environment for human pickers and picking robots. This chapter is motivated to enable a stepwise transformation from manual picking to highly automated picking processes by cooperative and learning robots. The main goal is to guarantee reliable order fulfilment by implementation of a feedback-loop between humans and robots for error handling and to gather data for machine learning algorithms to increase the performance of object detection. In this chapter a concept for measurement and evaluation of system performance is introduced ensuring successful processing of picking orders and training of picking robots to improve their ability for object detection. It is based on the amount of picking orders, the picking capacity of humans and robots, and the probability for successful automated order picking considering the training effort during system design. The proposed concept can be used for overall capacity planning as well as for operational control of picking processes.

1 Introduction

Modern supply chains are challenged by an increasing complexity and short product life cycles. Therefore, picking as central logistic process during order fulfilment must adapt to changing product ranges. Another rising challenge is the lack of personnel for manual picking processes. Therefore, automated picking systems handling steadily a changing product range become more and more important. In recent years, using technical progresses in robotics general concepts for automated picking are developed (Zou et al. 2019; Krug et al. 2016) or applied for specific use cases (Mester and Wahl

M. Rieder (✉) · R. Verbeet
Ulm University of Applied Sciences, Prittwitzstr. 10, 89075 Ulm, Germany
e-mail: mathias.rieder@thu.de

R. Verbeet
e-mail: richard.verbeet@thu.de

© The Author(s), under exclusive license to Springer Nature Switzerland AG 2020
P. Golinska-Dawson et al. (eds.), *Smart and Sustainable Supply Chain and Logistics – Trends, Challenges, Methods and Best Practices*, EcoProduction,
https://doi.org/10.1007/978-3-030-61947-3_3

2019). Thereby, the general concepts focus on automation of movement or manipulation of objects and not on an integrated handling of heterogenous and dynamic article ranges. Verbeet et al. (2019) propose a cooperative picking system to guarantee reliable automated picking by robots realized by a feedback-loop improving robots' ability for object detection by human support.

Classic methods for capacity planning and performance evaluation to control and design must be extended for partwise automated systems. The adaptive process model proposed by Verbeet et al. (2019) for a proposed cooperative picking system considers humans to support robots in addition to their normal workload to enable the learning process. In this system, not successful object detection during picking is the start of a learning process and therefore desired if overall order fulfilment is still guaranteed. The system is to be designed in such a way that all picking orders are completed, and the capacity of humans and robots is utilized. Furthermore, picking robots should reach their performance limits to trigger the proposed cooperative learning process to increase the picking performance of the whole system in the long term.

This chapter introduces a concept to calculate such an equilibrium of capacity for a partial automated picking system based on the adaptive process model. The calculation can be used to carry out a general capacity assessment based on available capacities and a pool of picking orders while allocating these picking orders to human pickers and robots by mathematical optimization. The overall system efficiency of picking robots' ability for object detection can be evaluated by an average probability for a successful object detection. For this, the threshold $\overline{P}_{Break}$ is calculated defining an average value for successful object detection of a single article.

The remainder of this chapter is organized as follows. In the second chapter, related work from the fields of existing robotic picking systems as well as planning and evaluation of order picking systems is discussed. This review shows the lack of a concept considering capacity during evaluation and design of cooperative partwise automated picking systems. The adaptive process model for an automated cooperative picking system is described in the third chapter. The fourth chapter proposes evaluation concepts for overall capacity planning, a preselection for order assignment and a calculation approach of a capacity-based working point. These concepts are discussed in fifth chapter and in the final chapter this paper is closed by a conclusion containing a brief summary and further research.

2 Related Work

Grosse et al. (2017) mention the following planning problems for picking systems: layout design (structure and dimension of shelfs), storage assignment (allocation of items to storage positions), zoning (assignment of working area to pickers), order batching (consolidation or splitting of picking orders), routing (sequence of picking positions and routing), and technical equipment (supporting equipment). In this paper, the problem of order assignment is added.

Many research activities focus on batching, sequencing and routing in order picking systems with multiple pickers (Scholz et al. 2017) and Chabot (2018) shows order picking is influenced most by facility layout, storage assignment policy and routing strategy. Jane and Laih (2005) improve utilization and completion time of a synchronized zone manual order picking system by a heuristic solving a natural cluster model. Yu and de Koster (2009) define order batching and zoning of the picking area using a queuing network approximation model. Bukchin et al. (2012) batch orders in a dynamic finite-horizon environment to minimize order tardiness and overtime costs of pickers using a Markov decision process-based approach to set an optimal decision-making policy. Lin et al. (2016) use Particle Swarm Optimization to solve the problems of order batching and picker routing. Zhang et al. (2017) present an approach to solve the on-line order batching and sequencing problem with multiple (manual) pickers using a hybrid rule-based algorithm minimizing turnover time (completion time of an order). In addition, they present a review about previous work about order batching. Pinto and Nagano (2020) solve the Optimized Billing Sequencing (order fulfilment) and Optimized Picking Sequence (batching, route planning) problems by combination of two genetic algorithms. Valle and Beasley (2019) discuss approaches using queueing theory, simulation, mathematical optimization and heuristics for system analysis, design optimization, and operations planning and control.

Henn (2015) is considering order assignment. The task of order assignment is strongly connected to workload balancing. In fast picking environments demand cannot be taken as known resulting in the requirement of shorter execution times of picking orders and a dynamic workload balancing (de Koster et al. 2007). Vanheusden et al. (2017) show a necessity to balance workload within a picking system not only in long-term range but also within a day or during a shift due to the steady rising requirement of flexibility. A reliable forecast is necessary to balance workload. van Gils et al. (2017) provide an overview of various time series forecasting models for predicting the workload within a picking system and indicators for measurement of accuracy of forecasting results. In van Gils (2019) different planning problems considering various real-life features to match demand and resource allocation are combined. Tu et al. (2019) focus on workload balancing within an order picking system by storage assignment. Merschformann et al. (2018) show that order assignment has the major impact on throughput of a picking system using a robotic transport system. Chen et al. (2017) combine different strategies of order sequencing, order release and storage assignment to balance workload and capacity.

Molnár (2004) suggests an integrated concept for planning a picking system by a genetic algorithm solving a constraint programming model followed by a simulation to estimate the number of pickers and picking schedule considering time constraints while minimizing total costs. Hwang and Cho (2006) plan a warehouse by minimizing costs considering throughput and storage space with a concept to measure travel time of transporters for manual picking and using a simulation model to define the necessary number of transporters. Seyedrezaei et al. (2012) present a dynamic mathematical model for the order picking planning problem maximizing order fulfilment considering product life, customer importance, probabilistic demand, and backorder strategy. Kłodawski and Jachimowski (2013) propose a concept for using an ant algorithm for planning a picking system considering various parameters but doesn't provide a definition for an evaluation function.

A mechanism to evaluate a picking system is the basis for planning and operational decisions. A qualitative approach for evaluation of a picking system is a Balanced Scorecard (Heine and Wenzel 2013). In contrast, VDI describes more than 350 KPIs for a quantitative evaluation of logistics processes (VDI 2007). Many quantitative concepts evaluate a picking system's performance by order fulfilment. Chabot (2018) uses order lead time for evaluation. Gong et al. (2010) define a framework to evaluate different storage and order picking policies by a DEA model considering total costs and service level. Brynzér et al. (1994) present an evaluation methodology using zero-based analysing of manual picking processes. Dallari et al. (2009) describe a design methodology for picking systems measuring performance by response time, picking rate and number of pickers. Pan and Wu (2012) evaluate the efficiency of a multi-picker system by estimation of the number of picking items per time to avoid inaccuracies during measurement of travel distance or travel time due to congestion. Yu and de Koster (2009) use mean throughput time of an arbitrary order as measurement of efficiency. Lamballais et al. (2017) evaluate the performance of a Robotic Mobile Fulfilment System (RMFS) that realizes a parts-to-picker environment by measuring maximum order throughput, robot utilization, and order cycle time. Hwang and Cho (2006) evaluate a system by transportation time of transporters for manual picking.

The mathematical concepts mentioned so far mainly minimize used time and travelled distance. Grosse et al. (2017) point out time to be still the most important indicator to evaluate the outcome of an order picking system. A review presented by Gu et al. (2010) shows amongst others the evaluation of performance by analytic models considering travel time or service time. Jane and Laih (2005) measure the improvement of completion time, Bukchin et al. (2012) use a measurement by slack, i.e. comparison of an order's picking time and its remaining time to supply. Zou et al. (2019) minimize the total time needed to pick items of an order. Manzini et al. (2007) evaluate performance of order picking by travel distance. Hsieh and Huang (2011) show how strategies of storage assignment, order batching and picker routing affect the overall performance also measured by travel distance Lin et al. (2016). measure the total picker routing distance. In Hernandez et al. (2017) the evaluated metrics are travel distance and travel time and Pinto and Nagano (2020) also combine these metrics by maximizing order portfolio billing and minimizing total picking time and travel distance. In Seyedrezaei et al. (2012) the degree of order fulfilment is maximized.

Hanson et al. (2018) and Jaghbeer (2019) mention the categories throughput, order lead time, availability, flexibility, quality, training time, resource utilization, costs, and ergonomics to evaluate performance of robotic picking systems. Jaghbeer (2019) states no studies using these categories for robot-to-parts picking systems exist. Even considering further technical and conceptual progress in automated picking, robots will depend on humans in order picking systems. Therefore, an efficient setup of an operational human–robot picking system needs a reliable human–machine-interaction (Azadeh et al. 2017). Bonini et al. (2019) propose a method to distribute various tasks among humans and robots within a warehouse to use synergies in human–machine-interaction. Hoffman (2019) describes the successful coordination of humans and robots as robot collaborative fluency measured by specific metrics for idle time for humans and robots each, concurrent activity, functional delay, and interaction between objectives. Within RMFS a lot of research about cooperation of transport robots delivering shelfs with articles to pickers located at static picking stations exists (Zou et al. 2019; Valle & Beasley 2019; Hanson et al. 2018). Implications for humans within a cooperative human–robot picking environment are discussed by Lee et al. (2017). An overview of different types of co-working (cell, coexistence, synchronized, cooperation, collaboration) can be found at Bender et al. (2016). RMFS can be implemented with approaches for navigation in a warehouse. Some research on navigation can be found in Nguyen et al. (2016) or Hernandez et al. (2017). Magazino realizes a picking robot capable of travelling to shelves, picking specific articles (shoe boxes) and delivering them to a transfer station (Mester and Wahl 2019). Within this system robots and humans work in parallel within a joint area. Bormann et al. (2019) show a buckling arm robot mounted on a mobile platform detecting objects by a camera system. They state the need for an adequate amount of training samples to enable a reliable object detection. The collection of these samples is automated by an object recording station collecting colored 3d point clouds. Furthermore, different systems for bin picking (Martinez et al. 2015) or shelf picking (Liang et al. 2015; Zhang et al. 2016; Zhu et al. 2016; Wahrmann et al. 2019) are proposed. Gripping of complex formed articles is discussed by Liu et al. (2019) and Kozai and Hashimoto (2018) calculate the risk for collision in case of different objects in a picking scene. Verbeet et al. (2019) describe a cooperative human–robot picking system using an integrated feedback-loop to improve the ability of object detection of robots. The following section explains this concept in detail.

3 Cooperative Picking System

Rieder and Verbeet (2019) present an adaptive process model to realize a cooperative picking system containing an Application-Phase and a Learning-Phase. This model was extended by Verbeet et al. (2019) by an Adjustment-Phase and a Cooperation-Phase as well as by a conceptual picking system describing its components and their interactions. The model is shown in Fig. 1. Within the Learning-Phase models for object detection are created and improved using image data recorded in a controlled environment as well as data from operational processes. This phase is decoupled from operational order picking within the Application-Phase where humans and robots work in parallel within a picking environment. A picking robot is supposed to successfully grip and withdraw from a storage location after a successful object detection. In case of an unsuccessful object detection it tries to find a solution on its

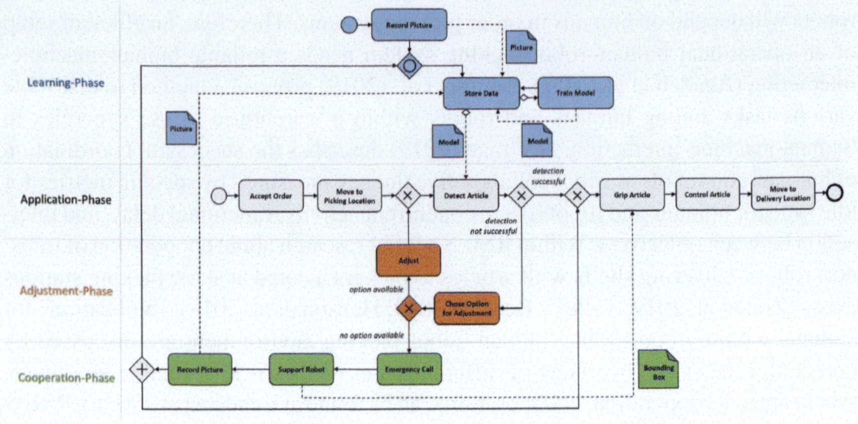

Fig. 1 Adaptive process model for picking-robots (Verbeet et al. 2019)

own by predefined options during the Adjustment-Phase, e.g. by moving its camera to a different position. If this is not successful, the Cooperation-Phase is triggered calling a human picker (Emergency Call) to support the robot by picking the article and generating feedback for an improvement of object detection. This feedback contains image data gathered from the operative situation at the shelf by the robot's camera and information added by the human picker (article-ID, position of article defined by a bounding box) to enable training.

The process model can be realized using an agent-based system architecture, whose components are shown in Fig. 2. A Warehouse-Managemen-System (WMS) is responsible for administration of inventory data and allocation of picking orders. Human pickers and picking robots cooperatively process assigned orders, whereby human pickers are interacting with IT systems and picking robots by wearables (Kong et al. 2019). Furthermore, a Picture Recording Machine is used for efficient and controlled image recording (Rieder and Verbeet 2019). These images are stored on a data server and are used for training of models for object detection by a computation cluster. Communication is realized by MQTT enabling topic controlled publishing and receiving of FIPA-conform Agent Communication Language (ACL) messages. Interaction patterns define the sequence of messages between components and embed it into the picking processes.

Each article can be successfully detected by a picking robot with a probability of P_{OD}. An average probability is introduced to evaluate picking robots' performance for successful object detection. The working points P_{Break} ($Effort_{Robot} = Benefit_{Robot}$), P_{Human} ($Error_{Robot} = Error_{Human}$), and $P_{Improve}$ ($Epoch\text{-}\Delta = \delta_{Limit}$) are defined. This probability describes the efficiency of object detection but not the overall performance of the system. Regarding the overall system performance, the capacity of human pickers and picking robots and the effort for Emergency Calls resulting from unsuccessful object detection must be considered. The assignment of orders to human pickers and picking robots during the interaction pattern "Picking Order" is of major

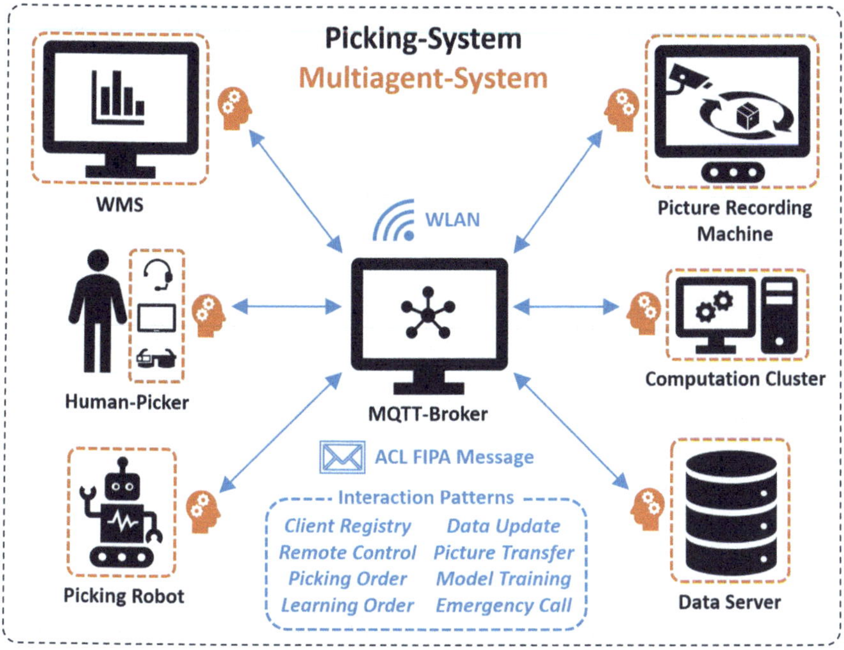

Fig. 2 Picking system to realize the adaptive process model according to Verbeet et al. (2019)

importance within this capacitive evaluation because it enables workload balancing. It is realized by a one-stage auction process arranged according to Contract Net Protocol (FIPA 2001) allocation orders depending on the effort for order fulfilment measured by time and using current workload, order lead time and duration of probably arising Emergency Calls.

The adaptive process model must reserve capacity to allow the feedback-loop to improve object detection. Therefore, an equilibrium between order fulfilment and improvement must be found, i.e. a working point must be defined at which fulfilment of all picking orders is guaranteed but robots are free to cause Emergency Calls to gather operational data. In the following chapter, a calculation for such an equilibrium is presented to setup picking capacity and control order assignment ensuring order fulfilment while maximizing robots' workload to trigger Emergency Calls.

4 System Evaluation

The general approach is a capacitive evaluation of the performance of a picking system. It can be used for proactive capacity planning or operational control of order assignment. Moreover, an approach for an overall performance evaluation of picking robots' ability for object detection is proposed according to the classification of

Verbeet et al. (2019) using a minimal probability of object detection $_{\text{Break}}$ to define an efficient working point for picking robots.

4.1 Capacity Planning of Picking System

According to the adaptive process model picking orders can be fulfilled by humans and robots in parallel. The overall picking performance of a system is reduced by the robots' dynamic learning process respectively by Emergency Calls. The following calculation enables the evaluation if a specific order assignment, i.e. an allocation of picking orders to human pickers and picking robots, allows the execution of all orders with the existing capacity. Basis for this calculation is a demand forecast for a time interval, e.g. one shift, containing articles and their number of picks, whereby this forecast is divided into two subsets for humans and robots each:

$$D_F = \sum_{\text{Article, Forecast}} \text{Picks}_{\text{Article, Forecast}} \tag{1}$$

$$\begin{aligned} D_F &= \text{Subset}_{\text{Robot}} + \text{Subset}_{\text{Human}} \\ &= \sum_{\text{Article, Robot}} \text{Picks}_{\text{Article, Robot}} + \sum_{\text{Article, Human}} \text{Picks}_{\text{Article, Human}} \end{aligned} \tag{2}$$

During the processing of orders from $\text{Subset}_{\text{Robot}}$, unsuccessful object detection can trigger an Emergency Call leading to a time effort for human ($L_{EC,H}$) and robot ($L_{EC,R}$). This effort is assumed to be constant for a picking system and can be evaluated from empirical data (Verbeet et al. 2019). The expected total effort for human pickers and picking robot is the sum of the effort for a single Emergency Call multiplied with the probability for an unsuccessful object detection weighted by the number of picks from $\text{Subset}_{\text{Robot}}$:

$$L_{EC,H,SR} = \sum_{\text{Article, Robot}} \left(\text{Picks}_{\text{Article, Robot}} \cdot L_{EC,H} \cdot \left(1 - P_{\text{OD, Article}}\right)\right) \tag{3}$$

$$L_{EC,R,SR} = \sum_{\text{Article, Robot}} \left(\text{Picks}_{\text{Article,Robot}} \cdot L_{EC,R} \cdot \left(1 - P_{\text{OD, Article}}\right)\right) \tag{4}$$

A picking capacity for human pickers (C_H) and picking robots (C_R) is calculated by multiplying an individual picking rate (picks per time unit) depending on warehouse organisation and picking environment with the number of humans respectively robots:

$$C_H = PickingRate_{SingleHuman} \cdot Number_{Human} \tag{5}$$

$$C_R = PickingRate_{SingleRobot} \cdot Number_{Robot} \tag{6}$$

In addition, an effective working time (WT) without breaks, charging or technical down time within the time interval of the forecast is defined. The two subsets must be defined in such a way that they can be fulfilled with the existing picking capacity:

$$C_H \cdot WT - \frac{L_{EC,H,SR}}{WT} \geq Subset_{Human} \tag{7}$$

$$C_R \cdot WT - \frac{L_{EC,R,SR}}{WT} \geq Subset_{Robot} \tag{8}$$

To enable a picking system to fulfil all picking orders the effective picking capacity of all humans and robots reduced by the capacity to handle Emergency Calls must be greater than or equal to the demand forecast. Therefore, the following equilibrium is defined:

$$C_H \cdot WT - \frac{L_{EC,H,SR}}{WT} + C_R \cdot WT - \frac{L_{EC,R,SR}}{WT} \geq D_F \tag{9}$$

The linear optimization model from Fig. 3 is based on the Eqs. (1)–(9) and calculates the subsets. The goal is to maximize the workload of robots and humans subject

```
execute { cplex.epgap = 0.0001; }

//********************** Variables ****************************************
int WT = ...;          float C_H = ...;            int numArticle = ...;
int L_EC_H = ...;      float C_R = ...;            range Articles = 1..numArticle;
int L_EC_R = ...;      int D_F[Articles] = ...;    float POD[Articles] = ...;

//********************** Decision Variables *******************************
dvar int SubsetRobot;    dvar float L_EC_H_SR;    dvar int+ PicksRobot[Articles];
dvar int SubsetHuman;    dvar float L_EC_R_SR;    dvar int+ PicksHuman[Articles];

//********************** Goal Function ************************************
maximize SubsetRobot;

//********************** Constraints **************************************
subject to {
    SubsetRobot == sum(i in Articles)(PicksRobot[i]);
    SubsetHuman == sum(i in Articles)(PicksHuman[i]);
    forall(i in Articles) D_F[i] == PicksRobot[i] + PicksHuman[i];
    L_EC_H_SR == sum(i in Articles)(PicksRobot[i] * L_EC_H * (1 - POD[i]));
    L_EC_R_SR == sum(i in Articles)(PicksRobot[i] * L_EC_R * (1 - POD[i]));
    SubsetRobot <= (C_R * WT) - (L_EC_R_SR / WT);
    SubsetHuman <= (C_H * WT) - (L_EC_H_SR / WT);
};
```

Fig. 3 Calculation of subsets by linear programming in OPL to capacity restrictions and total order fulfilment. Therefore, robots are assigned as many picking orders as possible to learn from resulting Emergency Calls. The model is programmed in OPL using OPL-Studio 12.8.0.0 (IBM 2020).

The picking system is modelled by the constant input parameters working time WT, theoretical picking capacity (C_R and C_H) and a set of articles each possessing a probability for successful object detection P_{OD}. Constraints are the capacity restrictions from Eqs. (7) to (8). The subsets are defined by the additional variables PicksRobot and PicksHuman defining the number of picks of an article within its corresponding subset. After introducing these variables, order fulfilment can also be defined as constraint by matching the sum of PicksRobot and PicksHuman against an article's picks within D_F.

4.2 Preselection for Order Assignment

The equilibrium defined by Eq. (5) can also be used for a preselection during order assignment. Two variants are introduced: the generation of a list of articles which may be assigned to a robot and the calculation of a threshold $\overline{P}_{OD}$ for the expected probability for successful object detection during order picking.

4.2.1 Preselection by Article List

If the articles of a picking order are part of $Subset_{Robot}$, robots are considered by the WMS during order assignment process. There is no need for a robot-specific assignment: Offering many human-only orders will stepwise increase their workload (order queue). When orders accessible by robots are offered, this workload will increase their effort values increasing the probability robots will win the auction. The effect of this mechanism is shown in Fig. 4.

It assumes the initial forecast D_F in total is reliable within working time WT but does not consider the variation of $Demand_{Real}$ over time. Therefore, it must be encountered by a feedback control during operation. At first, an initial calculation of subsets is using a linear smoothed total demand D_F for WT as $Demand_{Calculation,0}$. By a rolling recalculation which is based on the remaining working time and the difference of executed orders and original forecast new subsets are defined. In each recalculation a new P_{OD} for an article can also be considered. These recalculations are heavily affected by the rolling time span: the smaller the time span, the more effective is the matching of forecast and real demand. In contrary, the system is not allowed to trigger many Emergency Calls and learning is limited. Greater time spans allow higher delay during order fulfilment but enable a higher amount of Emergency Calls and thereby more input data for learning.

In Fig. 4 recalculations for the time values t_1, t_2 und t_3 are shown. Even if all existing orders are fulfilled at t_1, the number of executed orders C_1 is beneath the

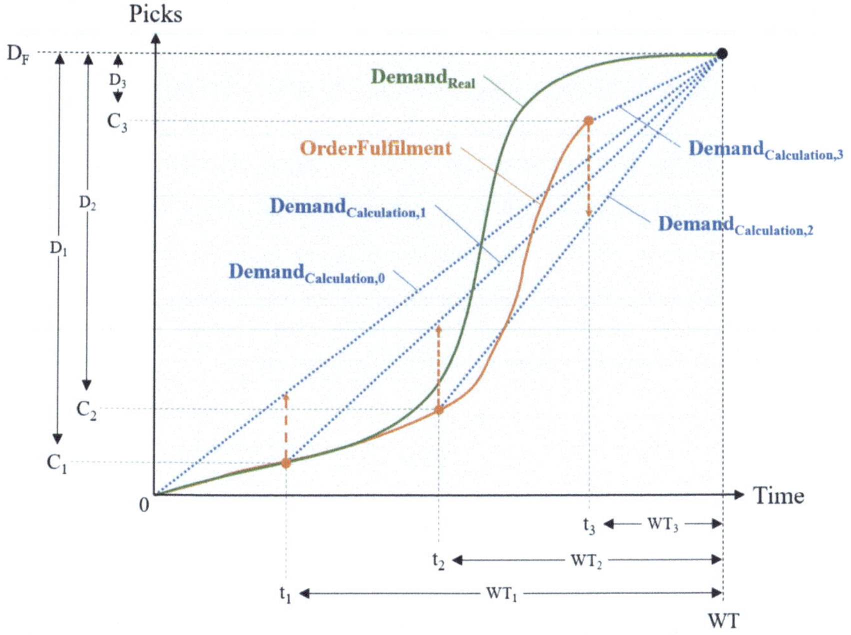

Fig. 4 Concept for a rolling calculation of subsets for a preselection before order assignment

calculated Demand$_{Calculation,0}$ resulting in lower capacity for Emergency Calls and "pulling" OrderFulfilment upwards. The same effect can be observed at t_2, whereby actually a real backlog of picking orders exists. At t_3, the number of executed orders is above the last recalculation Demand$_{Calculation,2}$ and more capacity for Emergency Calls is considered during the calculation of subsets, although the backlog still exists.

The question arises why a preselection is necessary during an order assignment by an auction which inherently should balance the workload. One reason is the variation of demand during WT. If an order's deadline for fulfilment is the end of WT, a statistical approach is sufficient because temporary backlog induced by Emergency Calls can be caught up over time. But if too much capacity is bound by Emergency Calls and due times for picking orders exist, there must be a regulating mechanism. Free capacity at the end of WT resulting from this mechanism can be used for learning orders described in Verbeet et al. (2019) to further improve object detection. A problem according to this mechanism can arise from the calculation of subsets by the linear program. Its solution is not unique, and articles can occur rarely in a subset. This effect could be avoided by further constraints or some higher meta control of the optimizing.

4.2.2 Preselection by Threshold

The forecast D_F is defined as a set of picks, i.e. no connection between picks and articles exists and the subsets are abstract volumes of picks. A threshold $\bar{P}_{Threshold}$ is calculated according to the expected probability of object detection for the articles of $Subset_{Robot}$ and is compared with $\bar{P}_{Order}$ before the initial call for bids during order assignment. However, the P_{OD} of a single article can be lower than $\bar{P}_{Threshold}$. In general, $\bar{P}$ is defined as the arithmetic mean of object detection P_{OD} weighted with the set of picks for a set of articles as shown in Eq. (10). Thereby, Eqs. (3) and (4) can be defined by using a constant value for object detection:

$$\bar{P} = \sum_{Article} \left(\frac{Picks_{Article}}{\sum_{Article} Picks_{Article}} \cdot P_{OD,\,Article} \right) \tag{10}$$

$$L_{EC,H,SR} = Subset_{Robot} \cdot L_{EC,H} \cdot \left(1 - \bar{P}_{Subset,\,Robot} \right) \tag{11}$$

$$L_{EC,R,SR} = Subset_{Robot} \cdot L_{EC,R} \cdot \left(1 - \bar{P}_{Subset,\,Robot} \right) \tag{12}$$

The linear program shown in Fig. 3 is modified by defining a subset as a set of picks. Using the Eqs. (11) and (12) the article-specific P_{OD} is replaced by a constant value. This value is minimized by the goal function. The equilibrium from Eq. (5) is still a constraint forcing the calculated P_{OD} to be the smallest value allowing an allocation of subsets. However, the goal function loses its linearity by this modification and therefore the optimization model is solved by Constraint Programming (using CP) in OPL-Studio. This also makes a modification by "dexpr" and a scaling-factor of 100 necessary because this method only accepts integers. The resulting optimization model is shown in Fig. 5.

At recalculation the set of executed picks and the remaining working time are updated. If there are more executed orders than expected, more capacity for Emergency Calls is released, $\bar{P}_{Threshold}$ is reduced, and robots are assigned more picking orders. In contrast, if execution is beneath the expected value, P: $^{-}_{Threshold}$ must be increased to raise picking performance by reducing the chance of triggering Emergency Calls. This approach is using a general P_{OD} enabling the handling of articles with a weak P_{OD} by picking robots. On the other hand, control of operational order fulfilment is weaker as picks are treated independent from articles.

```
using CP;

//*********************** Variables ********************************************
int WT = ...;          float C_H = ...;           int numArticle = ...;
int L_EC_H = ...;      float C_R = ...;           range Articles = 1..numArticle;
int L_EC_R = ...;      int D_F[Articles] = ...;

//*********************** Decision Variables ************************************
dvar int+ SubsetRobot;          dvar int+ SubsetHuman;
dvar int+ scalePOD;             dexpr float POD = scalePOD/100;
dvar int scaleL_EC_H_SR;        dexpr float L_EC_H_SR = scaleL_EC_H_SR/100;
dvar int scaleL_EC_R_SR;        dexpr float L_EC_R_SR = scaleL_EC_R_SR/100;

//*********************** Goal Function *****************************************
minimize POD;

//*********************** Constraints *******************************************
subject to {
    sum(i in Articles)(D_F[i]) == SubsetRobot + SubsetHuman;
    L_EC_H_SR == SubsetRobot * L_EC_H * (100 - POD);
    L_EC_R_SR == SubsetRobot * L_EC_R * (100 - POD);
    SubsetRobot <= (C_R * WT) - (L_EC_R_SR / WT);
    SubsetHuman <= (C_H * WT) - (L_EC_H_SR / WT);
};
```

Fig. 5 Calculation of $\bar{P}_{Threshold}$ by linear programming in OPL

4.3 Definition of Working Point for System Efficiency

Verbeet et al. (2019) introduce different thresholds to evaluate object detection of an article. P_{Break} is of major importance for real applications defining an equilibrium of effort for Emergency Calls and successful picks by robots. But to evaluate the efficiency of robots the overall system performance must be considered. A system can work in an efficient way even if the probabilities of object detection for single articles are less than P_{Break}. Consequently, an average probability of object detection for all articles is defined as well as an equilibrium between effort and benefit of using robots.

The Eqs. (13) and (14) calculate an expected effort for humans ($L_{EC,H,F}$) and robots ($L_{EC,R,F}$) for the forecast D_F. In Eq. (15) an equilibrium is defined equalizing the picking capacity of robots reduced by their effort due to Emergency Calls, i.e. their effective picking capacity and the effort of human pickers for handling Emergency Calls:

$$L_{EC,H,F} = D_F \cdot L_{EC,H} \cdot \left(1 - \bar{P}\right) \tag{13}$$

$$L_{EC,R,F} = D_F \cdot L_{EC,R} \cdot \left(1 - \bar{P}\right) \tag{14}$$

$$\frac{L_{EC,H,F}}{WT} = C_R \cdot WT - \frac{L_{EC,R,F}}{WT} \tag{15}$$

Therefore, by P; – the capacity human pickers must reserve for error handling is defined. However, this does not give any information about the ability of the picking system to fulfil all picking orders. This is guaranteed by meeting Eq. (9). In this case $\bar{P}$ matches $\bar{P}_{Break}$ and can be calculated by transforming Eq. (15):

$$\bar{P}_{Break} = 1 - \frac{C_R \cdot WT^2}{D_F \cdot (L_{EC,R} + L_{EC,H})} \tag{16}$$

Comparing $\bar{P}_{Break}$ and $\bar{P}_{Real}$ the efficiency of picking robots can be evaluated, whereby $\bar{P}_{Real}$ is calculated from real probabilities for object detection P_{OD}:

$\bar{P}_{Break} > \bar{P}_{Real}$: Human pickers must expend more capacity for handling Emergency Calls than robots can compensate by executing successful picks, i.e. the effective picking capacity of the system would be higher without robots.

$\bar{P}_{Break} = \bar{P}_{Real}$: The effort of human pickers for handling Emergency Calls and contribution of successful picks executed by robots equals each other. At this point, the system benefits from the collection of data to improve the ability of object detection of the picking robots resulting in an increasing $\bar{P}_{Real}$.

$\bar{P}_{Break} < \bar{P}_{Real}$: Picking robots are working efficient, meaning the effort of humans and robots for handling Emergency Calls is smaller than robots' contribution by executed picks. Consequently, the effective picking capacity is increased by the deployment of robots.

5 Discussion

The presented calculation approaches specify the mechanism for order assignment of the interaction pattern "Picking Order" and enable a capacitive evaluation of the picking system. However, the static input variables are problematic, i.e. the time requirements for calculating the effort value during order assignment and the empirical loss values for an Emergency Call $L_{EC,H}$ and $L_{EC,R}$. These assumptions make sense for a sufficiently large picking system that is evaluated over a longer period, so the real expectancy values match with the values of the assumptions. In systems with strongly fluctuating travel times, assuming constant effort can lead to an unacceptable weighting of P_{OD}. Therefore, the mentioned input variables should be calculated dynamically. For the calculation of the duration of an emergency call, standardized time assessments of processes such as MTM (Britzke 2010) can be used to calculate them context-dependent. Empirical data can be generated from an operative system using the approach described by Feldhorst (2018). Within robots, time values can be derived from calculations of their internal controller.

The preselection based on the capacitive equilibrium from Eq. (5) requires a static forecast for a time interval. This assumption is acceptable in systems with plannable or predictable demands. Without such a known lead a system can only be controlled reactively making the concept not reliable, because fluctuating demand must be covered by additional time buffer or resources. Alternatively, a forecast could be generated for each recalculation based on current order data and empirical values from previous periods. The threshold calculation is a statistical approach allowing a system to fluctuate to a certain degree and only slightly limits the auctioning process of order assignment. However, an empirical study must show whether the calculated threshold values are reliable. The approach tends to be more resilient the more complex the system is, i.e. the more articles exist, and the more robots and humans can process orders. The calculation of P_{Break} is an approach to evaluate the general efficiency of picking robots. In Verbeet et al. (2019), the additional working points P_{Human} and $P_{Improve}$ are also defined, whereby $P_{Improve}$ describes a very artificial value which can only be achieved in exceptional cases in operational systems. Most articles are expected to reach values between P_{Break} and P_{Human} to meet the capacity restrictions and ensure order processing.

The evaluations are based on a capacitive view of humans and robots and initially do not make any statements about economic aspects of the picking system. For such a consideration, an additional cost model would have to be integrated into the calculation. This cannot be formulated generally for the heterogeneous requirements and technical specifications of picking systems, which is why a capacitive approach was chosen in this paper.

6 Conclusions

The challenge of finding personnel for picking and handle continuously changing article ranges can be countered with partially automated picking systems creating a cooperative working environment for humans and picking robots. The motivation is to ensure reliable order fulfilment by implementation of a feedback-loop (Emergency Call) between humans and robots for error handling and data collection for machine learning algorithms in order to continuously improve object detection and thereby improve overall performance of picking robots. In this paper, a concept for measurement and evaluation of system performance to ensure the processing of picking orders and the training of picking robots is introduced.

The proposed approach is using a capacitive evaluation of a picking system to define equilibrium between the requirements of order processing, the picking performance of humans and robots and the effort for improving object detection of the robots. In a picking system this equilibrium can be used for strategic evaluation of the automated picking performance of robots (working point), for tactical resource planning (capacity planning) or for operational workload balancing (order assignment). Even if this evaluation mechanisms extend the adaptive process model and the conceptual picking system, there are still open questions for future research.

The presented calculation uses static assumptions for actually dynamic parameters making its equations only reliable for complex systems considering a sufficiently long runtime. However, the evaluation concept can be expanded by context-based calculations using MTM-based approaches. The actions of robots can be evaluated based on functions of their internal control, e.g. travel time between picking locations as outcome from path planning.

The components for a demonstrator, which is to validate the presented evaluation approach with empirical data, have been developed at Ulm University of Applied Sciences in recent months. The configuration of these components and the definition of suitable scenarios are still pending and will be completed soon. The evaluation approach will also be integrated into an intra-logistic scenario with real robots. These provide context-dependent estimations for the required time of their actions in order to enable a dynamic and context-based calculation of time effort and fulfilment time during order processing.

The model for object detection is also subject of current research. In current work, a singular neural network is used for the detection of all existing articles in a picking system. An alternative approach pursues the dynamic combination of several neural networks for one article each, which are compared for object detection with one another by an algorithm based on their storage locations provided by an overall WMS. This is intended to modularize object detection and shorten training times. The comparison "singular" versus "combined" can also be carried out by the demonstrator.

Acknowledgments This work is part of the project "ZAFH Intralogistik", funded by the European Regional Development Fund and the Ministry of Science, Research and Arts of Baden-Württemberg, Germany (F.No. 32-7545.24-17/3/1). It is also done within the post graduate school "Cognitive Computing in Socio-Technical Systems" of Ulm University of Applied Sciences and Ulm University, which is funded by Ministry for Science, Research and Arts of the State of Baden-Württemberg, Germany.

References

Azadeh K, de Koster RBM, Roy D (2017). Robotized and automated warehouse systems: review and recent developments. https://doi.org/10.2139/ssrn.2977779

Bender M, Braun M, Rally P, Scholtz O, Bauer W (2016) Leichtbauroboter in der manuellen Montage-Einfach anfangen: Erste Erfahrungen von Anwenderunternehmen. Fraunhofer IAO, Stuttgart

Bonini M, Urru A, Echelmeyer W (2019) The quality interaction function deployment for lean human-robot interaction. In: 24th international conference on methods and models in automation and robotics (MMAR). IEEE, pp 145–151. https://doi.org/10.1109/MMAR.2019.8864667

Bormann R, Brito BF, de Lindermayr J, Omainska M, Patel M (2019) Towards automated order picking robots for warehouses and retail. In: Tzovaras D, Giakoumis D, Vincze M (eds) Theoretical computer science and general issues, computer vision systems: 12th international conference, ICVS 2019, Thessaloniki, Greece, 23–25 September 2019, Proceedings, 1st edn. Springer International Publishing, pp 185–198

Britzke B (2010) MTM in einer globalisierten Wirtschaft: Arbeitsprozesse systematisch gestalten und optimieren (1. Aufl.). mi-Wirtschaftsbuch. FinanzBuch-Verl
Brynzér H, Johansson MI, Medbo L (1994) A methodology for evaluation of order picking systems as a base for system design and managerial decisions. Int J Oper Produc Manag 14(3):126–139. https://doi.org/10.1108/01443579410058595
Bukchin Y, Khmelnitsky E, Yakuel P (2012) Optimizing a dynamic order-picking process. Eur J Oper Res 219(2):335–346. https://doi.org/10.1016/j.ejor.2011.12.041
Chabot (2018) Material handling optimization in warehousing operations, Thesis, Université Laval, Quebec, Canada. https://www.semanticscholar.org/paper/Material-handling-optimization-in-warehousing-Chabot/ea2e64e1cdd15aee660e252883ca0fc022bc913c, https://corpus.ulaval.ca/jspui/bitstream/20.500.11794/36277/1/34371.pdf
Chen B, Qu T, Thurer M, Huang GQ, Li C, Xu S (2017) Warehouse workload control for production logistic. In 13th IEEE international conference on automation science and engineering (CASE). IEEE, pp 237–242. https://doi.org/10.1109/COASE.2017.8256109
Dallari F, Marchet G, Melacini M (2009) Design of order picking system. Int J Adv Manuf Technol 42(1–2):1–12. https://doi.org/10.1007/s00170-008-1571-9
de Koster R, Le-Duc T, Roodbergen KJ (2007) Design and control of warehouse order picking: a literature review. Eur J Oper Res 182(2):481–501. https://doi.org/10.1016/j.ejor.2006.07.009
Feldhorst S (2018) Automatische Aktivitäts- und Kontexterkennung zur Analyse des Kommissionierprozesses. Dissertation, Praxiswissen Service UG; Technische Universität Dortmund]. GBV Gemeinsamer Bibliotheksverbund
FIPA (2001) Contract net interaction protocol specification. https://www.fipa.org/specs/fipa00029/
Gong Y, de Koster RBM, van Nunen JAEE (2010) A flexible evaluative framework for order picking systems. Produc Oper Manag 19(1):70–82. https://doi.org/10.1111/j.1937-5956.2009.01047.x
Grosse EH, Glock CH, Neumann WP (2017) Human factors in order picking: a content analysis of the literature. Int J Prod Res 55(5):1260–1276. https://doi.org/10.1080/00207543.2016.1186296
Gu J, Goetschalckx M, McGinnis LF (2010) Research on warehouse design and performance evaluation: a comprehensive review. Eur J Oper Res 203(3):539–549. https://doi.org/10.1016/j.ejor.2009.07.031
Hanson R, Medbo L, Johansson MI (2018) Performance characteristics of robotic mobile fulfilment systems in order picking applications. IFAC-PapersOnLine 51(11):1493–1498. https://doi.org/10.1016/j.ifacol.2018.08.290
Heine F, Wenzel S (2013). Comparison of performance measurement-systems used for order-picking. https://doi.org/10.2195/lj_Rev_heine_en_201312_01
Henn S (2015) Order batching and sequencing for the minimization of the total tardiness in picker-to-part warehouses. Flex Serv Manuf J 27(1):86–114. https://doi.org/10.1007/s10696-012-9164-1
Hernandez K, Bacca B, Posso B (2017) Multi-goal path planning autonomous system for picking up and delivery tasks in mobile robotics. IEEE Latin Am Trans 15(2):232–238. https://doi.org/10.1109/tla.2017.7854617
Hoffman G (2019) Evaluating fluency in human-robot collaboration. IEEE Trans Hum-Mach Syst 49(3):209–218. https://doi.org/10.1109/thms.2019.2904558
Hsieh L-F, Huang Y-C (2011) New batch construction heuristics to optimise the performance of order picking systems. Int J Prod Econ 131(2):618–630. https://doi.org/10.1016/j.ijpe.2011.02.006
Hwang HS, Cho GS (2006) A performance evaluation model for order picking warehouse design. Comput Ind Eng 51(2):335–342. https://doi.org/10.1016/j.cie.2005.10.002
IBM (2020) IBM ILOG CPLEX optimization studio. https://www.ibm.com/de-de/products/ilog-cplex-optimization-studio
Jaghbeer (2019) On the performance of robotic parts-to-picker order picking systems, Thesis, ISSN1654-9732, Division of Supply and Operations Management, Department of Technology Management and Economics, Chalmers University of Technology, Gothenburg, Sweden. https://research.chalmers.se/publication/513777/file/513777_Fulltext.pdf
Jane C-C, Laih Y-W (2005) A clustering algorithm for item assignment in a synchronized zone order picking system. Eur J Oper Res 166(2):489–496. https://doi.org/10.1016/j.ejor.2004.01.042

Kłodawski M, Jachimowski R (2013) Ant algorithms for designing order picking systems. Prace Naukowe Politechniki Warszawskiej. Transport (z. 97): 259–269. https://www.infona.pl/resource/bwmeta1.element.baztech-4e4dac6b-1209-4d34-85a4-c4d66fbf8cf4

Kong XTR, Luo H, Huang GQ, Yang X (2019) Industrial wearable system: the human-centric empowering technology in Industry 4.0. J Intell Manuf 30(8): 2853–2869. https://doi.org/10.1007/s10845-018-1416-9

Kozai K, Hashimoto M (2018) Determining robot grasping-parameters by estimating "Picking Risk". In: International workshop on advanced image technology (IWAIT). IEEE. https://doi.org/10.1109/iwait.2018.8369707

Krug R, Stoyanov T, Tincani V, Andreasson H, Mosberger R, Fantoni G, Lilienthal AJ (2016) The next step in robot commissioning: autonomous picking and palletizing. IEEE Robot Autom Lett 1(1):546–553. https://doi.org/10.1109/lra.2016.2519944

Lamballais T, Roy D, de Koster RBM (2017) Estimating performance in a robotic mobile fulfillment system. Eur J Oper Res 256(3):976–990. https://doi.org/10.1016/j.ejor.2016.06.063

Lee JA, Chang YS, Choe YH (2017) Assessment and comparison of human-robot co-work order picking systems focused on ergonomic factors. In: International conference on applied human factors and ergonomics. Springer, Cham, pp 516–523. https://doi.org/10.1007/978-3-319-60525-8_53

Liang C, Chee KJ, Zou Y, Zhu H, Causo A, Vidas S, Teng T, Chen IM, Low KH, Cheah CC (2015) Automated robot picking system for e-commerce fulfillment warehouse application. In: 14th IFToMM world congress

Lin C-C, Kang J-R, Hou C-C, Cheng C-Y (2016) Joint order batching and picker Manhattan routing problem. Comput Ind Eng 95:164–174. https://doi.org/10.1016/j.cie.2016.03.009

Liu Y-R, Huang M-B, Huang H-P (2019) Automated grasp planning and path planning for a robot hand-arm system. In IEEE/SICE international symposium on system integration (SII). IEEE. https://doi.org/10.1109/sii.2019.8700433

Manzini R, Gamberi M, Persona A, Regattieri A (2007) Design of a class based storage picker to product order picking system. Int J Adv Manuf Technol 32(7–8):811–821. https://doi.org/10.1007/s00170-005-0377-2

Martinez C, Boca R, Zhang B, Chen H, Nidamarthi S (2015) Automated bin picking system for randomly located industrial parts. In IEEE international conference on technologies for practical robot applications (TePRA): Crowne Plaza Boston-Woburn 15 Middlesex Canal Park Woburn, Massachusetts, 01801 USA, 11–12 May 2013. IEEE, pp 1–6. https://doi.org/10.1109/TePRA.2015.7219656

Merschformann M, Lamballais T, de Koster R, Suhl L (2018) Decision rules for robotic mobile fulfillment systems. https://arxiv.org/pdf/1801.06703

Mester J, Wahl F (2019) Robotik in der Intralogistik-Ein Projekt der Unternehmen Fiege und Magazino. In Logistik der Zukunft—Logistics for the future. Springer, Gabler, Wiesbaden, pp 199–211. https://doi.org/10.1007/978-3-658-23805-6_7

Molnár B (2004) Planning of order picking processes using simulation and a genetic algorithm in multi-criteria scheduling optimisation. In: Proceedings 16th European simulation symposium. SCS Press.

Nguyen TH, Kim DH, Lee CH, Kim HK, Kim SB (2016) Mobile robot localization and path planning in a picking robot system using kinect camera in partially known environment. In: International conference on advanced engineering theory and applications. Springer, Cham, pp 686–701. https://doi.org/10.1007/978-3-319-50904-4_70

Pan JC-H, Wu M-H (2012) Throughput analysis for order picking system with multiple pickers and aisle congestion considerations. Comput Oper Res 39(7):1661–1672. https://doi.org/10.1016/j.cor.2011.09.022

Pinto ARF, Nagano MS (2020) Genetic algorithms applied to integration and optimization of billing and picking processes. J Intell Manuf 31(3):641–659. https://doi.org/10.1007/s10845-019-01470-3

Rieder M, Verbeet R (2019) Robot-human-learning for robotic picking processes. In: Kersten W, Blecker T, Ringle CM (eds) Proceedings of the hamburg international conference of logistics (HICL): Vol. 27. Proceedings of the Hamburg international conference of logistics (HICL)/Artificial intelligence and digital transformation in supply chain management: innovative approaches for supply chains, 10th edn., pp 87–114. https://www.econstor.eu/handle/10419/209370, https://doi.org/10.15480/882.2466

Scholz A, Schubert D, Wäscher G (2017) Order picking with multiple pickers and due dates—simultaneous solution of order batching, batch assignment and sequencing, and picker routing problems. Eur J Oper Res 263(2):461–478. https://doi.org/10.1016/j.ejor.2017.04.038

Seyedrezaei M, Najafi SE, Aghajani A, Bagherzadeh Valami H (2012) Designing a genetic algorithm to optimize fulfilled orders in order picking planning problem with probabilistic demand. Int J Res Indus Eng 1(2):40–57

Tu M, Shih PH, Yang MF, Lin CK, Kao SL (2019). Using multi-objective genetic algorithm on order picking system. In: Ao S-I (ed) Lecture notes in engineering and computer science. International multiconference of engineers and computer scientists 2019. International Association of Engineers, pp 35–39.

Valle CA, Beasley JE (2019) Order allocation, rack allocation and rack sequencing for pickers in a mobile rack environment. https://arxiv.org/pdf/1903.06702v3

van Gils T (2019) Designing efficient order picking systems: combining planning problems and integrating real-life features. 4OR 17(3):331–332. https://doi.org/10.1007/s10288-019-00405-1

van Gils T, Ramaekers K, Caris A, Cools M (2017) The use of time series forecasting in zone order picking systems to predict order pickers' workload. Int J Prod Res 55(21):6380–6393. https://doi.org/10.1080/00207543.2016.1216659

Vanheusden S, van Gils T, Ramaekers K, Caris A (2017) Reducing workload imbalance in parallel zone order picking systems. https://www.Semanticscholar.Org/paper/Reducing-Workload-Imbalance-in-Parallel-Zone-Order-Vanheusden-Gils/e4ab7b0c6ae339849731d5b60049c0d0e7894b81

VDI (2007) Operational logistics key figures from goods receiving to dispatch (Richtlinie, 4490). Beuth Verlag GmbH, Berlin

Verbeet R, Rieder M, Kies M (2019) Realization of a cooperative human-robot-picking by a learning multi-robot-system using BDI-agents. SSRN Electr J. Advance online publication. https://doi.org/10.2139/ssrn.3502934

Wahrmann D, Hildebrandt A-C, Schuetz C, Wittmann R, Rixen D (2019) An autonomous and flexible robotic framework for logistics applications. J Intell Rob Syst 93(3):419–431. https://doi.org/10.1007/s10846-017-0746-8

Yu M, de Koster RBM (2009) The impact of order batching and picking area zoning on order picking system performance. Eur J Oper Res 198(2):480–490. https://doi.org/10.1016/j.ejor.2008.09.011

Zhang H, Long P, Zhou D, Qian Z, Wang Z, Wan W, Manocha D, Park C, Hu T, Cao C, Chen Y, Chow M, Pan J (2016) Dorapicker: an autonomous picking system for general objects. In: IEEE international conference on automation science and engineering (CASE). IEEE, pp 721–726. https://doi.org/10.1109/COASE.2016.7743473

Zhang J, Wang X, Chan FTS, Ruan J (2017) On-line order batching and sequencing problem with multiple pickers: A hybrid rule-based algorithm. Appl Math Model 45:271–284. https://doi.org/10.1016/j.apm.2016.12.012

Zhu H, Kok YY, Causo A, Chee KJ, Zou Y, Al-Jufry SOK, Liang C, Chen I-M, Cheah CC, Low KH (2016) Strategy-based robotic item picking from shelves. In: IEEE/RSJ international conference on intelligent robots and systems. IEEE. https://doi.org/10.1109/iros.2016.7759354

Zou Y, Zhang D, Qi M (2019) Order picking system optimization based on picker-robot collaboration. In: Proceedings of the 2019 5th international conference on industrial and business engineering—ICIBE 2019. ACM Press, pp 1–6. https://doi.org/10.1145/3364335.3364386

Integrating RFID Signal with Scene for Automatic Identification in Logistics and Production

Seng Fat Wong and Weng Ian Ho

Abstract With the booming e-commerce development, advanced industrial logistics technique is necessary to enhance its working accuracy and efficiency. Meanwhile, the demand of Auto Guidance Vehicles (AGVs) is increasing while the technique can be popularized to apply in logistics and production. However, the AGV is necessary to work with precise technology for enhancing its performance. Therefore, the demands on the automatic identification and localization with RFID technology for supporting logistics and production are raising and it becomes the most important. In this chapter, the RFID technology integrating into the indoor positioning technology with LANDMARC methodology is studied. Moreover, the virtual reference elimination (VIRE) algorithm that is based on LANDMARC algorithm is applied to advance the performance of RFID localization. The VIRE positioning method is added with virtual reference tags to improve the accuracy of positioning in this study. This chapter summarizes three main contributions in this methodology. First, the RFID localization is more cost-efficiency in logistics and production, because it is no need to consider additional readers and tags. The hardware cost is the same as in the case of both LANDMARC and VIRE systems, so the accuracy can be easily improved in the comparison with LANDMARC. Second, the estimated position of target tags are more accurate because it works with virtual tags for localization. Third, the VIRE system can better adapt to dynamic indoor scenarios than the LANDMARC system in real environments.

Keywords RFID localization · LANDMARC · Virtual reference elimination (VIRE) · Auto guidance vehicles (AGVs)

S. F. Wong (✉) · W. I. Ho
Faculty of Science and Technology, University of Macau, Avenida da Universidade, Taipa, Macau 11-4081, China
e-mail: fstsfw@um.edu.mo

W. I. Ho
e-mail: edcoho@gmail.com

© The Author(s), under exclusive license to Springer Nature Switzerland AG 2020
P. Golinska-Dawson et al. (eds.), *Smart and Sustainable Supply Chain and Logistics – Trends, Challenges, Methods and Best Practices*, EcoProduction,
https://doi.org/10.1007/978-3-030-61947-3_4

1 Introduction

Online shopping is rapid development in the life and industries. Therefore, the logistic development is rapidly increasing into new generation. It is necessary to concern for enhancing its efficiency. Most of the logistic industries has been changed to use Auto Guidance Vehicles (AGVs) to instead of traditional human operation. However, the performance of AGV is being studied to more precise. Some researchers are considering to apply the Vision based Simultaneous Localization and Mapping (vSLAM) technique to optimize the transportation path for AGV (Wong and Yu 2019a). Meanwhile, the professional software is also supported to enhance the performance of logistics management (Wong and Yu 2019b). However, some logistic companies are not considered to use AGV with vSLAM technique because of cost-efficiency. Radio frequency identification (RFID) has been widely used in supporting the logistics management special in applying big data approach for logistics trajectory discovery (Zhong et al. 2015). Moreover, it is also applied in the RFID logistics system applicable to ubiquitous-city (Hong and So 2009). The general cargo handler logistics had studied to enhance the performance by the RFID technique with human error consequences (Giustia et al. 2019). Therefore, logistics and production are increasing the demand of AGVs with RFID technology. The LANDMARC methodology is recently referenced for the positioning algorithms with RFID technology (Ho and Wong 2012), the coordinates of readers and reference tags are known and the target tags can be calculated by different weights of them. The virtual reference elimination (VIRE) algorithm that is based on LANDMARC algorithm is applied to advance the performance of RFID localization, virtual tags are used to replace the reference tags in LANDMARC which can reduce the noise between tags and enhance the accuracy as more points can be used to calculate.

LANDMARC algorithm, utilizes Received Signal Strength Indicator (RSSI) to track moving objects, was the first attempt using active RFID for indoor location sensing with satisfactory result. In order to increase accuracy, it utilize the reference tags as reference points to assist readers in locating the unknown tracking tags, is one of the most classic algorithm using RFID for indoor location. LANDMARE contributed an idea of estimate the coordinates of tracking tags by comparing their RSSI values with those of k-nearest reference tag at known coordinates. Apparently, more reference tags means tracking tag located with greater precision. At the same time, excessive amount of reference will not only increase cost, but also introduce expected radio frequency interference, which led to inaccuracy of positioning.

For overcoming this defect of LANDMARC, VIRE introduce a concept of "virtual reference tags" to obtain more likely accurate positions without additional cost. The real reference tags in LANDMARC are properly placed to form a 2D regular grid. This real regular grid is further divided into n × n equal sized virtual grid cells, and each virtual grid cells are treated as covered by four virtual reference tags. The RSSI of virtual reference tags is calculated from those real reference tags by linear interpolation algorithm. The whole sensing area is divided into a number of location regions, which the centre of each region correspond to a virtual reference tag. Every reader has its own proximity map. If the difference between the RSS measurement of the unknown tag and the RSS measurement of a region is smaller than a threshold, the region is marked as 1. The fusion of all the n readers' maps provides a global proximity map for the tag.

In this paper, the VIRE positioning method is studied and further validated for supporting the advanced performance with RFID technology in logistics management. The RFID localization is cost-efficiency in logistics and production as no additional readers and tags is required. Meanwhile, the estimated position of target tags is more accurate because it works with virtual tags. The VIRE system can better adapt to dynamic indoor scenarios than the LANDMARC system in real environments. It is focused to discuss and compare the VIRE system with the LANDMARC system in the static and dynamic situation in this paper.

2 Methodology

The experiment of RFID localization system is applied UHF RFID technology. In this technology, the tags could be identified by the reader which shows by the identification code. The reader could also show the RSSI (Signal Strength Indication). It is decreased by the free path space loss. With the theory of the LANDMARC formula. The location of the tags could be found. The LANDMARC formula is using statistics to work out the probable location. It contrasts the reference resources and the test tags for providing the most probable location of the test tags. In this paper, it is combined with LANDMARC formula and image processing to match the test tags in the real scene with the identification code. In order to calculate the location more accuracy, it is put forward the VIRE system that matches the test tags in real scene.

The RFID Reader is RF-CODE 443 MHz M250 Reader that receives and reports the radio frequency messages emitted by RF Code tags. Its read range is around 45 m. The M100 asset tags have 2 s motion alert and 10 s beacon time. The dynamic part in the experiment is very important, so the M100 asset tags are chosen. The experiment needs a high-speed stimulant which near to 60 km/h. Therefore, the simulation car is applied and instead of AGV. The stimulation car is the simplest one with a detection tag. The traction engine is an electric engine which work in 24 V. Its rated speed could be 20,000 Rev. It's a powerful traction engine. The high speed of the engine makes the traction line in a horrible speed. The traction line is designed away from the engine, otherwise, the engine is easy to wound by the traction line. Thus, the roller and bearing system are designed to solve this problem. The holder is used to simulate the real situation of the reader. In the real situation, the reader on the warehouse will put upon the AGV. The Logitech C270 HD IPTV webcam is used to photographed and judge the location of the simulation car. In the software development is applied to Node.js, because it is an open-source, cross-platform runtime environment for developing server-side Web applications.

The real reference tags are properly placed to form a 2D regular grid. Objective tags can be placed anywhere within gird. In order to improve the precision of tracking tags within gird's position. Thus, the gird should be divided into a finer gird based on the concept of virtual reference tags. The core of VIRE approach is that four real reference tags as per physical grid and then divided into n x n equal sized virtual gird cells. The coordinate of virtual reference tags to be defined on the basis of the coordinate of the four real reference tags are known. On the RSSI values of virtual reference tag, the approach suggests the linear interpolation algorithm to obtain the RSSI values.

To obtain the RSSI value of each virtual reference tag to each reader, therefore, the reference tags' RSSI value and coordinate position are known in advance. Thus, the RSSI values of virtual tags are interpolated by the formula below:

$$S_k\left(T_{P,b}\right) = S_k(T_{a,b}) + \mathrm{p} + \frac{S_k(T_{a+n1,b+n2}) - S_k(T_{a,b})}{n+1}$$
$$= \frac{p \times S_k(T_{a+n1,b+n2}) + (n+1-p) \times S_k(T_{a,b})}{n+1} \tag{1}$$

where $S_k(T_{i,j})$ represents the RSSI value of the virtual reference tag located at the coordinate (i, j) for the kth reader. Assuming there are N × N virtual reference tags, the complexity of the interpolation algorithm is O (N^2).

Therefore, the coordinate position of all real reference tags and virtual reference tags are known. On the other hand, the signal strength of all real reference tags and virtual reference tags also are known. Meanwhile, it is important to choice the ideal threshold value kth in the LANDMARC algorithm to obtain the localization of objective tags by the ideology of LANDMARC algorithm:

$$E_j = \sqrt{\sum_{i=1}^{n} (\theta i - Si)^2} \tag{2}$$

where $j \in (1, \mathrm{m})$. The nearer reference tag to the tracking tag has a smaller E value. When there are m reference tags, an tracking tag has the vector $\mathrm{E} = (\mathrm{E}_1, \mathrm{E}_2, \ldots\ldots \mathrm{E}_\mathrm{m})$. These E values are used to reflect the relations of the tags, the smallest E_j means that the reference tag is the nearest reference tag surrounding the tracking tag. K-reference tags which have relative lower E values are selected from the m reference tags as the k neighbors. The tracking tag's coordinates are estimated by computing the weighted average of the k neighbors' coordinates (Fig. 1):

$$(x, y) = \sum_{i=1}^{k} wi(xi, yi) \tag{3}$$

$$\mathrm{wi} = \frac{1}{\mathrm{E}_\mathrm{i}^2} / \sum_{\mathrm{i}=1}^{\mathrm{k}} \frac{1}{\mathrm{E}_\mathrm{i}^2} \tag{4}$$

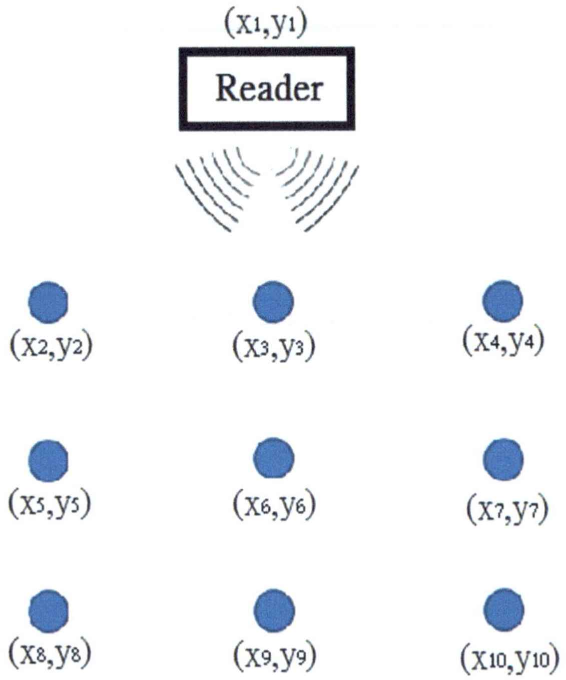

Fig. 1 The RFID localization system setup

3 Results and Discussion

The experiment of this research is included four parts which are the factor affecting the RSSI value, the database of reference tags, objective tags being matched at rest and objective tags being matched at dynamic. The experiments are done the same environment and condition. However, it is focused to discuss and compare the VIRE system with the LANDMARC system in the static and dynamic situation in this paper, because it can show the main contributions for logistics management. As a result, the objectives tags are matched by the VIRE system which can more accurate than the LANDMARC system and Log-path loss models.

It had been determined the RSSI data from the ten random object tags. In this discussion, it is put forward the VIRE system which matched ten object tags in the static situation and compared with LANDMARC system in the static situation. The virtual reference tags are obtained though the formula (Eq. 1) that includes interpolated concepts. On the other hand, to confirm the data based on VIRE system is regarding to the LANDMARC system. Nevertheless, the VIRE system is different for the LANDMARC system because the VIRE approach adopts much more virtual reference tags. In this two type of database, the main location method both are the weighting algorithm. The database of LANDMARC system has real 192 data. On the contrary, the database of VIRE System has 620 data which be divided into 192 real data and 428 virtual data.

$$\text{There are error rate: } \sigma = \frac{\text{VIRE coordinate} - \text{Real location coordinate}}{\text{Real location coordinate}} \quad (5)$$

As the results of Tables 1, 2, and 3, it can obtain that the error rate of VIRE system is lower than the error rate of LANDMARC system. In the X-axis, the average error of VIRE is 25.81% and the LANDMARC is 33.36% regarding Table 4. In the Y-axis, the average error of VIRE is 15.27% and the LANDMARC is 21.88% according to Table 4. Some results using VIRE positioning system has some error up to only 1% for X-axis. However, some particular data, the error rate has suddenly increased up to 106.21% for instance (0.45, 24 m) that may be affected by irregular factors, such as scattering or surrounding material. Meanwhile, the error rate of VIRE system has the error up to only 1.54% for Y-axis. Nevertheless, some particular data, the error

Table 1 The real location and VIRE location of ten random tags

Tag CODE	Real coordinate	RSSI	VIRE location (coordinate)	Error rate X (%)	Error rate Y (%)
LOCATE00365552	(2, 3)	(−62, −59)	(2.023, 3.073)	1.15	2.43
LOCATE00365554	(2, 4)	(−63, −62)	(1.937, 4.426)	3.15	10.65
LOCATE00365555	(2.5, 5)	(−65, −65)	(2.075, 5.143)	17.00	2.86
LOCATE00365569	(1.5, 5)	(−65, −68)	(2.001, 8.132)	27.96	62.64
LOCATE00365556	(2, 8)	(−66, −68)	(1.980, 8.512)	1.00	6.21
LOCATE00365562	(1.5, 12)	(−79, −75)	(2.776, 14.369)	85.03	5.74
LOCATE00365561	(2, 18)	(−73, −74)	(2.027, 13.514)	1.35	24.90
LOCATE00365559	(3, 24)	(−89, −88)	(2.585, 26.909)	13.80	4.62
LOCATE00365560	(1, 25)	(−89, 86)	(2.224, 25.387)	106.21	1.54
LOCATE00365557	(2, 28)	(−77, −81)	(1.935, 16.816)	1.42	31.07

Table 2 The real location and LANDMARC location of ten random tags

Tag CODE	Real coordinate	RSSI	LANDMARC location (coordinate)	Error rate X (%)	Error rate Y (%)
LOCATE00365552	(2, 3)	(−62, −59)	(2.150, 2.830)	7.50	5.67
LOCATE00365554	(2, 4)	(−63, −62)	(2.170, 5.480)	8.50	37.00
LOCATE00365555	(2.5, 5)	(−65, −65)	(1.960, 6.120)	21.60	22.40
LOCATE00365569	(1.5, 5)	(−65, −68)	(1.890, 8.140)	26.00	62.80
LOCATE00365556	(2, 8)	(−66, −68)	(2.070, 7.360)	3.50	8.00
LOCATE00365562	(1.5, 12)	(−79, −75)	(2.960 11.610)	97.33	3.25
LOCATE00365561	(2, 18)	(−73, −74)	(2.620, 12.320)	31.00	31.55
LOCATE00365559	(3, 24)	(−89, −88)	(2.410, 25.590)	19.67	6.62
LOCATE00365560	(1,25)	(−89, 86)	(2.170, 23.780)	117.00	4.88
LOCATE00365557	(2, 28)	(−77, −81)	(2.030, 17.730)	1.50	36.67

Table 3 Error rate comparison between LANDMARC and VIRE

Real coordinate	Error rate X VIRE (%)	Error rate Y VIRE (%)	Error rate X LANDMARC (%)	Error rate Y LANDMARC (%)
(2, 3)	1.15	2.43	7.50	5.67
(2, 4)	3.15	10.65	8.50	37.00
(2.5, 5)	17.00	2.86	21.60	22.40
(1,5, 5)	27.96	62.64	26.00	62.80
(2, 8)	1.00	6.21	3.50	8.00
(1.5, 12)	85.03	5.74	97.33	3.25
(2, 18)	1.35	24.90	31.00	31.55
(3, 24)	13.80	4.62	19.67	6.62
(1, 25)	106.21	1.54	117.00	4.88
(2, 28)	1.42	31.07	1.50	36.67

Table 4 The average error rate of VIRE and LANDMARC

	X-axis (%)	Y-axis (%)
VIRE	25.81	15.27
LANDMARC	33.36	21.88

rate of VIRE system has suddenly increased up to 62.64% for instance (0.95, 4 m) that may be affected by irregular factors too. Thus, the indoor environmental factor will be influenced with experimental error that will be studied in the further research as noise cancellation. It is shown that the VIRE system is more suitable for the real environment measurement comparing with LANDMARC system.

It had been obtained the empirical log-distance path loss formula in the VIRE database. In this discussion, it is analysed the method of VIRE System which compared with the empirical Log-distance path loss methods that are y express RSSI value and x express distance (Table 5).

From the Tables 6 and 7, the VIRE error rate is lower than the error rate of empirical Log-path loss models. The VIRE system can be matched the location of objective tags by the RSSI value, but the empirical Log-distance path loss only obtained the distance of Y-axis. Compared with the empirical Log-distance path loss models, the VIRE system is more suitable for locating in the real environment measurement.

In the experiment of dynamic identification in real scene, the RFID VIRE system is connected with the real scene to match the RFID signal with the image signal. The camera catches the image signal with the probable location. With the probable location and the RFID signal, the VIRE location matches with the probable location. The image signal can be matched with the RFID signal. For simulating the real scene in the high speed logistic operation, the testing simulation car should be assigned to different velocities which may be even to 60 km/h. Assume the range of the RFID is 20 m, in consideration of the fastest detection interval of the RFID reader is 3 s. The

Table 5 The experiment data with Log-distance path loss method of VIRE database

Location	Channel A		Channel B	
0.45 m X-axis	$y = -7.225\ln(x) - 55.064$	$x = e^{\frac{y+55.064}{-7.225}}$	$y = -7.312\ln(x) - 52.805$	$x = e^{\frac{y+52.805}{-7.312}}$
0.95 m X-axis	$y = -7.297\ln(x) - 53.484$	$x = e^{\frac{y+53.484}{-7.297}}$	$y = -7.041\ln(x) - 52.732$	$x = e^{\frac{y+52.732}{-7.041}}$
1.45 m X-axis	$y = -7.565\ln(x) - 50.305$	$x = e^{\frac{y+50.305}{-7.565}}$	$y = -6.675\ln(x) - 52.268$	$x = e^{\frac{y+52.268}{-6.675}}$
1.95 m X-axis	$y = -8.088\ln(x) - 50.624$	$x = e^{\frac{y+50.624}{-8.088}}$	$y = -6.814\ln(x) - 53.237$	$x = e^{\frac{y+53.237}{-6.814}}$
2.45 m X-axis	$y = -8.669\ln(x) - 52.278$	$x = e^{\frac{y+52.278}{-8.669}}$	$y = -6.892\ln(x) - 56.247$	$x = e^{\frac{y+56.247}{-6.892}}$

Table 6 Calculated path loss at different locations

Real location (m)	RSSI	The distance of log path loss, Channel A		The distance of log path loss, Channel B	
(1.45, 2)	(−62, −59)	$x = e^{\frac{y+50.305}{-7.565}}$	4.69 m	$x = e^{\frac{y+52.268}{-6.675}}$	2.74 m
(1.45, 3)	(−63, −62)	$x = e^{\frac{y+50.305}{-7.565}}$	5.35 m	$x = e^{\frac{y+52.268}{-6.675}}$	4.30 m
(1.95, 4)	(−65, −65)	$x = e^{\frac{y+50.624}{-8.088}}$	5.91 m	$x = e^{\frac{y+53.237}{-6.814}}$	5.62 m
(0.95, 4)	(−65, −68)	$x = e^{\frac{y+53.484}{-7.297}}$	4.85 m	$x = e^{\frac{y+52.732}{-7.041}}$	8.74 m
(1.45, 7)	(−66, −68)	$x = e^{\frac{y+50.305}{-7.565}}$	7.96 m	$x = e^{\frac{y+52.268}{-6.675}}$	10.56 m
(0.95,11)	(−79, −75)	$x = e^{\frac{y+53.484}{-7.297}}$	33.01 m	$x = e^{\frac{y+52.732}{-7.041}}$	23.63 m
(1.45, 17)	(−73, −74)	$x = e^{\frac{y+50.305}{-7.565}}$	20.09 m	$x = e^{\frac{y+52.268}{-6.675}}$	25.94 m
(2.45, 23)	(−89, −88)	$x = e^{\frac{y+52.278}{-8.669}}$	69.13 m	$x = e^{\frac{y+56.247}{-6.892}}$	100 m
(0.45, 24)	(−89, −86)	$x = e^{\frac{y+55.064}{-7.225}}$	109.62 m	$x = e^{\frac{y+52.805}{-7.312}}$	93.67 m
(1.45, 27)	(−77, −81)	$x = e^{\frac{y+50.305}{-7.565}}$	34.08 m	$x = e^{\frac{y+52.268}{-6.675}}$	57.83 m

highest velocity which must be detected to connect with this formula: $S = Vt$. The S is the range of the RFID, and the t is the detection interval. The fastest velocity $V = 6.67$ m/s which is 24 km/h. It means in this experiment the fastest velocity is 24 km/h.

The Simmah AR925 tachometer will be used to test the velocity of the simulation car. The Logitech C270 HD IPTV webcam instead of the camera on the warehouse.

The first experiment the tachometer shows that the velocity is 67.52 m/min which is 4.0512 km/h. The detection program is shown that:

Table 7 Error rate comparison between VIRE and the empirical log-distance path loss model

Real location, coordinate	Error rate X of VIRE (%)	Error rate Y of VIRE (%)	Error rate Y of Log path loss (Channel A) (%)	Error rate Y of Log path loss (Channel B) (%)
(1.45, 2 m), (2,3)	1.15	2.43	134.5	37
(1.45, 3 m), (2,4)	3.15	10.65	78.3	43
(1.95, 4 m), (2.5,5)	17.00	2.86	47.8	40.5
(0.95,,4 m), (1,5,5)	27.96	62.64	21.25	118.5
(1.45, 7 m), (2,8)	1.00	6.21	13.71	50.86
(0.95, 11 m), (1.5,12)	85.03	5.74	200	114.82
(1.45, 17 m), (2,18)	1.35	24.90	18.18	52.59
(2.45, 23 m), (3,24)	13.80	4.62	200.57	334.78
(0.45, 24 m), (1,25)	106.21	1.54	356.75	290.29
(1.45, 27 m), (2,28)	1.42	31.07	26.22	114.19

LOCATE00365566 1f0 760 −83 −90, 6:18:09 pm
LOCATE00365566 1f0 760 −77 −87, 6:18:12 pm
LOCATE00365566 1f0 760 −75 −79, 6:18:15 pm
LOCATE00365566 1f0 760 −54 −68, 6:19:18 pm

With the VIRE location shows that in the 6:18:12 pm the simulation car is in the (18.329, 2.029) which means the car is 17 m from the camera. In the camera the simulation car is near 17 m from the camera. The match test is successful.

The second experiment the tachometer shows that the velocity is 141.62 m/min which is 8.4972 km/h. The detection program is shown that:

LOCATE00365566 1f0 760 −85 −85, 6:37:06 pm
LOCATE00365566 1f0 760 −78 −84, 6:37:09 pm
LOCATE00365566 1f0 760 −50 −50, 6:37:12 pm

With the VIRE location shows that in the 6:37:09 pm the simulation car is in the (18.3014, 2.0252) which means the car is 17 m from the camera.

The third experiment the tachometer shows that the velocity is 555.24 m/min which is 33.3144 km/h. The detection program is shown that:

LOCATE00365566 1f0 760 −88 −79, 11:52:09 am

LOCATE00365566 1f0 760 −82 −84, 11:52:12 am
LOCATE00365566 1f0 760 −50 −50, 11:52:16 am
LOCATE00365566 1f0 760 −51 −53, 11:52:20 am

With the camera it is found that the simulation car starts at 11:52:13 am., and the simulation car reaches destination before 11:52:16 am. The VIRE location can only detect the location on the start and the end. It means that if the velocity is too high than 24 km/h, so it could not match the moving car on the RIFD range. In this experiment, it found out the limitation of dynamic situation.

The fourth experiment the tachometer shows that the velocity is 253.2 m/min which is 15.192 km/h. The detection program is shown that:

LOCATE00365566 1f0 760 −83 −88, 12:26:24 pm
LOCATE00365566 1f0 760 −83 −88, 12:26:27 pm
LOCATE00365566 1f0 760 −57 −68, 12:26:30 pm
LOCATE00365566 1f0 760 −51 −50, 12:26:34 pm

With the VIRE location shows that in the 12:26:30 pm the simulation car is in the (8.0747, 2.1636) which means the car is 7 m from the camera.

The VIRE system can calculate the coordinate of object tags through the RFID Signal. Combining the location of object tags in the scene, it can match the location of the moving object tags. If the velocity of car over 24 km/h, the system could not match the moving car on the RIFD range (20 m).

4 Conclusions

In this paper, it integrated VIRE with LANDMARC system to locate the position of the simulation car that instead of AGV. In reality, an open area on the high speed could greatly reduce the effect of the electromagnetic reflection, so as to enhance the accuracy of the positioning. When applied in practical, more readers can also be used to ensure the accuracy of the RFID signal and thus the direction and position of the target tags. Therefore in order to guarantee a better result when using this system on the road, readers with less restriction, i.e. faster detection speed and wider detection range should be used. However in the open area the weather becomes another vital factor influencing the behaviour of the electromagnetic wave, for instance tempest, mist, thunderstorm etc. Because of the environment and technical restriction it was not able to simulate this kind of variety.

Another route to achieve a better reliability and accuracy is to combine image analysis with the RFID signal. This methodology requires a prerequisite database which records the information of the simulation cars in the respective tags. In this way the system would filter incompatible matching pair when a tag is detected within the range. As a result, the objectives tags are matched by the VIRE system which can more accurate than the LANDMARC system and Log-path loss models. Moreover, the hardware cost is same as the LANDMARC system and VIRE system can improve the accuracy under the same situation as LANDMARC. Meanwhile, the estimated position of objective tags is more accurate because it works with virtual tags to define the more unknown position.

Acknowledgments The authors acknowledge the funding support by University of Macau under conference grant and The Science and Technology Development Fund, Macau SAR (File no. 0027/2018/ASC and 0041/2018/AIR). The authors also acknowledge Mr. Lok Ka Iat, Mr. Wang Jun Feng, and Mr. Wu Bi Xin for their support in data collection.

References

Andersen JB, Rappaport TS, Yoshida S (1995) Propagation measurements and models for wireless communications channels. IEEE Commun. Mag. 33(1):42–49

Giustia I, Cepolina EM, Cangialosi E, Aquaro D, Caroti G, Piemonte A (2019) Mitigation of human error consequences in general cargo handler logistics: impact of RFID implementation. Comput. Indus Eng 137:1–11 (106038)

Haykin S (2005) Cognitive radio: brain-empowered wireless communications. IEEE J Sel Areas Commun 23:201–220

Ho WI, Wong SF (2012) Noise factor of RFID-aided positioning. Int J Electr Commun Eng 6(7):684–689

Hong SK, So YS (2009) Cost of ownership model for the RFID logistics system applicable to u-city. Eur J Oper Res 194:406–417

Iskander MF, Yun Z (2002) Propagation prediction models for wireless communication systems. IEEE Trans Microw Theory Techn 50:662–673

Jain S, Sabharwal A, Chandra S (2010) An improvised localization scheme using active RFID for accurate tracking in smart homes. In: 12th international conference on computer modelling and simulation, pp 51–56

Jiang XJ, Liu Y, Wang XL (2009) An enhanced approach of indoor location sensing using active RFID. In: International conference on information engineering, pp 169–172

Kirch M, Poenicke O, Richter K (2017) RFID in Logistics and production—applications, research and visions for smart logistics zones. Procedia Eng 178:526–533

Li JH, Zhang GM, Yu L, Wang ZX, Zhang J (2013) An advanced RFID localization algorithm based on region division and error compensation. KSII Trans Internet Inform Syst 7(4):670–691 (2013)

Liu BH, Otis B, Challa S, Axon P, Chou CT, Jha S (2006) On the fading and shadowing effects for wireless sensor networks. In 2006 IEEE international conference on mobile adhoc and sensor systems (MASS), pp 51–60

Oztekin A, Pajouh FM, Delen D, Swim LK (2010) An RFID network design methodology for asset tracking in healthcare. Decis Support Syst 49:100–109

Sharma PK, Singh R (2010) Comparative analysis of propagation path loss models with field measured data. Int J Eng Sci Technol 2:2008–2013

Ukkonen L, Sydanheimo L, Kivikoski M (2005) Effects of metallic plate size on the performance of microstrip patch-type tag antennas for passive RFID. IEEE Antennas Wirel Propag Lett 4:410–413

Wong SF, Yu Z (2019a) The mobile robot anti-disturbance vSLAM navigation algorithm based on RBF neural network. Procedia Manuf 38:400–407

Wong SF, Yu Z (2019b) Flexsim-based modulation and simulation for logistics batch picking. In: Proceedings of 2019 the 3rd international conference on software and e-business (ICSEB 2019), pp 141–145

Zhong RY, Huang GQ, Lan SL, Dai QY, Xu C, Zhang T (2015) A big data approach for logistics trajectory discovery from RFID-enabled production data. Int J Prod Econ 165:260–272

Zhou WH, Jiang ND, Yan CC (2019) Research on anti-collision algorithm of RFID tags in logistics system. Procedia Comput Sci 154:460–467

Logic for Management of Vehicles at Warehouse

Kamila Kluska and Patrycja Hoffa-Dąbrowska

Abstract The main goal of this chapter is to present a mechanism for managing the movement of vehicles in a warehouse. The logic for managing two means of warehouse transportation is part of a research project for Zrembud Cieszyn, located in Poland. The company needs simulation to help design and verify rules for coordinating vehicle movement and communication. The logic was written for warehouse transport using forklifts and a specialized, new type of automatic guided vehicle (named Transfer Unit) in two different warehouse systems. The approach continuously determines vehicle positions based on controlling markers, traffic restrictions, communication, order assignment, and safety requirements in many different warehouse configurations (including rack parameters and inbound/outbound buffer locations). The method considers a new type of storage, referred to as drawer racks. The research work uses actual data, rules and technical information from the company.

1 Introduction

This chapter addresses one stage of simulation modelling of storage systems and supports the innovative design of a new semi-automatic storage system. The system consists of racks placed on metal frames (called drawers, platforms) that are turned 90° relative to the transport corridor (Pawlewski and Kunc 2019). Static platforms with single racks are placed at the end of each row with double platforms between them on which two racks are placed. Due to the impeded access to storage spaces in the double rack, an AGV vehicle (Transfer Unit, which slides in and out of the platforms perpendicular to the corridor) is implemented into the system. In order to reach the rack, the AGV moves under metal frames where the racks stand. In order to access the racks, an AGV moves under the platform, lifts and slides it out. After loading/unloading of containers, the platform with racks is pushed back into its

K. Kluska (✉) · P. Hoffa-Dąbrowska
Poznan University of Technology, Jacka Rychlewskiego St. 2, 60-965 Poznan, Poland
e-mail: kamila.kluska@put.poznan.pl

© The Author(s), under exclusive license to Springer Nature Switzerland AG 2020
P. Golinska-Dawson et al. (eds.), *Smart and Sustainable Supply Chain and Logistics – Trends, Challenges, Methods and Best Practices*, EcoProduction,
https://doi.org/10.1007/978-3-030-61947-3_5

Fig. 1 The view of prototype of drawer rack system (*Source* https://zrembud.com.pl/nowe-systemy-regalowe/)

default position in the rack row and is lowered. Implementation of described concept is shown in Fig. 1.

The most important difference between the SRS (drawer rack system) and SRR (row racking system) systems results from the dynamic structure of the SRS system and the static structure of the SRR system. The applied structures directly result in differences in the availability of loads in both systems and in the structure of racks. Loads are permanently available in SRR as racks have a fixed position on the warehouse layout. In the SRS system, loads are not available due to the perpendicular positioning of racks relative to the corridor and their placement next to each other. Access to loads is only possible if the rack slides out towards the corridor. This is possible through the use of a new storage structure—placing racks on movable metal platforms. Changing the position of the rack also causes the change of the position of stopping points (located on the sides of the rack), reduction of the corridor capacity and even its complete blocking. This creates new challenges related to SKU allocation, replenishment, picking, determining how to manage forklifts and AGVs, and the principles of their cooperation.

The methodology considers both traditional row racking systems (SRR and automatic drawer racking systems (SRS), as well as handling orders related to warehouse stocking (PZ) and shipments (WZ). The warehouse structure can be freely modified depending on the users' requirements. Algorithms for generating routes for forklifts and AGVs are adapted to consider operation in any type of warehouse structure.

The stages for developing the methodology are presented in Fig. 2.

The first two stages are for defining the project and preparing information and structures needed to build an automatic warehouse generation mechanism. The next three steps are the process of building this mechanism.

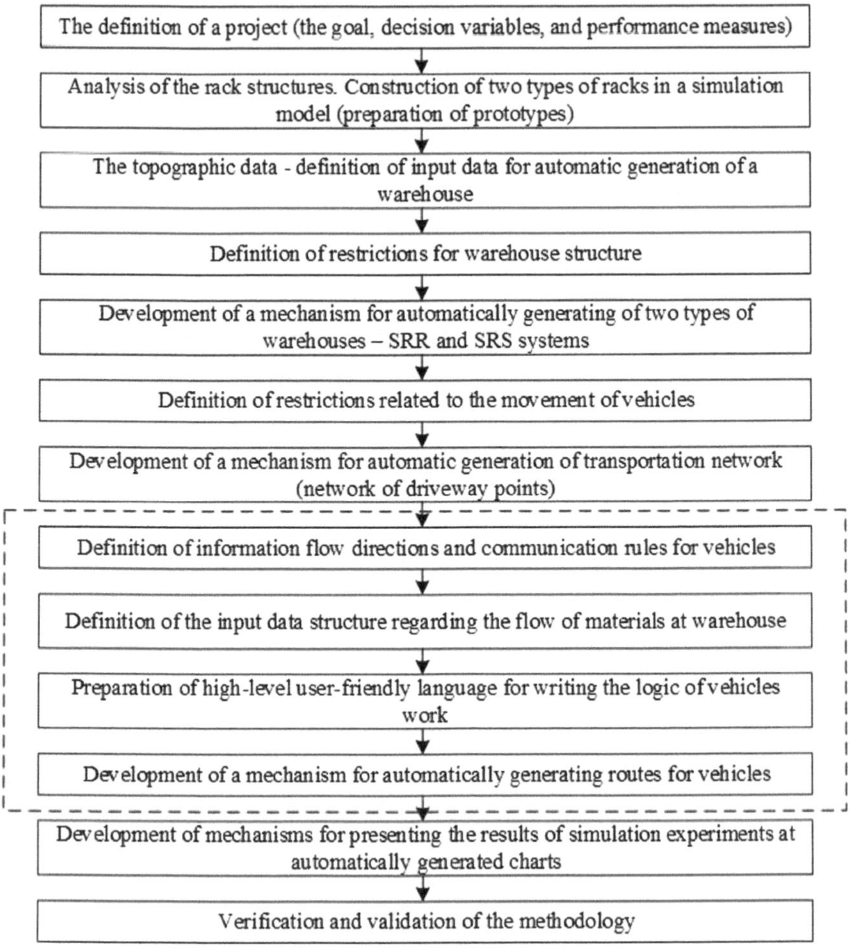

Fig. 2 The stages of building the methodology of simulation modeling of storage systems. *Source* Own work

The next two steps include the construction of a mechanism for automatically generating a transport network consisting of various kinds of checkpoints (driveway points). These points guarantee traffic safety and enable its management. Checkpoints designate:

- forklifts paths between corridors,
- places of forklift entry and exit to/from the corridor,
- forklifts paths between rows of racks,
- AGV paths under rows of shelves,

- points in the corridor to which AGVs slide out the platforms with racks to allow forklifts access to the storage locations and points that determine the path of the AGV to return to the row of racks when the platform is pulled back,
- places in front of the racks where forklifts stop to perform loading and unloading operations,
- stopping places for the forklifts while they are waiting for an AGV to stop moving,
- detour path around the extended AGV (with platform) for a forklift.

The next steps in building the methodology (marked with a blue frame in Fig. 1) relate to the definition of the principles of communication and cooperation between AGV and forklifts. This is the basis for the automatic management of their work. This mechanism is the subject of this chapter. The last stages of building the methodology include preparing mechanisms for generating charts (analysis of results) and testing the built tool.

Every simulation project requires a clear definition of goals. It also requires the definition of decision variables and performance measures. Through these activities, developers know how to evaluate the alternative solutions, what is being assessed in the alternatives, and what factors to modify to obtain the best result.

The objective of the project is to achieve maximum efficiency of the warehouse system by having fork trucks travel the smallest distance and requiring the smallest area of the warehouse.

The most important decision variable is the warehouse structure, i.e., its layout. A second key decision variable is the number of forklifts operating in the system.

The main performance measures that are used to assess alternatives and to drive the simulation experiments are:

- System performance: for a given list of transport orders, the time it takes to complete this list.
- Distance travelled by forklifts and transfer units (AGVs).

Another measure of performance, that is evaluated outside of the simulation experiments is the total warehouse area that is calculated as the warehouse layout changes.

The main goal of this chapter is to present a mechanism that manages the movement of vehicles in semi-automated and traditional warehouse system that adapts to the warehouse layout and the locations of inbound and outbound buffers.

The chapter is organized in five sections. Section 2 provides a literature review about planning and managing automated vehicles. Section 3 discusses the methodology in the context of simulating storage systems. Section 4 describes the logic for managing vehicles in warehouses that use SRS and SRR systems. Section 5 describes the simulation experiments and their results. Section 6 provides conclusions and discusses future work.

2 Literature Review

There are many different means of transport in storage areas, i.e. cranes, conveyors, transport trucks (Raczyk 2009), hand trolleys, forklifts, stacker cranes, and AGVs. Depending on the type of stored goods, type of warehouse, owned storage facilities and the level of automation, various means of transport are used. Globalization, the pace of production, and the level of technology and automation all impact the means of transport that are used in operating storage areas. As process automation increases, there is also an increase in the use of autonomous industrial trucks (Polten and Emde 2020; Saffar et al. 2017). The literature confirms the growing trend in AGV applications in storage areas. Searching the SCOPUS database for the words "automated guided vehicle" and "warehouse" in tittle/abstract/keyword, the number of articles was 200. Figure 3 presents the number of articles authored by year until 2019.

In recent years, the popularity of using drones in warehouse-picking area has increased, as indicated by Derpich et al. (2018), Perussi et al. (2019), Betti Sorbelli et al. (2019).

The automation of warehouse processes is associated with automated storage and retrieval system (AS/RS). A discussion of AS/RS systems is found in (Van den Berg and Gademann (2000). and a comparison of AS/RS and AVS/RS systems is provided in Ekren and Heragu (2011).

Various methods for managing the work of AGVs and planning their routes are found in the following:

- a quadratic optimization method for coordinating AGVs (Digani et al. 2019),
- a time-efficient approach to solve conflicts and deadlocks for scheduling AGVs (Failed 2018),

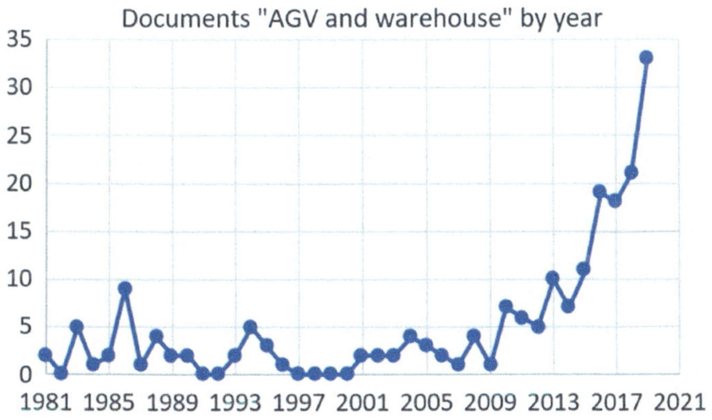

Fig. 3 The number of articles about automated guided vehicles in warehouses by year in SCOPUS base. *Source* SCOPUS base (13 March 2020)

- multi-AGVs conflict-free routing and dynamic dispatching strategies for automated warehouses (Li et al. 2019),
- using two adaptive genetic algorithms and a multi-adaptive genetic algorithm to optimize the task scheduling of AGVs (Liu et al. 2019),
- a neural network-based algorithm with genetic training for a combined job and energy management for AGVs (Pagani et al. 2018),
- fuzzy-set qualitative comparative analysis applied to the design of a network flow of AGVs (Llopis-Albert et al. 2019),
- a novel tabu search algorithm for routing problems (Xing et al. 2020),
- a time–space network model for path planning (Yin et al. 2019),
- collision-free route planning for multiple AGVs in automated warehouses (Zhang et al. 2018).

In case of AGV, many articles focus on routing, network planning, and collision avoidance.

3 Methodology

The mechanism for managing forklifts and AGVs that is described in this paper is a part of a larger simulation modelling methodology for storage systems that is contained within the LogABS software. LogABS is a 3D-based simulation software that is dedicated to the design and redesign of factory intralogistics. Production planning, workstation and production line balancing, analysis of work ergonomics and yamazumi analysis are also feasible using LogABS (https://logabs.com/). The most important reasons for choosing this software are: ease of use, access to set of tools supporting the creation and modification of the factory layouts, built-in mechanisms for assessing value-added and performing ergonomic analyses, and it uses a language that is understood by production engineers.

The use of the mechanism for automatic management of forklifts and AGVs is the stage of the methodology that is highlighted in Fig. 4.

The first four stages of the methodology relate to the procedures necessary to automatically generate a warehouse. The next two steps marked in Fig. 3 relate to the mechanism for managing the operation of vehicles in a simulation model. These steps concern the input of material flow data, the automatic processing of this data for the current generation of forklift and AGV truck routes during the simulation experiment (Pawlewski 2018b, 2019). Routes are generated by taking into account the spatial relationships of vehicles and their communication. The methodology also includes generating graphs that display the outcome of simulation experiments. The implementation of the stages related to the automatic generation of warehouses and the development of charts for the presentation of results, is associated with the stages of methodology construction presented in Fig. 1 and is the subject of other papers.

The use of a mechanism for automatic management of forklifts and AGVs in a warehouse requires.

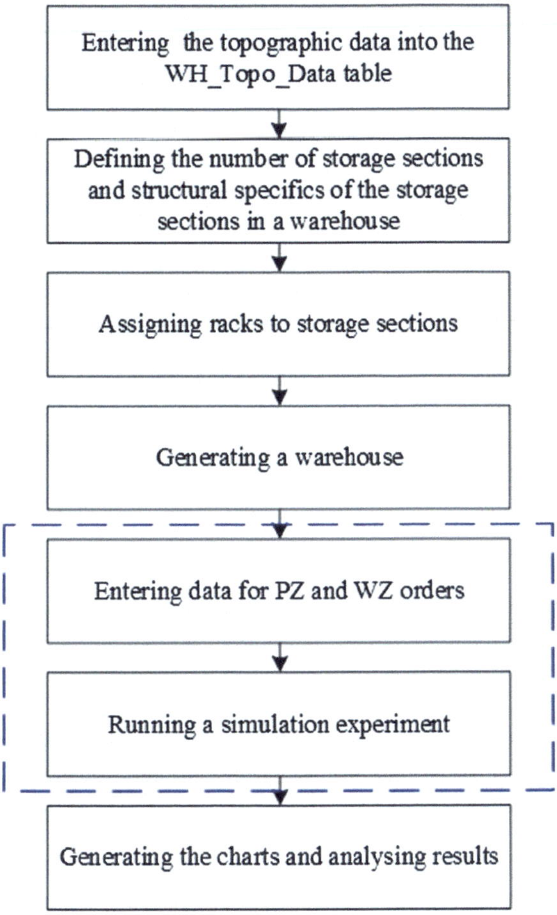

Fig. 4 The stages of methodology for automatically building simulation models of storage systems. *Source* Own work

3.1 Entering a Series of Input Data into the WZ_PZ_SRS and WZ_PZ_SRR Tables, Which Have the Same Structure

The tables are used to determine the different types of flows in warehouses that can be simulated. An example of the WZ_PZ_SRS table is shown in Fig. 5.

The type of transport order is specified in the first column of the table. PZ means the order related to accepting the load into the warehouse, i.e. transporting the load from the input buffer to the rack. WZ means a release order from the warehouse, i.e. transport of load from the rack to the output buffer.

The SKU column designates the load index to be transported. This is information is for the user and does not affect the operation of the mechanism.

Model | WZ_PZ_SRR | WZ_PZ_SRS

	PZ/WZ	SKU	Address	Address LogABS	Buffer	Status
Row 1	PZ	GL2279303	2B_21_e_2	S2B_11/P_38	IN_SRS/P_01	0
Row 2	WZ	GL263484	1A_01_a_2	S1A_01/P_02	OUT_SRS/P_01	0
Row 3	WZ	GL3757303	3B_12_c_1	S3B_07/P_09	OUT_SRS/P_01	0
Row 4	PZ	GL3798603	1A_22_b_2	S1A_12/P_06	IN_SRS/P_01	0
Row 5	PZ	GL428584	3B_16_f_1	S3B_09/P_21	IN_SRS/P_01	0
Row 6	WZ	GL6560603	1A_11_e_1	S3B_09/P_03	OUT_SRS/P_01	0
Row 7	WZ	GL4581503	3B_17_c_3	S3B_09/P_35	OUT_SRS/P_01	0
Row 8	PZ	GL4660603	3A_20_d_2	S3A_11/P_14	IN_SRS/P_01	0
Row 9	WZ	GL4560603	1A_02_d_4	S3A_11/P_01	OUT_SRS/P_01	0
Row 10	PZ	GL6581503	2A_21_f_3	S2A_11/P_47	IN_SRS/P_01	0
Row 11	WZ	GL6760603	2A_22_d_2	S2A_12/P_14	OUT_SRS/P_01	0
Row 12	PZ	GL6959103	2B_01_e_4	S2B_01/P_20	IN_SRS/P_01	0
Row 13	PZ	GL2279303	3B_17_c_1	S3B_09/P_33	IN_SRS/P_01	0
Row 14	WZ	GL263484	2B_14_d_2	S3B_09/P_04	OUT_SRS/P_01	0

Fig. 5 WZ_PZ_SRS data table for drawer racking systems. *Source* Own work

Address is the name of the storage location using the customer's notation. Address LogABS is an automatically generated name for an object in the simulation model; it is in the notation of the LogABS simulation program (Pawlewski 2018a). This address is based on the Address column that uses the customer's notation.

The Buffer column is the name of the buffer (along with the specification of the storage place number) that forms the link during the flow implementation. Loads are picked from this buffer during the execution of a PZ order or are dropped to this buffer during the execution of a WZ order.

The Status column specifies the status of an order. If an order is not completed, the status is 0 (i.e. the order has not been executed). After completing the order, the status changes to a value of 1. Based on this, the user can see what orders are completed as a simulation runs. After completing a simulation experiment, the user can also check whether any of the orders have been missed, e.g. as a result of defining an address that does not exist in the generated warehouse.

3.2 Determining the Variant of the Inbound and Outbound Buffers Arrangement Relative to the Storage Racks

The setting options currently available in the mechanism are shown in Table 1. In Table 1 the four locations of in and out buffers are presented. The green rectangles in the figures labelled "IN" denote an inbound buffer. Orange rectangles denoted with "OUT" represent outbound buffers. The blue areas are rows of racks.

Table 1 The variants of the inbound and outbound buffer arrangement relative to the storage racks

Variant 1	Variant 2
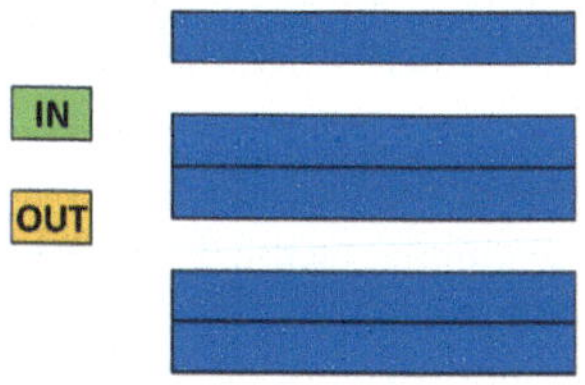	
Variant 3	Variant 4

Source Own work

Buffers can be located on the left side of the racks (variant 1), on the right side of the racks (variant 3) or separately on the left and right side (variant 2 and 4). Establishing the variant of setting buffers relative to the storage area determines the correct operation of the algorithm, i.e. the faultless indication of forklift trucks' paths through the warehouse during the execution of transport orders.

4 Logic of the Vehicle Management Mechanism

The most important assumptions for the algorithm for the SRS system:

1. A maximum of one forklift in a corridor. When there is a forklift in the corridor, no other forklift can enter the corridor to process an order.
2. Allocation of transport orders to forklifts is based on the availability of forklifts and corridors.
3. A list of orders is analyzed in turn, and a WZ order is selected for each PZ order (and vice versa), whose execution will take place from: (1) the same rack, (2) another rack located on the same platform, (3) another a rack in the same corridor in the same row, or (4), another rack in the same corridor but on the other side (in a different row). The search order is carried out according to priorities.
4. If a transport order cannot be made, the forklift leaves the corridor and waits in a parking area.
5. The system is operated by forklifts and AGVs.

6. Only one AGV truck can move under one row of racks, which is permanently assigned to it.
7. AGVs have priority over forklifts. When an AGV moves, a forklift must not enter within a safety range of one meter from its edges. If necessary, the forklift waits for the AGV to end its movement.
8. Forklift have the ability to communicate (send signals) to AGVs and vice versa.
9. Vehicles move according to control markers (driveway points) on the floor. Only one vehicle can stay on a given marker at a time.
10. The maximum capacity of a forklift is one transport unit (one container).
11. The maximum capacity of an AGV is one platform.
12. All forklifts have the same motion parameters and capabilities.
13. AGV trucks have the same motion parameters.
14. Each vehicle can carry out only one transport order at a time.
15. The capacity of inbound and outbound buffers is unlimited.
16. The capacity of one storage place on a rack is one transport unit (one container).
17. Loads are always available on racks. It is not possible for a transport order not to be carried out due to the lack of a load on a rack.
18. After a forklift has finished working with the racks on a double platform, the AGV must slide it into the rack row to its default location.

The most important assumptions for the algorithm for the SRR system:

1. The maximum number of forklifts that can be in a corridor at the same time is set by the user. When the maximum number of forklifts are in a corridor, no additional forklifts can enter the corridor to complete an order.
2. Allocation of transport orders to forklifts is based on the availability of forklifts and corridors.
3. A list of orders is analyzed in turn, and a WZ order is selected for each PZ order (and vice versa), whose execution will take place from (1) the same rack, (2) another a rack in the same corridor in the same row, or (3) another rack in the same corridor but on the other side but in a different row. The search order is carried out according to priority values.
4. In the absence of a transport order that can be made, the forklift truck leaves the corridor and waits in a parking area.
5. The system is only operated by forklifts.
6. Vehicles move according to control markers (driveway points) on the floor. Only one vehicle can stay on a given marker at a time.
7. The maximum capacity of a forklift is one transport unit (one container).
8. All forklifts have the same motion parameters and capabilities.
9. Each vehicle can carry out only one transport order at a time.
10. The capacity of inbound and outbound buffers is unlimited.
11. The capacity of one storage place on a rack is one transport unit (one container).
12. Loads are always available on racks. It is not possible for a transport order not to be carried out due to the lack of a load on a rack.

The most important part of the vehicle management mechanism are algorithms that:

- load and process data concerning transport orders,
- observe the availability of forklifts and AGVs,
- manage the allocation of transport orders (by analyzing and selecting orders according to priorities and allocating them for execution depending on the availability of transport resources) for forklifts and AGVs,
- generate vehicle work routes that enable proper vehicle movement and communication, meet the assumptions, and take into account the work coordination,
- record the execution of individual transport orders and the entire list of orders (completion of vehicle operations),
- manage the movement of vehicles to their default locations when idle.

4.1 The Algorithm for Managing Forklifts and AGVs in the SRS System

The algorithm is shown in Fig. 6 and is divided into six modules.

The first module is responsible for activating the algorithm, loading and processing the data concerning transport orders, as well as checking the number of available forklifts and AGVs. This module also includes a system preventing activation of the mechanism in the event when the drawer warehouse was not generated by the user.

The second module is a continuation of the first module. This part of the algorithm manages transport orders—their selection and assignment to available forklifts and AGVs. After assigning the transport order, the transition to the third module follows. In the event of completion of all transport orders from the list, the transition to module six follows.

In the third module analysis of the assigned task is taken place. The algorithm checks in turn:

- variant of space organization in the warehouse,
- is the address in the transport order on the single or double racking platform,
- from which side of the double rack the load will be picked up,
- how far the double rack will slide out (how many storage places in relation to the depth of the platform), as well as
- what is the type of order (PZ or WZ).

All this information is necessary to specify in the fourth module:

- whether it is necessary to use (activate) the AGV truck,
- what access path to set for vehicles,
- how routes should be generated,
- what number of routes is needed,
- at which moment vehicles will exchange signals initiating the movement.

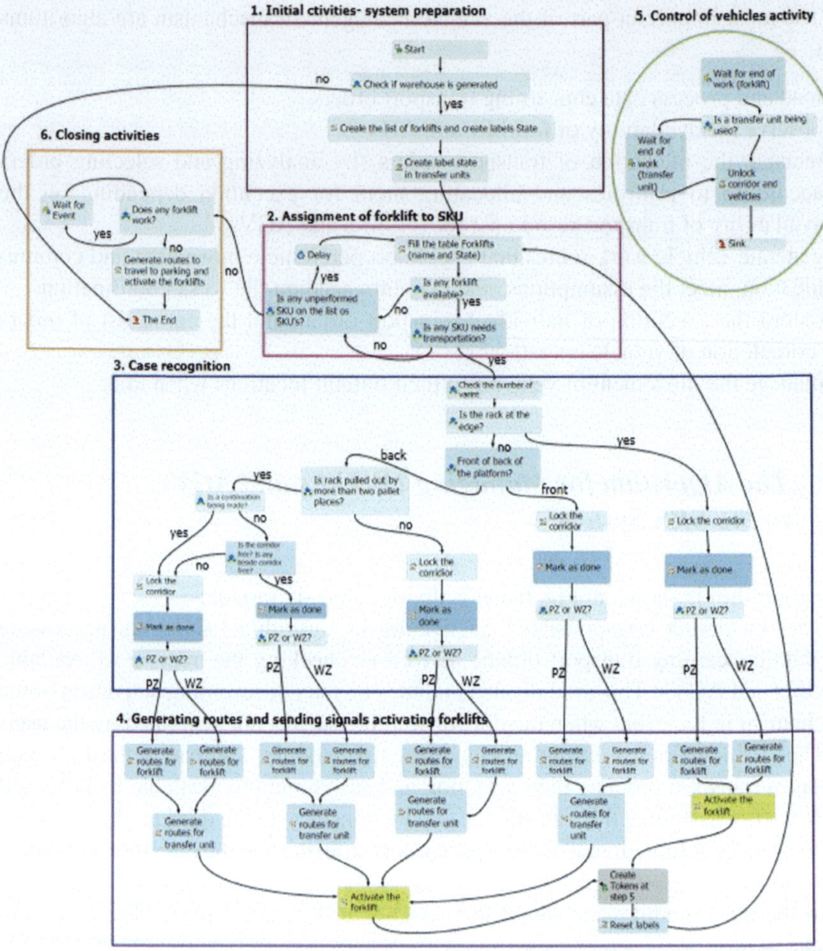

Fig. 6 The algorithm for managing forklifts and AGVs in the SRS system. *Source* Own study with use LogABS program

The fourth module generates routes for forklifts and AGVs, activate forklifts and also the mechanism embedded in the fifth module.

In the fifth module observation of the execution of the transport order by the vehicles is taking place. When the vehicles complete the generated routes (they finish carrying out activities related to the implementation of a given transport order), the mechanism will record it and set them as available. Availability means that the algorithm can resume the implementation of the second module and thus allocate another order for implementation.

The sixth module is activated when all orders from the list are completed. This module checks if all vehicles have completed their movement. If yes, it generates routes that allow vehicles to return to parking lots and activates them to execute these routes.

4.2 The Algorithm for Managing Forklifts in the SRR System

The SRR system does not use AGV trucks, but only forklifts. The algorithm works analogously to the algorithm for the SRS system and has been divided into six modules.

The first and second modules works analogously to the algorithm for the SRS system, without consideration of AGV's availability. After assigning the transport order, the transition to the third module follows.

In the third module analysis of the assigned task is taking place. The algorithm checks in turn:

- variant of space organization in the warehouse,
- what is the type of order (PZ or WZ).

All this information is necessary to specify in module four:

- what access path to set for vehicles,
- how routes should be generated,
- what number of routes is needed.

In the fourth module, routes for forklifts are generated, forklifts are activated and the mechanism embedded in the fifth module is activated.

The fifth and sixth modules work in a similar way to the modules included in the algorithm for the SRS system, but they omit the observation of AGVs activity.

4.3 High-Level Language for Vehicles Routes

High-level language is used to record routes performed by forklifts and AGVs. This language must make it possible to model the logic of forklift operation in a way that correctly recreates the procedure of loading and unloading the rack with a forklift. The recording of activities on the route must enable:

- moving the vehicle in accordance with the relevant rules and the right path,
- finding addresses in model space,
- picking and putting away containers at designated addresses,
- communicate with other traffic contractors in a given space.

Such conditions are met by the LogABS (Pawlewski 2018a, b) language in which logic is written using high-level instructions. LogABS uses a set of 64 different instructions that, written one after the other, form a work itinerary.

The structure of the LogABS instruction is as follows (Pawlewski 2019):

- Address (location of the object to which the instruction relates),
- Instruction (activity name),
- Parameter (field in which duration or quantity is defined depending on the type of instruction).

The following LogABS instructions are used to write the logic of vehicles:

- Travel—go to the designated place in the model space (without load),
- TravelLoaded—go to the designated place in the model space (with load),
- Load—load a specified number of parts or containers directly from the location,
- Unload—unload a specified number of parts/containers directly to the location,
- Call—activates the route saved in the table,
- ReadyForTask—set the readiness status. It means that the operator has finished his work and can accept new order,
- StartOperator—initiate the work of the operator,
- RepTime—save time. Based on the entries, work time charts are generated,
- PickUp—lift the racks located in the drawer—lift the drawer,
- Lower—put the drawer on the floor.

Logic of transport vehicles work is built from the LogABS instructions listed above. A fragment of the forklift route is shown in Fig. 7, while a fragment of the AGV route is shown in Fig. 8.

	ID	Where	Activity	Duration [s] / How many
Row 1	1		RepTime	24
Row 2	2	GG_002	Travel	0
Row 3	3	Transport_1A/Ou_01/S1A_02/N_06	Travel	0
Row 4	4	Transport_1A/Ou_01/S1A_02/P_06	Load	1
Row 5	5	Transport_1A/Ou_01	StartOperator	2
Row 6	6	Buffer/I_01	Travel	0
Row 7	7	Buffer/P_01	Unload	1
Row 8	8		RepTime	24
Row 9	9		ReadyForTask	0

Fig. 7 A fragment of the forklift route recorded with use LogABS language. *Source* Own study with use LogABS program

	ID	Where	Activity	Duration [s] / How many
Row 1	1	GG_041	Travel	0
Row 2	2	/S1A_05	PickUp	1
Row 3	3	GG_042	Travel	0
Row 4	4	ForkLifts/Op_01	StartOperator	3
Row 5	5		ReadyForTask	0

Fig. 8 A fragment of the AGV route recorded with use LogABS language. *Source* Own study with use LogABS program

5 Simulation Experiments

A number of computer simulations were carried out for the described simulation model in order to compare the SRS and SRR systems. For the purposes of the paper, three experiments are compared. In each experiment, the arrangement of racks is different, while the total storage capacity for the SRS and SRR systems is the same. The rack settings for each experiment are shown in Table 2.

In respective simulation variants, the number of rows and the number of racks in a row were changed. Basic information is presented in Table 3.

For each experiment, the distance travelled by all vehicles in a given simulation was compiled. The results are summarized in Table 4. It can be noticed that each time the distance travelled in the SRR system was lower by 218 m, 1,156 m and 706 m respectively.

In addition, the time of execution of 70 PZ and WZ orders was measured. The results are shown in Table 5. In all cases this time was shorter for the SRR system.

Comparing the SRS and SRR systems based on the three experiments, it can be seen that:

- the area occupied by a certain number of locations is smaller in the case of the SRS system, which is based on condensed racking in the case of the SRS system,
- the distance travelled by all vehicles is lower in the case of the SRR system,
- the time of completing 70 PZ/WZ orders is shorter in the SRR system.

Table 2 Rack setting for the SRS and SRR systems in each (respective) experiment

E. no	SRS	SRR
1	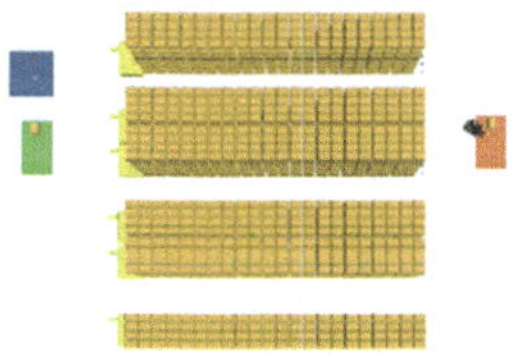	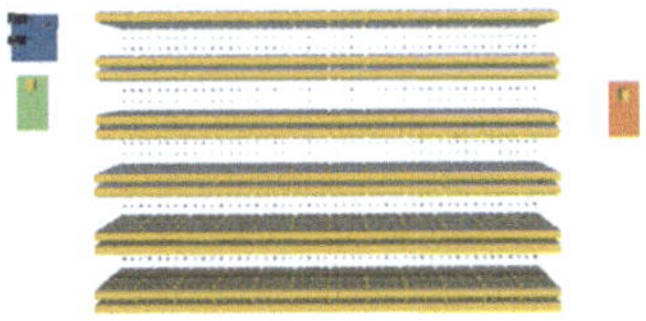
2	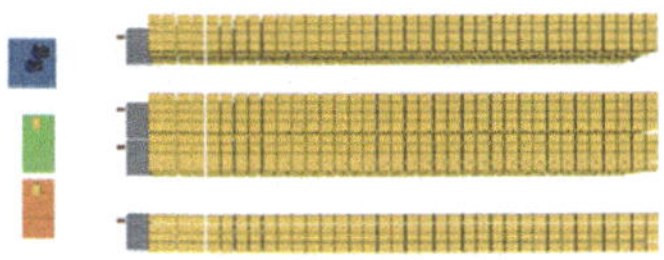	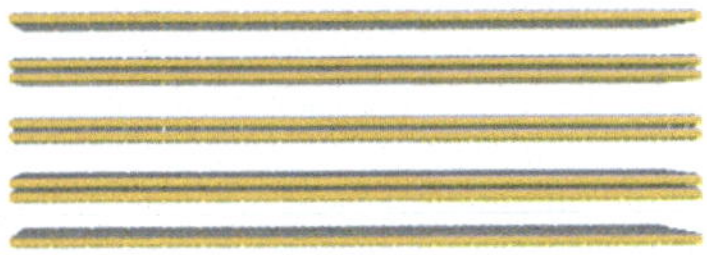
3	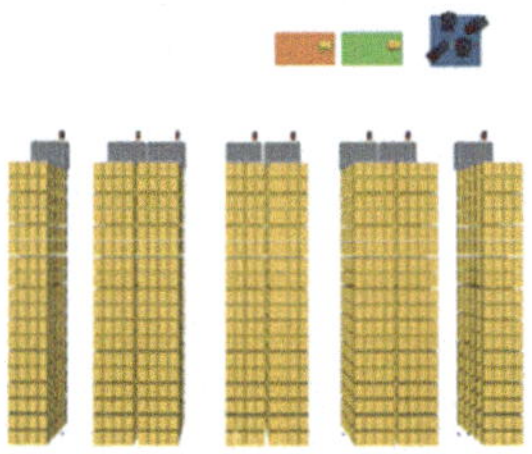	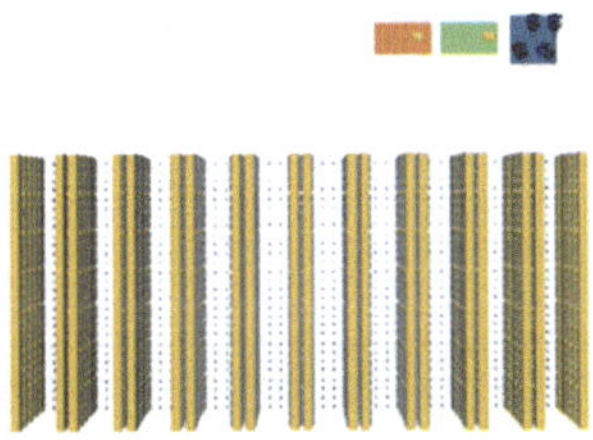

Source Own work

Table 3 Basic SRR and SRS system settings for each experiment

	Experiment 1		Experiment 2		Experiment 3	
	SRS	SRR	SRS	SRR	SRS	SRR
Number of rows of racks	6	11	4	8	8	20
Number of racks within a single row	12	12	19	18	10	7
Total capacity of warehouse (no. of locations)	2640	2640	2880	2880	2880	2800
Width of storage area (m)	32	44	48	66	25	25,5
Length of storage area (m)	34	32	22,5	22,5	44,86	56,6
Storage area (m^2)	1088	1408	1080	1485	1121,5	1443,3
Number of forklifts	2	2	2	2	4	4
Number of AGV vehicles	6	0	4	0	8	0

Source Own work

Table 4 Distance travelled by vehicles in experiments

	Experiment 1		Experiment 2		Experiment 3	
	SRS	SRR	SRS	SRR	SRS	SRR
Distance traveled by forklifts and transfer units (m)	5276,851	5052,86	5972,205	4816,567	4318,776	3613,194

Source Own work

Table 5 The time of execution of 70 PZ and WZ orders

	Experiment 1		Experiment 2		Experiment 3	
	SRS	SRR	SRS	SRR	SRS	SRR
Time of completing the list of 70 tasks PZ/WZ (min)	34,32	29,61	38,27	28,38	17,73	11,94

Source Own work

6 Conclusions and Future Directions

The paper describes the mechanism of managing forklifts and AGVs work, which is an important part of the simulation modelling methodology for storage systems. The mechanism enables the user without programming skills to build a simulation model of two storage systems and support advanced algorithms from the level of two tables with a simple structure. Algorithms automatically adjust the way forklifts move to modelled storage system structures. The tool allows to enter data on the flow of materials and automatically processes them to model transport processes and vehicle operation logic, including their communication and complex decision-making processes.

The result of building the mechanism is the execution of orders in simulation models. To sum up the results of simulation experiments, in terms of time and distance, the SRR system prevails over the SRS system. However, it is worth considering both cases. Depending on whether the speed of order execution or area occupied by the same number of locations is more important, different rack storage system will be chosen. Both the SRS and SRR systems are characterized by their specific advantages and functionalities, as well as limitations. Therefore, it is worth using simulation tools that allow comparing both concepts in the virtual world, and then select the most suitable variant for the enterprise.

Further development of the research involves building a mechanism for collecting, processing and presenting results in the form of charts. To increase the usability of the methodology, it is planned to expand the mechanism of managing the operation of trucks by including other types of vehicles and enabling the operation of other storage systems. An important stage of the planned methodology evolution is the development of mechanisms enabling the implementation of resource and process cost accounting.

References

Betti Sorbelli F, Coro F, Pinotti CM, Shende A (2019) Automated picking system employing a drone. In: Proceedings—15th annual international conference on distributed computing in sensor systems, DCOSS 2019, May 2019, Article number 8804800, pp 633–640. https://doi.org/10.1109/DCOSS.2019.00115

Derpich I, Miranda D, Sepulveda J (2018) Using drones in a warehouse with minimum energy consumption. In: 2018 7th international conference on computers communications and control (ICCCC). IEEE. https://doi.org/10.1109/ICCCC.2018.8390444

Digani V, Hsieh MA, Sabattini L et al (2019) Coordination of multiple AGVs: a quadratic optimization method. Auton Robot 43:539–555. https://doi.org/10.1007/s10514-018-9730-9

Ekren BY, Heragu SS (2011) Simulation based performance comparison of AVS/RS and AS/RS. In: IIE annual conference proceedings, pp 1–7

https://logabs.com/. Accessed 6 May 2020

Li X, Zhang C, Yang W, Qi M (2019) Multi-AGVs conflict-free routing and dynamic dispatching strategies for automated warehouses. In: Kim K, Kim H (eds) Mobile and wireless technology 2018, ICMWT 2018. Lecture notes in electrical engineering, vol 513. Springer, Singapore

Liu Y, Ji S, Su Z, Guo D (2019) Multi-objective AGV scheduling in an automatic sorting system of an unmanned (intelligent) warehouse by using two adaptive genetic algorithms and a multi-adaptive genetic algorithm. PloS One 14(12)

Llopis-Albert C, Rubio F, Valero F (2019) Fuzzy-set qualitative comparative analysis applied to the design of a network flow of automated guided vehicles for improving business productivity. J Bus Res 101, August 2019, 737–742

Pagani P, Colling D, Furmans K (2018) A neural network-based algorithm with genetic training for a combined job and energy management for AGVs. Logist J: Proc 2018. https://doi.org/10.2195/lj_Proc_pagani_en_201811_01

Pawlewski P (2018a) Script language to describe agent's behaviors. In: Highlights of practical applications of agents, multi-agent systems, and complexity: the PAAMS collection: international workshops of PAAMS 2018, Toledo, Spain, 20–22 June 2018, Proceedings, red. Javier Bajo, Patrycja Hoffa-Dąbrowska (WIZ). Springer, 2018, pp 137–148

Pawlewski P (2018b) Methodology for layout and intralogistics redesign using simulation. In: Rabe M, Juan AA, Mustafee N, Skoogh A, Jain S, Johansson B (eds) Proceedings of the 2018 winter simulation conference. IEEE Press, 2018, pp 3193–3204

Pawlewski P (2019) Built-in lean management tools in simulation modelling. In: 2019 winter simulation conference (WSC), National Harbor, MD, USA, pp 2665–2676

Pawlewski P, Kunc T (2019) Using agent base simulation to model operations in semi-automated warehouse. In: Highlights of practical applications of survivable agents and multi-agent systems: the PAAMS collection: international workshops of PAAMS 2019 Ávila, Spain, 26–28 June 2019, Proceedings, red. Fernando De La Prieta, Alfonso González-Briones, Paweł Pawlewski (WIZ), Davide Calvaresi, Elena Del Val, Fernando Lopes, Vincente Julian, Eneko Osaba, Ramón Sánchez-Iborra. Springer Nature Switzerland AG, Cham, Switzerland, pp 50–61

Perussi JB, Gressler F, Seleme R (2019) Supply chain 4.0: autonomous vehicles and equipment to meet demand. Int J Supply Chain Manag 8(4), August 2019, 33–41. ISSN 20513771

Polten L, Emde S (2020) Scheduling automated guided vehicles in very narrow aisle warehouses, Omega. Available online 23 January 2020, 102204, In Press, Corrected Proof, Elsevier.https://doi.org/10.1016/j.omega.2020.102204

Raczyk R (2009) środki transportu bliskiego i magazynowania, Wydawnictwo Politechniki Poznańskiej, Poznań. ISBN 978-83-7143-828-8

Saffar S, Jamaludin Z, Jafar F (2017) Investigating the influences of automated guided vehicles (AGVs) as material transportation for automotive assembly process. J Mech Eng SI 4(1), 1 August 2017, 47–60. ISSN 18235514

Tai R, Wang J, Tian W, Chen W, Wang H, Zhou Y (2018) A time-efficient approach to solve conflicts and deadlocks for scheduling AGVs in warehousing applications. In: 2018 IEEE international conference on real-time computing and robotics (RCAR), Kandima, Maldives, pp 166–171

Van den Berg JP, Gademann AJRM (2000) Simulation study of an automated storage/retrieval system. Int J Prod Res 38(6):1339–1356

Xing L, Liu Y, Li H, Wu C-C, Lin W-C, Chen X (2020) A novel tabu search algorithm for multi-AGV routing problem. Mathematics 8(2), 1 February 2020, Article number 279.

Yin S, Xin J (2019) Path planning of multiple agvs using a time-space network model. In: 2019 34rd youth academic annual conference of chinese association of automation (YAC), Jinzhou, China, pp 73–78

Zhang Z, Guo Q, Chen J, Yuan P (2018) Collision-free route planning for multiple agvs in an automated warehouse based on collision classification. IEEE Access 6:26022–26035

The Possibilities of Digitizing the Preparation Process for Shipping Batteries in a Distribution Warehouse—A Case Study

Łukasz Brzeziński and Magdalena Krystyna Wyrwicka

Abstract Modern technologies are gaining more and more importance in supply chain management, leading to the transformation of its three elements: network structure, business processes and management components. Enterprises use the potential of digital technologies as sources of value and transform supply chains to varying degrees and in various ways. Certainly, the development of digital technologies is one of the sources of building competitive advantage of enterprises and supply chains in the twenty-first century and is an important subject of research. Digitalization of storage related processes can contribute to the organization receiving multiple benefits—in terms of time and cost. This type of solution will be discussed as a part of the case study in this article. The aim of chapter was the analysis of the effects of digitisation of the sub-process of preparing goods for shipment in the distribution warehouse (including picking, labelling, quantity and quality control, packaging and loading). The analysis of the research material was based on the use of the following research methods: Exide case study, key performance indicators (KPI's), as well as elements of investment profitability assessment (ROI).

Keywords Distribution warehouse · Digitization of processes · Digitization in storage · Improvement of storage processes

1 Introduction

New business circumstances are related to openness to way of operating that are different from the traditional ones (Barton and Thomas 2009; Aztori et al. 2010; Bughin et al. 2011; Columbus 2015). The ability to efficiently use the potential of digital technologies and the digital competences of organisations are becoming the

Ł. Brzeziński (✉)
Faculty of Management and Logistics, Poznan School of Logistics, 61755 Poznan, Poland
e-mail: lukasz.brzezinski@wsl.com.pl

M. K. Wyrwicka
Faculty of Engineering Management, Poznan University of Technology, 60965 Poznan, Poland

© The Author(s), under exclusive license to Springer Nature Switzerland AG 2020

P. Golinska-Dawson et al. (eds.), *Smart and Sustainable Supply Chain and Logistics – Trends, Challenges, Methods and Best Practices*, EcoProduction,
https://doi.org/10.1007/978-3-030-61947-3_6

source of competitive edge of entire sectors and economies. In 2020, the global economy saturation level will reach 25%, whereas in 2005 that was only 15% (Concoran and Datta 2016).

Contemporary logistics (an in particular the functioning of distribution warehouses), which has already become global in its nature, must be flexibility and capable quick identification of business chances brought about by new tech, including the ability to use all available information (Ballou 2007; Carmichael et al. 2011). The ever-growing scope, complexity and dynamism of logistics processes fosters the need to implement new solutions in logistics. Logistics is affecting, directly and indirectly, the economy to a higher and higher degree, and the changes within it (Barton and Thomas 2009; Detlor 2010; Coyle et al. 2010; Boyes 2016) create new challenges for logistics to face.

The growth of informational technologies and digitisation of warehousing processes have a massive impact on the way enterprises and entire economy sectors function. Among enterprises, these relations can be seen on three operating levels: (a) within the scope of creating value in new business relations; (b) within the scope of creating value in processes that use the concept of managing the customer's experience and (c) within the scope of creating the enterprise's fundamental capabilities (Dörner and Edelman 2015). In every sphere of business activity, new tech can support its growth, on one hand, and change this specific sector and the prevailing balance of forces, on the other one. Some enterprises are thus forced to expand their activity to another sector, some others to find new business models creating value in their present sectors.

Such approach aims to bring benefits not only within the area of improving production processes, but also in logistics and transport. Increasingly more often, both directly and indirectly, the new tech impact logistics within logistics processes, driving its never stopping evolution and new challenges (Wodnicka 2019).

The phenomena affecting the transformation of logistics include, among else, the ones related to economic development in global scale (globalisation and internationalisation of activity) and the digitisation era combined with the fourth industrial revolution (Industry 4.0), which bring about transformations in the way enterprises cooperate and the control of processes in the supply chain in the scale of industries and sectors (Chopra and Meindl 2007; Johnson et al. 2008; Wood 2010; Wu et al. 2016; Uton 2017). It is therefore worth listing such phenomena as:

- the use of ICT solutions, including satellite and radio identification system that enable the coordination of the flows of products and information in time and space;
- standardisation and automation of economic processes, their processes require increased discipline of deliveries and continuity of stock optimisation processes;
- product customisation, allowing the personalisation of products, requires that a higher quantity of them must be in turnover, there are more selection options, the quality and shopping convenience are accelerated, all of which implicates production flexibility, as evidenced by signals from the market, as well as the search for and development of distribution channels;

- the higher number of large area shops that keep lower stocks and require better service from their suppliers;
- the change of nature of distribution processes which, going beyond the current limits of enterprises, require the collection and analysis of data, the expansion of the chain of information resources for integrating, in real time, the data from external business partners, suppliers or clients;
- the expansion of forms of ties among manufacturers, the implementation of new business model concepts, translating into the rising tendency to outsource logistics services (the increase in the share of external service costs in the business cost structure, the alteration of the material cost structure and stock maintenance, the reduction of investment outlays and the increase of variable costs). This results from the pressure to reduce costs as an element of the growth of competitiveness of businesses (Szukalski and Wodnicka 2016);
- the reduction of the market life cycle of products which imposes temporal discipline on the production process and the process of implementation of new products, because clients do not want to wait long for new or improved products.

Based on the above, a conclusion can be made that the challenges faced by logistics are not just limited to transport or warehousing, but are also related to processes of managing and coordinating activities among companies within the supply chain, in real time—the ability to react to the rapidly changing demand.

Industry 4.0, mentioned before, means the integration of smart solutions covering the equipment and IT resources, new work methods and new roles of human resources (Schwab 2016; Word Economic Forum 2016a, b). The diffusion of Industry 4.0 is connected to three phenomena (Paprocki 2016):

- the common digitisation and maintaining continuous communication between individuals themselves, between individuals and devices and between devices themselves;
- the increasingly more often implemented disruptive innovations that allow incremental increase of efficiency and effectiveness of functioning of the social and economic system;
- achieving such a level of development of machinery that they gain the ability to behave autonomously thanks to the use of artificial intelligence in their control processes.

The rapid acceleration of developments in IT and the transposition of their use lead to the formulation of the fourth industrial revolution thesis, although it is vocally disputed whether to consider it evolution or transformation (Neubauer 2011; Cohen and Kietzmann 2014). For quite some time, ICT have been reaching a more mature form, combining physical systems with the Internet of Things, Artificial Intelligence, Big Data and cloud computing.

The awareness of the ground-breaking nature of technological solutions and the vision of application of new possibilities become a challenge for many businesses and encourages to change the paradigm of conducting business.

The following are listed among the fundamental premises (Still 2018):

- global networks connecting production plants, machinery and warehouse management systems;
- autonomous exchange of information in cyber-physical systems, within which the equipment and data bases may operate jointly and control each other;
- identifiable, real time localized "smart products" that know their own history, status and paths for reaching their destinations.

Digitization makes information and communication available anywhere, anytime, within any context, and for any user using any device and type of access (Rappa 2004; Manyika et al. 2013; Kayikci 2018; Pluralsight 2018). The functioning of modern organisation within global economy requires the use of management methods and development strategies that are adequate to the new management conditions in the stage of digital transformation. Put shortly, numerous solutions and tools are being implemented in organisations. Especially, if they function in global scale. This article presents an example of digitisation of processes in Exide Technologies S.A. It should be noted that currently, most businesses, including SME's and large businesses operating locally, regionally or internationally, support their processes with IT solutions. Although the entire process is now commonly managed with electronic tools, there is still potential, or room, for further digitisation at the level of sub-processes or even particular activities. This also applies to entities that are leaders in their respective sectors (in global view). The analysed improvement is an example of this sort of approach—namely, the sub-process of preparation of goods for shipment (a part of the process of warehousing in the organisation's Distribution Warehouse). The solution's originality and added value results from showing the identified organisational problems, the implementation of the solution and its effects in preparing shipments to around 20 European states, within the context of the global entity's experience and know-how. The article aims to evaluate the effectiveness of digitisation within the sub-process of preparing goods for shipment from the distribution warehouse.

1.1 Methodology

The article's research purpose is the analysis of the effects of digitisation of the sub-process of preparing goods for shipment in the distribution warehouse (including picking, labelling, quantity and quality control, packaging and loading).

The applied study methods included a case study and a documentation method. The case study, or analysis and description of a single, most usually real case, that allows drawing conclusions on the causes and results of its course and, in a broader sense, a given business model, market specifics, technical, cultural and social conditions, etc. This means drawing conclusions on the basis of a single case. It is a study method consisting in a comprehensive description of a certain group or a unit, approached without any initial hypotheses. The subject of exploration is individual in nature. The case study's most important element, looking from the perspective of credibility, is the quality of its substantial content. The quality is manifested, among else, by the accuracy of the raised problems, formulated diagnoses and the adequacy of solutions proposed for a given situation (Langley 1999). The documentation method consists in the use of factual information for examination purposes, collected, in advance, for economic practice purposes and recorded in relevant documents.

The improvement project was designed, implemented and evaluated in 2019 (it took six months). The examined entity is Exide Technologies S.A.—Distribution Warehouse in Swarzędz (Wielkopolska province, Poland).

It was exploratory case study data collection method accompanied by additional data collection method such as interviews, tests etc. Advantages of case study method included data collection and analysis within the context of phenomenon, integration of qualitative and quantitative data in data analysis. Due to this approach it was possible to capture complexities of real-life situations so that the phenomenon was studied in greater levels of depth. Case studies involved both qualitative and quantitive research methods.

On one hand things were measurable and were expressed in numbers, on the other survey questions were in most cases closed-ended and created in accordance with the research goals, thus making the answers easily transformable into numbers. Used methods to collecting data: direct observation, archival records and personal interviews. The research was conducted by project team.

The improvement project is the outcome of the research. Based on the research we discovered different possibilities how to optimise the operations, how to reduce time to pick orders, but also how to eliminate activities which does not add any value.

Exide Technologies is an American multinational lead-acid batteries manufacturing company. It manufactures automotive batteries and industrial batteries. Exide's predecessor corporation was the Electric Storage Battery Company, founded by Gibbs (in 1888 year). Exide produces batteries and accessories for the Transportation markets with applications in the original-equipment and aftermarket channels for Auto/Truck/SUV, Heavy Duty, Lawn and Garden, Marine/RV, Golfcarts and Powersport, using Absorbed Glass Mat (AGM), Flooded, Enhanced Flooded Battery and Gel (VRLA) technologies. Exide also markets lithium-ion batteries for motorbikes in Europe (Exide 2020a). Exide has production plants in 89 countries and is one of the largest producers of automotive batteries in the world (Exide 2020b).

In Poland, Exide supplies batteries as the so-called first equipment to manufacturers such as: FCA Poland, VW Poznań, CNH in Płock, MAN in Niepołomice and Same Deutz-Fahr near Lublin. The Poznań-made batteries are also exported and used in the newest models of cars manufactured: by Jaguar and Land Rover in England and Slovakia (Nitra), TPCA in Kolin (Czechia), VW in Bratislava, Škoda (plant in Mlada Boleslav and Kvasiny), Volvo Cars in Belgium and Sweden, Audi (Ingolstad, Germany). Additionally, Exide makes batteries for Scania, MAN and Volvo Trucks, all made in Europe. The company's aftermarket brands are Centra and Exide. The Poznań-based producer's most important buyers are from Western Europe and the Commonwealth of Independent States (Ukraine, Moldova, Kazakhstan, Azerbaijan, Georgia and Armenia) (Exide 2020b).

1.2 Results

The analysed improvement project covered the process of preparation of goods for shipment in the distribution warehouse. It should be noted that, prior to digitization, there were multiple problems related, among else, to the use of various pallet labels, the circulation of printed documentation, determining responsibility for inconsistencies (for goods for which clients filed complaints). The result of the described change was the introduction of scanners in the picking process, standardization of the supplementation and picking of batteries, creation of a set of KPI's and monitoring of pallet movements of load units within the warehouse (Table 1).

The analysed improvement project contributed to improved parameters of the evaluation of the sub-process of preparation of goods for shipment in the organization's distribution warehouse. Most importantly, a positive change was noted with regard to the number of quantitative errors (decrease from 5.8 to 3.2%), generic errors (decrease from 3.8 to 2.1%) or the preparation of incorrect pallets (reduction to 0%). Digitization also allowed to reduce the number of employees involved in the process (from 54 to 49). Additionally, the savings for 12 months were EUR 90,000 (the cost of implementation was around EUR 35,000) (Table 2).

Referring to the analyzes related to the use of digitization in the broadly understood logistics processes, enterprises use the potential of digital technologies as a source of value and transform supply chains in various ways and in various ways. At the same time, it is pointed out that the implementation of digital technologies requires strategic changes in both the business model and the operational model of the company (Bock et al. 2017). The main challenge for the development of applications of these technologies are threats in global cyberspace, which significantly increase the negative risk regarding data security or transmitted information (Boyes 2016). Barriers to the implementation of digital technologies also include: a lack of sufficient knowledge about the nature and importance of digitization, an incorrect assessment of its potential, a lack of capital for investments in technologies and a lack of confidence preventing the exchange of information (Sherman and Chauhan 2016).

It is indicated that currently 76% of enterprises use social media, in addition 63% of entrepreneurs say that analytical tools allow them to gain a competitive advantage. As many as 92% of entrepreneurs are satisfied with the services used in the cloud and plan to increase their use, and 54% of mobile phones are smartphones (Słowik 2015).

The results of a global survey conducted by Ernst & Young indicate that managers in the area of information and communication technologies (ITC) currently spend over 25% of the budget on innovation in the field of modern technologies such as social media, mobile technologies, advanced data analytics and cloud computing (Ernst and Young 2015). These technologies contribute to improving the quality of operations and better meeting customer expectations.

Table 1 Comparison of work organization before and after the change

Work organization before the change	Work organization after the change
There were different pallet labels used in the warehouse	Implementation of scanners for picking
Workers prepared batteries using only paper documents	Standardization of replenishment and battery picking processes
Some information on pallet labels were missing e.g. formation codes	Standardization of the process of picking from the storage
The planning of operations in the warehouse was based on paper documents/Excel files	Standardization of the label preparation process
The correctness of labelling/finishing of batteries was controlled only visually (without any support from the WMS system)	Crating sets of KPI's for monitoring productivity
Voltage control results were registered on paper	Creating an application for displaying results
The productivity report was prepared in an Excel file	Different roles and tasks assigned to employees in the application
The information on the current status of the productivity of the finishing line was not displayed on the screen/seen by the workers	Tracking and tracing movement in the warehouse
The complaints (regardless of the reason) were not easy to investigate for the root cause	Tracking and tracing of preparation of customers' orders
There were difficulties in implementing an action plan to reduce the number of mistakes	Monitoring of tasks assigned to particular users
Not enough information on the pallet labels	
Unclear information on who prepared the pallet (illegible signatures)	
Waste of time due to unnecessary travel within the warehouse (incorrect voltage in the racks, labels not available)	
Some ready pallets stored in incorrect places	
Searching for the ready pallets in the warehouse due to missing information in the documents	
Difficulties in assessment of the workload of a shift	
Difficulties in assessment and execution of productivity (poor planning based on paper documents, Excel files)	
Difficulties in executing correct planning and transfer of information at the turn of the shift	
Different responsibilities of employees	

2 Conclusions

The role and importance of digitisation in business management and improving the functioning of supply chains is a topic widely discussed in the literature. The originality of the article is primarily due to the indication of a practical example of improvement in one of the leading battery manufacturers—Exide Technologies.

Table 2 Evaluation of digitization of the sub-process of preparation of goods for shipment

KPI's	Before the change	After implementation of scanners
Average number of batteries picked by a picker	428	467
Average number of batteries labelled per finishing line	890	997
% of complaints (errors in qty)	0.058	0.032
% of complaints (errors in type)	0.038	0.021
% of complaints (errors in loading wrong pallet)	0.0001	0
No. of mistakes when preparing reports	12	3
Reduction in paper consumption	60	55
No. of employees needed for the process	54	49
Savings in EUR (12 months)		90,000
Money spent (in EUR) on the project	35,000	
ROI	1.57	

The distribution warehouse serves customers from over 20 European countries. The indicated picking solutions have contributed to shortening the time of preparation of goods for shipment, reducing the number of process errors and reducing the costs.

A conclusion can be made, on the basis of the conducted analyses, that the digitisation of processes leads to a multi-dimensional transformation of the nature of the entire supply chain that is subject to virtual imaging through the continued collecting, processing and monitoring of the data from all cells.

Economic changes within the process of digital transformation and the evolution of business relations mean that companies wanting to compete effectively on global markets, thanks to the organization of supply chains, must give crucial importance to their flexibility and its ability to implement innovative business models along with the reorganization of processes. This will ultimately allow for achieving higher levels of digital maturity, which will translate into greater efficiency in the functioning of supply chains during the period of digital transformation—significant changes are already visible in the reconfiguration of logistics processes and business communication.

Referring to the benefits of research, it is worth noting that digitization can contribute to improving the functioning of the organization. This is of significant importance especially at the operational level, where it is possible to study in detail the functioning of individual work stations. Which is also to some extent related to the improvement of the organizational structure from the bottom, as IT tools are usually implemented at the strategic level. However, when it comes to certain limitations of research analyzes, they are primarily associated with the study of a single organization, which also means that the results of similar activities in other enterprises can generate completely different results. In the aspect of future research, the authors will focus on increasing the research sample of business entities to obtain a broader perspective of potential benefits and threats from the digitization of logistics processes.

It should be noted that even global enterprises have a large potential to introduce improvements with the application of digital tools within sub-processes and operations (parts of the warehousing process in this example). The described solution contributed to improvements both in quality (reduction of errors) and in costs—reduction of time needed to prepare the goods for shipment or the involvement of employees. In the view of global crisis situations (COVID-19 pandemics), this may play a big role for the organisation's survival.

Acknowledgments Poznan University of Technology 11/142/SBAD/1005.

References

Atzori L, Iera A, Morabito G (2010) The Internet of things: a survey. Comput Netw 54
Ballou RH (2007) The evolution and future of logistics and supply chain management. Europ Bus Rev 19(4):332–348
Barton R, Thomas A (2009) Implementation of intelligent systems, enabling integration of SMEs to high-value supply chain networks. Eng Appl Artif Intell 22
Bock R, Iansiti M, Lakhani KR (2017) What the companies on the right side of the digital business divide have in common. Harv Bus Rev 31 January 2017
Boyes H (2016) Cybersecurity and cyber-resilient supply chains. Technol Innov Manag Rev, April 2016
Bughin J, Livingstone J, Marwaha S (2011) Seizing the potential of big data. McKinsey Quart 4
Carmichael F, Palacios-Marques D, Gil-Pechuan I (2011) How to create information management capabilities through web 2.0. Serv Indus J 31
Chopra S, Meindl P (2007) Supply chain management. strategy, planning & operation. In: Boersch C, Elschen R (eds) Das Summa Summarum des Management. Gabler
Cohen B, Kietzmann J (2014) Ride on! Mobility business models for the sharing economy. Organ Environ 27(3):279–296
Columbus L (2015) 10 ways mobility is revolutionizing manufacturing. Forbes 4
Corcoran P, Datta SK (2016) Mobile-edge computing and the internet of things for consumers: extending cloud computing and services to the edge of the network. IEEE Consum Electr Mag 5(5):73
Coyle JJ, Bardi EJ, Langley CJ (2010) Zarządzanie logistyczne. Polskie Wydawnictwo, Warszawa
Detlor B (2010) Information management. Int J Inform Manag 30
Dörner K, Edelman D (2015) What 'digital' really means [2018-02-15]. McKinsey Digital. https://www.mckinsey.com/industries/high-tech/our-insights/what-digital-really-means.Ekonomiczne
Ernst & Young (2015) SMAC 3.0: digital is here. Enterprise IT trends and investments 2015 [2020-05-05]. https://pingpdf.com/pdf-smac-30-digital-is-here-ey.html
Exide (2020a). https://www.exide.com/en/about-us
Exide (2020b). https://www.akumulator.pl/artykuly/exide-konczy-130-lat-w-poznaniu-obecny-jest-od-ponad-20-lat/
Fourth Industrial Revolution. World Economic Forum, Cologny/Geneva

Johnson MW, Christensen CM, Kagermann H (2008) Reinventing your business model. Harv Buss Rev 12
Kayikci Y (2018) Sustainability impact of digitalization in logistics. Procedia Manuf 21:782–789
Langley A (1999) Strategies for theorizing from process data. Acad Manag Rev 4
Manyika J, Chui M, Bughin J, Dobbs R, Bisson P, Marrs A (2013) Disruptive technologies: advances that will transform life, business, and the global economy. Raport McKinsey Global Institute. McKinsey & Company
Neubauer NM (2011) Business models in the area of logistics: in search of hidden champions, their business principles and common industry misperceptions. Gabler Verlag. Wiesbaden: Springer Science & Business Media
Paprocki W (2016) Koncepcja Przemysł 4.0 i jej zastosowanie w warunkach gospodarki cyfrowej. W: J. Gajewski, W. Paprocki, J. Pieriegud (red.). Cyfryzacja gospodarki i społeczeństwa. Szanse i wyzwania dla sektorów infrastrukturalnych (s. 39–57). Gdańsk: Instytut Badań nad Gospodarką Rynkową; Gdańska Akademia Bankowa
Pluralsight. (2018). Technology in 2025: Prepare for the fourth industrial revolution [2018-03-28]. Pluralsigh, https://www.pluralsight.com/blog/career/tech-in-2025.
Rappa MA (2004) The utility business model and the future of computing services. IBM Syst J 43(1):32–42
Schwab K (2016) The fourth industrial revolution. World Economic Forum, Cologny/Geneva
Sherman R, Chauhan V (2016) Just my (re-)imagination. Supply Chain Manag Rev, March/April
Słowik J (2015) Nadaj SMAC swojemu biznesowi. Nowoczesne Zarządzanie 1:48–49
Still (2018) Przemysł 4.0 – przyszłość czy mrzonka? [2018-04-04]. Still Polska. https://www.still.pl/28657.0.0.html
Szukalski SM, Wodnicka M (2016) Outsourcing. Metodyka przygotowywania procesów i ocena efektywności. Difin, Warszawa
Uton J (2017) Setting sights on the smart supply chain. Pharmaceutical Executive—Oper Manag 3
Wodnicka M (2019) Technologie blockchain przyszłością logistyki. Zeszyty Naukowe Małopolskiej Wyższej Szkoły Ekonomicznej W Tarnowie, t. 41(1):43–53
Wood R (2010) Creating the smart supply chain. MHD Supply Chain Solut 7–8
World Economic Forum (2016a) The future of jobs: employment, skills and workforce strategy for the Fourth Industrial Revolution, Cologny/Geneva
World Economic Forum (2016b) Top 10 emerging technologies of 2016. World Economic Forum, Cologny/Geneva
Wu L, Yue X, Jin A, Yen DC (2016) Smart supply chain management: a review and implications for future research. Int J Logist Manag 27

Stakeholder Involvement Added Value Indicators in IT Systems Design for Industry 4.0 Digital Innovation Hubs

Adam Olszewski and Pawel Pawlewski

Abstract This chapter is addressing the creation of added value by means of concurrent engineering in the early stages of IT systems design for Industry 4.0 enterprises in the context of Digital Innovation Hubs. The recent phenomenon of Digital Innovation Hubs poses new opportunities and threats in introducing advanced collaborative IT services co-developed by a number of complementary subjects, such as ISVs, infrastructure providers, consulting firms, research entities and others. Unlike other business networks or chambers of commerce, Digital Innovation Hubs place research entities at the core of innovation ecosystems. Such approach may result in transversal involvement of external stakeholders in research and development IT projects. While consequent opportunities are many, so are the threats and uncertainties. This article analyses related pros and cons based on research results published thus far and selected ongoing use cases. Furthermore, this article proposes new investment risk factors resulting from broadening the pool of IT system design stakeholders.

Keywords Industry 4.0 · Concurrent engineering · Digital innovation hub

1 Introduction

The Information System (IS) industry has to provide added value, or value proposition, desired by their customers in order to spur profits. This article is addressing the creation of added value by means of external stakeholder engagement and concurrent engineering in the early stages of IT systems design for 4.0 (Industry/Logistics) enterprises in the context of Digital Innovation Hubs. The recent phenomenon of Digital Innovation Hubs poses new opportunities and threats in introducing advanced

A. Olszewski (✉)
IBCh PAS-Poznan Supercomputing and Networking Center(PSNC), Poznan, Poland
e-mail: adol@man.poznan.pl

P. Pawlewski
Faculty of Engineering Management, Poznan University of Technology, J. Rychlewskiego 2, 60-965 Poznań, Poland
e-mail: pawel.pawlewski@put.poznan.pl

© The Author(s), under exclusive license to Springer Nature Switzerland AG 2020
P. Golinska-Dawson et al. (eds.), *Smart and Sustainable Supply Chain and Logistics – Trends, Challenges, Methods and Best Practices*, EcoProduction,
https://doi.org/10.1007/978-3-030-61947-3_7

collaborative IT services co-developed by a number of complementary subjects, such as ISVs, infrastructure providers, consulting firms, research entities and others. Unlike other business networks or chambers of commerce, Digital Innovation Hubs place research entities at the core of innovation ecosystems. Such approach may result in transversal involvement of external stakeholders in research and development IT projects. While consequent opportunities are many, so are the threats and uncertainties. This article analyses related pros and cons based on research results published thus far and selected ongoing use cases. Furthermore, this article proposes new investment risk factors resulting from broadening the pool of IT system design stakeholders.

The goal of this chapter is to verify by means of qualitative comparison analysis if there is a need to conduct action research in the field of value-driven/concurrent engineering, which would enable comparing and balancing the benefits and risks resulting from an early inclusion of external entities (members of I4.0 manufacturing companies' ecosystems) in the early stages of information system (IS) design.

While the vast majority of research focuses on applied IT system design done by corporates with clear predefined targets, this article analyses the tools and approaches research organizations take at the early stages, up to Technology Readiness Level 3 (TRL3), of technology integration and information system development. Unlike corporates, whose main objectives are to grow profits and cut costs, research organizations focus on expanding the boundaries of knowledge. On this account the perspective of research entities on concurrent engineering may differ significantly from the corporate or industrial perspectives and analysing this phenomenon shall bring to light new, unobvious values, underrated in corporate-driven ecosystems. Also, this same analysis may initiate taking a broader perspective on the value creation issue—the one of R&D investment risk mitigating tools, potentially useful for both corporate and academic research environments.

The highlights of the chapter are:

- analysis of information system development perceived added values shared by academic and business ecosystems,
- comparison of closed and open innovation paradigms
- proposition of new unobvious R&D investment risk factors
- subjective prioritization of R&D investment risk factors
- analysis of investment risk management tools.

The main contribution is the development of a collaborative DIH based service development approach which enables an intentional management of uncertainty as a critical R&D investment risk factor.

The chapter is organized as follows: first, in Sect. 2 academic vs industrial values and indicators are contrasted, thence presenting the different approaches to R&D and IS design. Section 3 defines the problem—what is the unobvious added value of collaborative IS/service design and how can the key factors be managed in industrial settings. The following Sect. 4 proposes a DIH based approach to managing IS design uncertainty by providing a subjective ranking of unobvious values shared by industry and academia, listing their success factors and assessing the impact of the digital innovation hubs approach on these factors. Conclusions of the final Sect. 5 are followed by further research recommendations.

2 Literature Background

Literature hardly addresses the fairly new phenomenon of Digital Innovation Hubs in view of value-driven Information System (IS) design. Although it has not yet been thoroughly examined, a number of related studies have been published, mainly addressing the corporate perspective on IS design in closed environments, which can serve as a reference platform for our analysis.

On the one hand, closed IS design environments have been thoroughly studied—by focusing on R&D investment risk, Bahli and Rivard (2005) list the factors having biggest impact on the efficiency and results of outsourced IS design. On the other hand, design environments following the open innovation paradigm focus on the methods rather than efficiency and economic results. Some analyse the very concepts of design, showing how new, collaborative methods can be co-created either by in-house employees or by involving external professionals. For this purpose novel event formulas are tested, such as ideation contests or codesign workshops. IS design, being one of many areas in such studies, appears interchangeably with service design. Consequently, these analyses and the underlying recommendations referring to team-building, market exploration or process management refer to codesign as the primary subject, while IS design plays the role of a reference point rather than the primary subject. Although references to IS design across codesign studies are rather mild, selected conclusions can be quite relevant and trend sensitive. Like in any codesign endeavour, also in IS design, the general migration from granting a sole expert individual the full responsibility for design outcomes to enabling customers or users join IS design teams and play the roles of experts of their experiences (Visser et al. 2005).

The process of involving external stakeholders in service design has been divided into stages. Its application from stages from TRL1 to TRL3 as ideation has been analysed, querying whether or not such approach can benefit an organization's innovation outcomes. From the process perspective two approaches to research have been proposed (Edvardsson et al. 2005), as a "category of market offerings" or a "perspective on value creation." Little attention has been paid however, to economic benefits of the novel process, resulting in limited usability of these studies in real-world decision-making, including added value or investment risk measurements. The existing research is based on a service centred approach, assuming a logic that is always uniquely and phenomenologically determined by the beneficiary (Vargo and Lusch 2016, p. 8).

In terms of types of stakeholder involved, the present research is based on B2B relations, rather than science-to-business or business-to-customer, which results in the neglecting of the two latter types. Moreover, the analysed cases most often focus on local, near proximity entities, leaving the broader context of international markets aside (Gemser and Perks 2015). Whereas, the ability to consciously involve specific extreme customers or partners who have particular need ahead of market with highly estimated expectations (von Hippel 1986) can be very beneficial for IS design and innovation (Mahr et al. 2014). One of repeated hypotheses across literature, is that by involving customers a company can become more resilient to investment risk by avoiding the problem of user needs being sticky, difficult to transfer, and articulate (von Hippel 2001; Witell et al. 2011).

It is worth mentioning that already the traditional design studies, unlike IS approaches, implied that participatory approach required the "exchange between people who experience products, interfaces, systems and spaces and people who design for experiencing" (Sanders et al. 1999).

The lack of proper participatory IS design tools has also been covered from public services perspective, such as medical software for patients and caregivers. While it is evident that IS shall support and not hinder the experiences of caregivers and patients, little or no tools actually enable gathering requirements or insights from these target groups. The consequent misalignment of medical software results in highly qualified staff spending more of their time with the computer screen rather than with the patient. In this case the key factor or added value resulting from involving patients and users in the IS design would be the shortening of time doctors, physicians and other caregivers must spent at their computers. Based on specific medical workflows, involving all key stakeholders, concrete steps and related insights are introduced, playing the role of a showcase of tools for healthcare innovation projects (Vollmer 2019). Similarly to industrial contexts, the medical software fails to take a holistic approach to the addressed process, to embrace the off-screen experiences and finally to deliver tools that solve existing real-life problems without causing new ones. The present studies of specific use cases in medical area may bridge the conceptual gap between the industrial and academic contexts. Some of these cases have been introduced as role models for novel methodologies, such as EVOKE (Early Value Oriented design exploration with Knowledge maturity), meant to facilitate the selection of the newly appearing IS design possibilities with a focus on added value related information.

Engineers' tendency to avoid redesign during development, especially at IS component level, is yet another added value hindrance addressed in recent studies. All too often IS developers go back to initial setting of value, agreed for the IS with the product owner early in the project and before external, non-technical stakeholders were even introduced to the planning. This phenomenon may give grounds to either earlier involvement of the user or to extending the IS concept definition phase. Either way, study shows that selection of the right moment when users' needs are introduced might be critical for IS development. Similarly, the knowledge [...] where user needs originate and mature becomes critical to understand which sub-system performances have to be sacrificed to optimize the overall system behaviour. This makes systems engineers to go back and refer to the original construct of 'value' to orient their early stage design decisions (Monceaux and Kossmann 2012). Both aspects are of critical impact on investment risk certainty factor.

However, when moving from the macro level to the micro, this 'value' notion becomes blurrier, and contextual understanding gets lost when requirements are communicated down the supply network (Monceaux et al. 2014). There, the actual struggle of component oriented engineers takes place, depriving the IS of the user-driven added value by following the original, presumably unnegotiable, IS specifications. Although more resilient to additional costs and multiplying iterations, such conservative approach can hardly maximize the desired added value (Isaksson et al. 2013).

The established IS supply processes, although effective in coordinating a definite pool of tasks and resources towards the predefined objective, cannot handle the early stages of innovative IS development, which requires continuous sharing of knowledge and negotiations embracing a range of interdisciplinary skills, experiences and tools originating at different, often distanced, entities. Narrowing down targets to local perspectives causes design teams to fail creating solutions configured in the most valuable way. Empirical observations show that when system-level requirements are not available or not mature enough, engineers dealing with the development of long lead-time sub-systems tend to target local optima, rather than opening up the design space (Bertoni et al. 2018). These 'local optima' seldom embody the best possible result for the overall system. Most likely they hinder the possibility of identifying solutions that would work even better and that maximize value (Collopy and Hollingsworth 2011).

Some authors underline the importance and poor results of gathering specific requirements for IS engineering. On the one hand, well defined requirements for an IS are known to be prerequisite to avoid customer disappointment. On the other hand, all too often the process of defining IS requirements is neglected, delayed or distributed across engineering teams. Many claim that failure to involve stakeholders or clients at this stage leads directly to uncontrolled cost increase. Conversely, to improve clarity, awareness and understanding of what should be included in a system design, and hence to minimize development time and later rework, iteration and negotiation with customers and stakeholders must be established since the earliest design phases (Jiao and Chen 2006; Withanage et al. 2010). Nonetheless, all too often the relation with customers is not managed in an intentional manner nor carried out consistently, resulting in overlapping or missing requirements for specific IS features, modules, layers, documentation. Research in established software engineering methods shows that, requirements elicitation is far from being a linear, monolithic process; rather, it follows a more concurrent process (Prasad 1999). Consequently, unlike the open innovation paradigm involving external real-life stakeholders, the established research focuses on empowering individual engineers like they are owners of the processes addressed by the systems they develop. It is claimed a significant part of the concurrent design method, that the individual engineer, not an external stakeholder, customer, nor future user, is given much more say in the overall design process due to the collaborative nature of concurrent engineering. Giving the engineer ownership is claimed to improve the productivity of the employee and quality of the product, based on the assumption that people who are given a sense of gratification and ownership over their work tend to work harder and design a more robust product, as opposed to an employee that is assigned a task with little say in the general process (Kusiak 1992).

3 Research Problem

The key problem analysed in this article is whether or not there are any unobvious values behind recent academic paradigms, such as the open innovation paradigm, and its implementations, such as digital innovation hubs, that could bring new value or increase the existing values for industrial R&D clients in the development of IT systems. From a business theory perspective, customer value refers to customers' perceptions of what they receive, in return for what they sacrifice (Zeithaml 1988). There are two aspects to customer value: desired value and perceived value. In order to assess value from industry 4.0 perspective one needs to dig deeper the notions of customer and value. This article focuses on the latter with minor references to the former.

Being an organization rather than individual, a manufacturing plant may employ thousands of professionals, all potentially having different perspectives on the value of delivered IT solutions. Additionally, partners of the manufacturing plant, such as service providers, hardware maintenance firms, external consultants and last but not least the logistics all assess the functionalities and features of the plant's IT layer from another angle. On this account, before starting a study, it takes to define and classify the notions of the key user, value, IT system.

In MIDIH project a collaborative approach to IT service development is proposed, by forming a network of digital innovation hubs, i.e. collaborative networks of research and business entities with complementary offerings. This approach calls for new value measurement tools and may discover novel values, absent in closed single provider settings. The collaborative MIDIH approach also redefines key notions, including values and their indicators from shared, intersubjective perspectives.

- Better use of internal transport,
- Better organization of work in the warehouse,
- Less consumption of internal transport by reducing storage space, Less electricity consumption in forklift trucks—by eliminating unnecessary movements on handling.

4 Problem Solution and Methodology

For sake of this article we propose an alignment of notions used within two analysed use cases—projects, MIDIH—Manufacturing Industry Digital Innovation Hubs and SymbIoTe—Symbiosis of Smart Objects across IoT Environment. While the latter project has been concluded and provides a full overview of IT value creation, the former MIDIH is in progress while writing this article, which enables us to continue observations and studies proposed in the final remarks of this article. Both projects focus on developing advanced IT tools that may enable new or more efficient processes within industrial environments. Moreover, both projects involve numerous R&D organizations, along with their perspectives and goals and a number of industrial partners, playing roles of pilot adopters of the developed IT tools. Therefore, by analysing these cases, we add a real-life layer to our otherwise theoretical considerations.

MIDIH project proposes a novel approach to the creation of technology based value proposition, influencing the investment risk in R&D. By experimenting with IS component development in DIH environments, MIDIH redefines the provider and the value and consequently proposes new tools for creating the value proposition. Unlike in Osterwalder's business model canvass, normally applied to a single product provided by a single vendor, MIDIH takes a DIH based collaborative provider approach, where the value proposition does not exist unless complementary resources are combined and integrated in a collaborative manner. Such multiple vendor value requires a more advanced consideration of ecosystem relations and value flows in order to properly analyse, plan and manage the value network across multiple stakeholders. For this purpose MIDIH redesigns the Osterwalder's canvass, combines it with the value network analysis tool and complements with the project's new DIH service portfolio analysis tool, covering as many as 34 service development factors embracing the actual collaborative service aspects in more detail than the classical single provider tools do.

In industrial IT research the customer is the business entity who orders IT R&D or development jobs and expects these jobs to be performed as planned, which in case of research is not always the case. From industrial client perspective, the overall value of IT R&D boils down to ensuring a positive balance between costs and benefits which is determined by the notion of IT R&D investment risk. Insights from transaction costs theory suggest that there exist three major sources of risk factors for IT outsourcing: the transaction, the client and the supplier (Bahli and Rivard 2005) divide the three risk sources into seven risk factors—Table 1. These are the factors that industrial ecosystems are familiar with and hence these factors are intentionally managed by businesses.

However, apart from those well-known key factors, there may be other factors and related values, largely ignored by the industrial environments, yet having significant impact on the overall IS design risk. The hypothesis we are analysing here is that

Table 1 Risk factors in IT outsourcing operation. (*Source* own study)

Source of risk	Risk factors
Transaction	Asset specificity
	Small number of suppliers
	Uncertainty
	Relatedness
	Measurement problems
Client	Expertise with the IT operation
	Expertise with outsourcing
Supplier	Expertise with the IT operation
	Expertise with outsourcing

by intentionally managing these new, unobvious risk factors, industry could lower their investment risks and consequently boost their innovation and research activity (Table 2).

MIDIH project has made a new source of risk, ecosystem, evident and covering an array of risk factors. However in this article we focus only on a single, high impact factor—uncertainty—potentially including values going beyond the obvious business criteria. In the context of IT outsourcing, uncertainty may be present because [...] the transacting parties have incomplete or imperfect information, or because there are numerous unimaginable possibilities, which may arise during the course of the transaction. This means that, in the face of uncertainty, contracts are unavoidably incomplete, and may require renegotiation and frequent adjustments when unexpected contingencies occur. This renegotiation adds to ex ante costs and postpones the realization of outsourcing's perspective value. Ultimately, the resolution of ex ante uncertainty must wait ex post reality (Pilling et al. 1994).

Table 2 Risk factors largely ignored by the industrial environments. (*Source* own study)

Source of risk	Risk factors	Related added values from DIH approach	Level of impact on risk
Transaction	Asset specificity	Improved understanding	Low
	Small number of suppliers	Collaborative supply options	Mid
	Uncertainty	Early and thorough validation of assumed desired features	High
	Relatedness	Multiple interdependence	Low
	Measurement problems	Broad access to academic measurement tools	Low
Client	Expertise with the IT operation	Extended area-specific pool of IT expertise	Mid
	Expertise with outsourcing	Improved access to shared services and consultancy firms	Low
Supplier	Expertise with the IT operation	Extended area-specific pool of IT expertise	Mid
	Expertise with outsourcing	Improved access to shared services and consultancy firms	Low
Ecosystem	IPR management	Alternative IPR options	High
	Technology lifecycle management	Transparent distribution of responsibilities and costs	High

Fig. 1 View of of living laboratory—SymbIoTe project

The data for this part of analysis have been gathered in a living laboratory mode concentrating at three workshops, carried out every six months within SymbIoTe project—Fig. 1.

During workshops the same value-related questions were asked to check the correctness and level of certainty about values expected by industrial partners. Initially, R&D partners, in this case suppliers of IS, listed the values they had assumed key for their clients—industrial partners. Later, those same categories of values were collected from industrial partners. Finally, after the initial failed validation process, values listed by both sources were once more gathered, combined and completed. The study reveals that prior to the involvement of industrial partners the levels of certainty about assumed values were high, even though very generic and in some cases incorrect. Only after listing values by industrial partners the values assumed by R&D could have been validated.

The revealed discrepancy between values assumed by suppliers and expected by clients had an impact on the definition and execution of the developed technology components and further IS functionalities. For instance, interoperability of sensor data across modules and levels was one of the core system assumptions. However, it was only after DIH workshops that camera entered the subject matter pool of sensors as video recordings were indicated by consumers the type of data needed in such systems, which had never been considered before. Without repeatedly guiding both suppliers and clients through the IS usage scenarios, the resulting IS prototype would have been construed accordingly with the suppliers' initial assumptions. Consequently, the resulting proof of concept IS tools would not have met the expectations of clients to the level it finally did.

In MIDIH project the mutual sharing of resources (data and infrastructures) and complementing competences is observed to bring the critical added value for industrial partners. The usage of FIWARE architecture and its Arrowhead components, built by a numerous R&D, in combination with middleware and hardware elements coming from MIDIH partners gave birth to a robotic arm demonstrator, operating in a flat structure in factories giving superior managerial and analytical qualities surpassing those of the pyramid PLC based structure. Another interesting result is the logistical monitoring systems, developed by FIAT Research Centre in collaboration with Cefriel and Engineering. Here, CPS/IoT Technologies have been adopted, and, leveraging on MIDIH Open Platform and on the methodologies that have been developed within the project, it has been possible to enable the optimization both of Inbound Logistics Processes (Smart Supply Chain scenario) and Industrial Processes (Smart Factory scenario) in FCA. Consequently, a scenario has been developed where international logistics can monitor a container condition live and from historical data, indicating cases of free fall, side falling and crossing parameters such as humidity or temperature. Likewise in SymbIoTe, this and other MIDIH project industrial solutions would have missed critical requirements had it not been for the collaborative open approach at early development stages, involving multiple stakeholders on both provider and customer sides.

5 Conclusions and Further Works

Values sought for by academics pushing state-of-the-art resulting from the open innovation paradigm, i.e. opening up for external ideas and allowing internal ideas outside or early involvement of industrial partners in IT R&D, may add substantially to the values desired by industrial clients, in it to lowering R&D investment risk, by raising transactional certainty and enabling new unobvious industrially desired benefits. At present, the industrial benefits grow unintentionally in science-to-business consortia and their economic potential lingers largely undisclosed. It takes further research to verify if intentional open innovation could bring substantial increase of the desired value for industry and consequently raise the uptake of R&D projects results.

An array of interdisciplinary research would be needed, involving disciplines such as IT engineering, design and economy, to identify and propose respective IS design risk management tools. The issue analysed in this article merely touches the surface of a broader problem of added value generation and maximization across the whole innovation development up to TRL9 and lifecycle. An extended research is needed to assess more collaborative approaches and tools and to analyse their specific implementations.

Acknowledgements The work was carried out as part of the Project no. 767498 "Manufacturing Industry Digital Innvation Hubs", financed by European Union Framework Programme for Research and Innovation Horizon 2020.

References

Bahli B, Rivard S (2005) Validating measures of information technology outsourcing risk factors. Omega 33(2):175–187
Bertoni M, Bertoni A, Isaksson O (2018) EVOKE: a value-driven concept selection method for early system design. J Syst Sci Syst Eng 27:46–77. https://doi.org/10.1007/s11518-016-5324-2
Collopy PD, Hollingsworth PM (2011) Value-driven design. J Aircraft 48(3):749–759
Edvardsson B, Gustafsson A, Roos I (2005) Service portraits in service research: a critical review. Int J Serv Ind Manag 16(1):107–121
Gemser G, Perks H (2015) Co-Creation with Customers: an Evolving Innovation Research Field. J Prod Innov Manag 32(5):660–665
Isaksson O, Kossmann M, Bertoni M, Eres H, Monceaux A, Bertoni A, Wiseall S, Zhang X (2013). Value-driven design: a methodology to link expectations to technical requirements in the extended enterprise. In: Proceedings of the 23rd INCOSE international symposium, 23 (1): 803–819, Philadelphia, PA
Jiao J, Chen C-H (2006) Customer requirement management in product development: a review of research issues. Concur Eng Res Appl 14(3):173–185
Kusiak A (1992) Concurrent engineering: automation, tools, and techniques. Wiley-Interscience; 1 edn
Mahr D, Lievens A, Blazevic V (2014) The value of customer co-created knowledge during the innovation process. J Prod Innov Manag 31(3):599–615
Monceaux A, Kossmann M (2012) Towards a value-Driven design methodology enhancing traditional requirements management enterprise. Proceedings of the 22nd INCOSE symposium, 22 (1):910–925, Rome, Italy
Monceaux A, Kossman M, Wiseall S, Bertoni M, Isaksson O, Eres MH, Bertoni A, Ndrianarilala R (2014) Overview of value driven design research: methods, applications and relevance for conceptual design. Insight 17(4):37–39
Pilling B, Crosby L, Jackson D (1994) Relational bonds in industrial exchange: an experimental test of the transaction cost economic framework. J Bus Res 30(3):237–251
Prasad B (1999) Enabling principles of concurrency and simultaneity in concurrent engineering. Artif Intell Eng Des Anal Manuf 13(3):185–204
Sanders BE, Dandavate U (1999) Design for experiencing: new tools. 1st International Conference on Design and Emotion, Delft, the Netherlands
Vargo L Stephen, Lusch RF (2016) Institutions and axioms: an extension and update of service-dominant logic. J Acad Mark Sci 44(1):5–23
Visser S Froukje, Stappers PJ, Van der Lugt R, Sanders EB (2005) Contextmapping: experiences from practice. CoDesign 1(2):119–149
Witell L, Kristensson P, Gustafsson A, Löfgren M (2011) Idea generation: customer co-creation versus traditional market research techniques. J Serv Manag 22(2):140–159
Vollmer F (2019) 4D Wireframing as a tool for integrating digital with physical touchpoints for an elevated patient experience. In: Pfannstiel M, Rasche C (eds) Service design and service thinking in healthcare and hospital management. Springer, Cham
von Hippel E (1986) Lead users: a source of novel product concepts. Manage Sci 32(7):791–805
von Hippel E (2001) User toolkits for innovation. J Prod Innov Manag 18(4):247–257
Withanage C, Park T, Choi H-J (2010) A concept evaluation method for strategic product design with concurrent consideration of future customer requirements. Concur Eng Res Appl 18(4):275–289
Zeithaml VA (1988) Consumer perceptions of price, quality, and value: a means-end model and synthesis of evidence. J Market 52(3):2–22

References

[illegible]

Comparison of CRM Systems Dedicated to SMEs in Terms of the Omnichannel Concept

Roman Domański and Hubert Wojciechowski

Abstract The new form of multi-channel sales—omnichannel—entails the need for companies to adapt to provide the best possible customer experience. Dedicated CRM systems are solutions that enterprises use to manage customer relationships. They should evolve from classic solutions, currently existing, towards customer relationship management systems that meet the requirements of omnichannel—i.e. Omnichannel Customer Relationship Management (OCRM) systems. Due to the large number of solutions, only CRM systems dedicated to small and medium enterprises (SME) is studied in this chapter. This chapter aims to develop best possible configuration of CRM system dedicated to omnichannel and, in the next step, to compare existing CRM systems for SME with the best possible one. State-of-the art solutions in modern forms of distribution suggest that there is a research gap in ranking of key factors for CRM systems that use the omnichannel approach. The originality of this article is based on: undertaking a very niche topic—SME-dedicated CRM systems enabling the implementation of the omnichannel concept; a pioneering solution to the problem based on an innovative research tool—the Grey Incidence Analysis (GIA) method from the Grey System Theory (GST) family.

Keywords Omnichannel · Customer relationship management (CRM) systems · Small and medium enterprises (SMEs) · Grey incidence analysis (GIA) · Grey system theory (GST)

R. Domański · H. Wojciechowski (✉)
Faculty of Engineering Management, Institute of Logistics, Division of Production Engineering and Logistics, Poznan University of Technology, ul. Jacka Rychlewskiego 2, 60-965 Poznań, Poland
e-mail: hubert.wojciechowski@put.poznan.pl

R. Domański
e-mail: roman.domanski@put.poznan.pl

© The Author(s), under exclusive license to Springer Nature Switzerland AG 2020
P. Golinska-Dawson et al. (eds.), *Smart and Sustainable Supply Chain and Logistics – Trends, Challenges, Methods and Best Practices*, EcoProduction,
https://doi.org/10.1007/978-3-030-61947-3_8

The chapter used the interdependence of analysis of qualitative and quantitative elements. In the theoretical part, a systematic literature review was used to analyze secondary sources listed in the Scopus database. In the practical part, GIA method, in primary study, was used as a tool for comparing CRM systems.

1 Introduction

A new form of distribution called omnichannel is defined as universal retail or a universal channel. It offers the possibility of making purchase/sale transactions by switching from one channel (e.g. conclusion of a contract through a call centre) to another (e.g. collecting goods from parcel pick-up stations) at any stage of the transaction, including parallel operation in several channels at the same time (e.g. comparison of the price level in an online shop while visiting a brick-and-mortar shop). For customers, this solution means shopping comfort. For enterprises, this is a challenge related to shaping distribution logistics systems adequate to customers' expectations.

What can be of use hers is modern technology. Customer Relationship Management (CRM) systems are used to maintain good shopping experience. CRM systems support enterprises in implementing operational, analytical and contact (interactive) functions. Therefore, this category of software was selected as the object of the study in the article. However, due to the emergence of new forms of retail, the existing (classic) CRM systems should evolve towards omnichannel (modern) OCRM systems. This transformation is now taking place.

In each country, the dominant fraction (although numbers are different) is small and medium enterprises (SMEs)—e.g. in 2018 in Poland they made up 95% (LEs made up just 5%). They largely create the economic position of a given country. In this group of enterprises, the effects of changes (organizational and technological transformations) have the greatest potential to affect the economy of each country. That is why SMEs were chosen in the article as the research subject.

A promising executive concept of the postulates outlined above is the Grey System Theory (GST). Due to the rapid rate of changes in business and the long time it takes to obtain statistical data, conducting objective comprehensive research is very difficult to carry out and is also economically unviable. GST is an alternative. Using this theory, in the lean language terms, one can come to the same conclusions using less time and fewer resources.

One of the GST tools is Grey Incidence Analysis (GIA). It enables the study of interrelationships (degree of similarity) between systems. The idea of this article is based on formulating the functionality of an ideal OCRM system and then comparing it with the list of CRM systems available on the market for SMEs in order to assess the extent to which they meet the omnichannel retail requirements. That is why GIA was chosen in the article as a research implementation tool.

The article aims to develop the best possible configuration of the CRM system dedicated to omnichannel and then to compare the existing CRM systems for SMEs with the best possible one. The article consists of the following sections: literature search in the Scopus database—quantitative and qualitative analysis of articles; description of a four-stage original methodology of testing CRM systems dedicated to SMEs in the context of omnichannel; presentation of the sequence results of research on the omnichannel of CRM systems dedicated to SMEs using the GIA method, results analysis and final conclusions.

2 Literature Search in the Scopus Database

2.1 Quantity Analysis

The only source of literature search is the Scopus database. This is only seemingly a research limitation. The authors' previous numerous experiences allow to conclude that in the majority of cases the same list of publications is the result of a search in the Web of Science database. Hence, there is no need to duplicate literature search. Literature search in the Scopus database was conducted on 26 February 2020. In each case, interesting phrases were sought in the following places: "article title, abstract, keywords"—identification of potential articles.

Literature search starts with the recognition of interest in individual concepts. The evaluation criterion in this respect is the number of publications on a given topic—individually (Table 1). As part of the search, various possibilities of recording a given term were taken into account.

Table 1 Number of publications on a given topic—individually

Term number	Phrase	Number of articles
1	CRM system	4,978
2	Customer relationship management system	5,715
3	SME	34,739
4	Small and medium enterprise	27,307
5	Omni-channel	261
6	Omni-channel	315
7	Omni channel	1,043
8	GST	31,462
9	Grey system theory	9,697
10	GIA	2,956
11	Grey incidence analysis	2,157

Source Own work based on the Scopus database

Table 2 Number of publications on a given topic—collectively

Term number	Phrase	Metaphrase	Number of articles
1	CRM system	CRM	8,971
2	Customer relationship management system		
3	SME	SME	46,060
4	Small and medium enterprise		
5	Omnichannel	OMNI	1,276
6	Omni-channel		
7	Omni channel		
8	GST	GST	41,090
9	Grey system theory		
10	GIA	GIA	5,096
11	Grey incidence analysis		

Source Own work based on the Scopus database

It was then decided to aggregate the same terms, recorded in different ways, into groups. To this end, phrases were searched for in pairs, searching for the sum of both sets (OR operator). Table 2 presents a term-related aggregated number of publications.

Comparing the results in both tables, it can be concluded that within a given article a given phrase is sometimes recorded in several ways. Therefore, the results in Table 2 are not the exact sum of the results from Table 1. From the perspective of the object and subject of research, by far the most numerous groups in Table 2 are publications devoted to various aspects of SME. CRM and omnichannel systems, as narrowly specialized terms, constitute smaller groups. The advantage of CRM publications over omnichannel should not come as a surprise. After all, the former term is historically older. Hence, it already has significant output. From the perspective of the research implementation tool, the smaller number of publications dedicated to GIA, as one of the detailed solutions of GST (a larger number of articles), is very natural.

However, the most interesting was the search for interrelationships between individual phrases. Based on Table 2, a phrase search was made in pairs, looking for the common part of both sets (AND operator). The interrelationships of terms through the prism of the number of publications are presented in Table 3 (explanation of the meaning of numbers in parentheses in the part devoted to qualitative analysis—Sect. 2.2).

The final results of the quantitative analysis of the publications are very interesting. The subject of CRM in the context of SME has already been of considerable interest to scientists (247 articles). However, omnichannel issues in CRM systems are just beginning to be discussed—just 11 articles. In the SME fraction, omnichannel is characterized by an even greater niche of interest (only 4 articles). GST has already been used as part of CRM (14 articles) and SME (34 articles). However, its specific

Table 3 Number of publications on a given topic—reports

	CRM	SME	OMNI	GST	GIA
CRM	x	247	11 (10)	14 (4)	–
SME		x	4		3 (2)
OMNI			x	–	–
GST				x	259
GIA					x

Source Own work based on the Scopus database

tool—GIA—has so far been used only in a pilot form and only on the basis of SME (only 3 articles), no use of GIA in the context of CRM systems and omnichannel. GST has never been used as a research method or tool in the context of omnichannel.

3 Quality Analysis

Out of 11 publications in the CRM—OMNI relation, 1 will be intentionally omitted—a conference report. Table 4 presents a brief description of each publication.

The publications in Table 4 span the last 5 years. In this case the bottom limit is the moment the concept of omnichannel appears. The number of publications is increasing. At the moment, there are no clearly dominant authors or scientific centres (individual articles). France and the United Kingdom show a slight dominance. The discussed issues are most often placed in the areas of: Business, Management and Accounting; Computer science. From the citations' point of view, particular attention should be paid to the following articles: Picot-Coupey et al. (2016), Kung et al. (2008), Khan and Faisal (2015).

Only 4 publications are dedicated to the SME—OMNI relationship. Table 5 presents a brief description of each of them.

The publications in Table 5 also span the last 5 years. The number of publications is also increasing. At the moment, there are no clearly dominant authors or scientific centres (individual articles). The issues in question are most often placed in the area of Business, Management and Accounting. From the citations' point of view, attention should be paid especially to the article by Khan and Faisal (2015).

SME—GIA relationships were identified only in 3 (de facto 2) publications. Table 6 presents a brief description of each of them.

The publications in Table 6 span only 4 years. Quite puzzling is the lack of publications after 2014, which may indicate the lack of GIA's interest in the SME area. An interesting observation is the fact that both publications come from China, from which the grey system theory originates. They are placed in the areas of: Computer Science; Engineering. It is also peculiar that none of the publications have yet been cited.

Table 4 List of articles about CRM and omnichannel systems

Author	Year	Scope of interest
Vasiliev and Serov	2019	An economic—mathematical omnichannel model of sales management system in banking
Prodanova and Van Looy	2019	The evolution of business process management through different social media tools as a means to achieve a transition toward the recommended omnichannel management approach
Ieva and Ziliani	2018	The explosion in the number of touchpoints is putting pressure on companies to design omnichannel customers' experiences aimed at achieving long-term customer loyalty
Církovský and Maryška	2018	To provide a concept of customer relationship management systems usage for complex campaigns in the healthcare area
Won	2018	A case study of Lotte Shopping—the company is trying to reinforce the omni-channel strategy, which can create synergy among various distribution channels based on its core competences
Critchley	2018	Description of set up the Dynamics 365 Online system for sales, customer service, marketing
Park and Lee	2017	The retailers recognize the importance of the mobile channel as an efficient sales channel and as a tool for CRM—to explain channel choices in the omni-channel environment
Taufique Hossain et al.	2017	Additional research is required to know more novel outcomes of the channel integration within omnichannel services marketing
Picot-Coupey et al.	2016	The challenges faced in shifting to omni-channel strategy are so numerous and so engaging that, de facto, it is impossible to evolve directly from a multi-channel, siloed strategy to an omni-channel strategy without any transition
Hutchinson et al.	2015	Illustrates the value of a structured, formal CRM system which helps SME retailers compete in a complex, competitive and omni-channel marketplace

Source Own work based on the Scopus database

Out of 14 publications in the CRM—GST relation, 10 will be intentionally omitted—conference reports (Feng and Mei 2010), accidental hits (Chang et al. 2017). Table 7 presents a brief description of each publication.

The publications in Table 7 span the period of 9 years. The number of publications has a fixed tendency. At the moment, there are no clearly dominant authors or scientific centres (individual articles). China—the homeland of grey system theory—shows a slight dominance. The considered issues are most often placed in the area of: Computer Science. From the citations' point of view, particular attention should be paid to articles such as: Govindan et al. (2016), Xiong et al. (2016).

Table 5 List of articles about SME and omnichannel

Author	Year	Scope of interest
Calderón et al.	2019	Analyses the development of exploitation and exploration capabilities, and the role of ambidexterity, in the evolution of small Spanish wineries toward the multi-channel distribution system
Kim et al.	2018	The omni-channel platform represents the most sustainable approach for small business owners undergoing difficulties such as technological and organizational changes
Heidekröger et al.	2018	Omni-channel management is one trend that is increasingly gaining attraction, that profound understanding of influencing factors for sales and service to small and medium enterprises (SME) is missing
Hutchinson et al.	2015	Illustrates the value of a structured, formal CRM system to help SME retailers compete in a complex, competitive and omni-channel marketplace (repeating—see: Table 4—last row)

Source Own work based on the Scopus database

Table 6 List of articles about SME and GIA

Author	Year	Scope of interest
Tang and Wang	2014	Applying the grey incidence analysis method to study the GEM listed companies internal control effectiveness (repeating—see: —Tang and Wang 2013 row below)
Tang and Wang	2013	Applying the grey incidence analysis method to study the GEM listed companies internal control effectiveness
Yong	2011	A detailed analysis on the structure and environment of China's e-commerce development by means of grey incidence analysis

Source Own work based on the Scopus database

Table 7 List of articles about CRM and omnichannel systems

Author	Year	Scope of interest
Govindan et al.	2016	Proposes an integrated grey DEMATEL method to consider interdependent relationships among the 3PL provider selection criteria
Orzan et al.	2014	Some elements taken from grey systems are used in the relationships between the advertising campaigns and the buyers' decisions
Hu	2010	Based on the grey-fuzzy theory, it constructs a grey-fuzzy comprehensive performance evaluation model for the CRM system, and verifies the model's scientificalness and feasibility through empirical study
Xiong et al.	2008	Identifies customer behaviour using a grey correlation model to evaluate proposed segmented customers in CRM

Source Own work based on the Scopus database

Out of 34 publications in the SME—GST relation, 17 will be intentionally omitted—conference reports (Církovský and Maryška 2018), accidental hits (Hutchinson et al. 2015). Table 8 presents a brief description of each publication.

The publications in Table 8 span the period of 14 years. They concern the aspects of: modelling (programming), forecasting, decision making or controlling. For explanatory issues, authors usually use grey incidence analysis (GIA), grey model (GM) for predictive issues, and grey decision making (GDM) for design issues. The number of publications has a fixed-increasing tendency. At present, the only author with more than 1 publication (two) is Kung. As for nationality, it is definitely dominated by China (13 articles)—that is where GST comes from—Jiangsu University (2 publications), and also Taiwan (3 articles)—Chaoyang University of Technology (also 2 publications). The considered issues are most often placed in the

Table 8 List of articles about SME and GIA

Author	Year	Scope of interest
Chang et al.	2017	Data smoothing index—forecasting
Zeng et al.	2017	Grey relational analysis—selection of SME-specific ERP systems
Khan and Faisal	2015	Grey-based model—ERP vendor selection
Zhang and Chen	2014	Grey prediction model—predicts the monetization ratio and financial interrelations ratio
Tang and Wang	2014	Grey incidence analysis—study the GEM listed companies internal control effectiveness (repeating)
Tang and Wang	2013	Grey incidence analysis—study the GEM listed companies internal control effectiveness
Zhang et al.	2012	Grey prediction model—forecast the sales for the next decade
Yong	2011	Grey incidence analysis—analysis on the structure and environment of China's e-commerce development
Li et al.	2011	Grey-relational theory—system of technological SMEs' financing capability
Feng and Mei	2010	Grey relativity analysis—the electronic commerce application decision-making in SMEs
Lingyu et al.	2009	Grey relation—fuzzy multi-criteria decision making (FMCDM)
Ma and Fan	2009	Grey-fuzzy theory—risk evaluation of networked SME cluster
Cheng and Wang	2009	A grey hierarchy evaluation model—assess destructive innovation risks quantitatively
Kung et al.	2008	Grey statistic method—evaluate the optimal distribution for strategic resources in medium and small enterprises
Sun et al.	2007	Grey correlative degrees of the factors—marketing mix effectiveness
Kung	2005	Grey system method—assess the enterprise's decision on marketing resource distribution and its performance
Chen et al.	2004	Fuzzy-grey comprehensive evaluation method—safety management of small and medium enterprise

Source Own work based on the Scopus database

areas of: Decision Sciences; Computer Science; Engineering; Business, Management and Accounting. From the citations' point of view, particular attention should be paid to articles such as: Li et al. (2011), Khan and Faisal (2015), as well as: Církovský and Maryška (2018), Cheng and Wang (2009), Chen et al. (2004).

4 Summary of Literature Analysis

The qualitative analysis of publications leads to the conclusion that the number of publications in the context of the research interests of this article is even smaller than it initially resulted from the quantitative analysis of the publications. The results of literature research show a clear gap in the issue of omnichannel CRM systems (only 10 publications). Out of 247 CRM publications dedicated to SMEs, only 4 deal with omnichannel issues—therefore, the niche deepens even more. GST, or its specific tool—GIA, have never been used as an instrument in omnichannel testing (double lack of publication). An incidental use of GST in the context of CRM has been reported (4 publications). Therefore, GIA is a new research tool not yet used by any of the researchers. This broader literature research is a development of earlier authors' pilot studies regarding a comparative analysis of the functionality of CRM systems, carried out on a case study—the software service provider (Domański and Filipiak 2019).

To sum up, the originality of this article is based on: 1. undertaking a very niche topic—SME-dedicated CRM systems enabling the implementation of the omnichannel concept; 2. a pioneering solution to the problem based on an innovative research tool—the GIA method from the GST family. As part of the practical part, the CRM system specification dedicated to SMEs will be proposed for companies that use the omnichannel approach. Then, CRM systems for SMEs available on the market will be assessed in terms of meeting their omnichannel requirements.

5 Methodology for Testing CRM Systems Dedicated to SMEs in Terms of Omnichannel

The author's research methodology consists of 4 main steps (Fig. 1). First, research objects (CRM systems) were determined. To this end, research was conducted on the Internet, based on CRM system rankings published on the web—including https://crm.financesonline.com/, https://www.sellwise.pl/ranking-najlepych-systemow-crm-2019/, https://www.g2.com/categories/crm, https://www.capterra.com/customer-relationship-management-software/, 07.03.2020) as well as websites of manufacturers of individual CRM systems. Based on the available lists of CRM systems on the market, all those that were not dedicated to small and medium

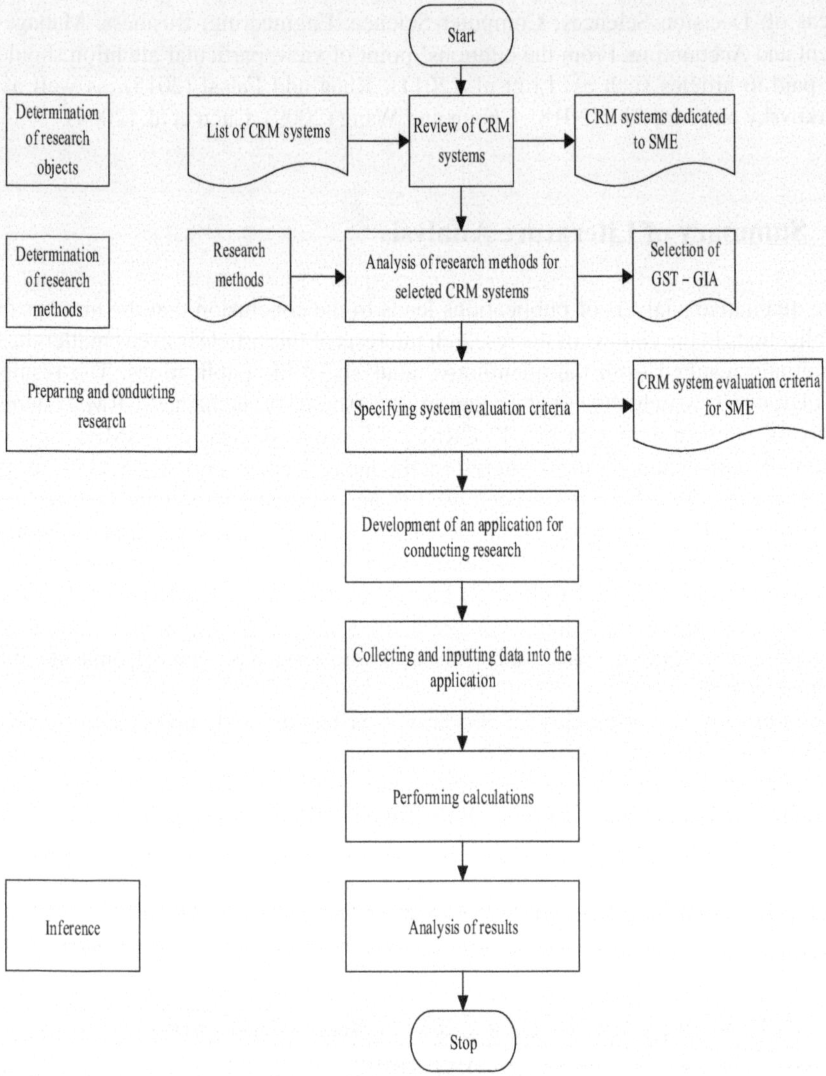

Fig. 1 Original research methodology. *Source* Own work

enterprises were rejected (systems for large enterprises were omitted from the beginning). The selection identified 84 CRM systems for small and 53 for medium-sized enterprises. Among the selected CRM systems, only 48 of them are dedicated to SMEs at the same time. By rejecting strictly specialized systems, the authors identified 31 common, universal CRM systems that will be covered by the study. These are: Act!, Agile CRM, Avochato, Close, Creatio, EngageBay, eWay-CRM, Freshsales, HubSpot, InfoFlo, InStream, Microsoft Dynamics 365 Sales, NetHunt, NetSuite, Onpipeline, Oracle EBS CRM, Pipedrive, Pipeliner, Prophet CRM, Really Simple

Systems, Salesflare, Salesforce, SAP CRM, Shape Legal, treak, SuiteCRM, Sumac, Upsales Sales and Marketing Platform, vCita, Zengine, Zoho.

The next step in preparing the study was to review the research methods that can be used (step 2). Based on authors' earlier experience (Wojciechowski and Hadaś 2020), it was decided to use the GIA method from the grey system theory, because it perfectly fits into the research assumptions, i.e. a description of reality, a small research sample required, unknown distribution of variables. In addition, as demonstrated in the light of literature analysis (Sect. 2), the GIA method is a pioneering research tool that the authors undertake.

Then the systems that went on to the next stage were evaluated by current users (y1)—the presence and scope of operation of individual modules was examined (step 3). The assessment by system users included, among others, possibilities of managing contacts and relations with business partners, task and schedule management in terms of marketing channels, technical support, reporting and analyses, integration with other platforms, possibilities for mobile devices and social media, awards and certificates possessed, costs of purchasing a given system. The final step in collecting data was to define criteria related to omnichannel in CRM systems. On this basis, it was decided that CRM systems will be compared in terms of integration of customer service channels (x1) and the possibility of using these channels. Selected channels in CRM systems included: email (x2), an enterprise or store website (x3), social media (x4), chat with employees (x5) and telephone contact (x6). For the operational needs of the study, the authors developed a simple IT application.

The final stage of the research methodology—analysis of results and inference (step 4) will be described later in the article as a separate fragment (Sect. 5).

6 Implementation of the Study of Omnichannel CRM Systems Dedicated to SMEs

The first step is to collect data and fill in Table 9 where the numbers in the top row correspond to consecutive enterprises: y1 is the characteristics of the CRM system, i.e. in this case the overall rating of the system by users (it can take values from 0 to 10; 10 maximum rating), while all other factors from x1 to x6 determine the impact on this characteristic (factor x1 can have values: 0—no integration, 0.5—integration of not all channels of the CRM system, 1—full integration of all channels; factors from x2 to x6 adopt binary values: 0—the channel is not present in the system, 1—the channel is present in the system).

The next step compared the results of 31 CRM systems (IDs from 2 to 32 in Table 9) against the ideal CRM system (column 1)—representing the maximum values in each row that exists only hypothetically and was determined artificially for the purposes of the study. According to the assumptions of the grey system theory, Si and Sj coefficients are used for comparisons. The following formulas were used for calculating Si, Sj and Sj—Si (Liu et al. 2017).

Table 9 Collected input data for calculations using the GIA method

Id	1	2	3	4	5	6	7	8	9	10	11	12	13	14	15	16
y1	10.0	8.0	8.4	8.8	9.4	9.3	8.9	8.5	9.6	9.8	9.1	9.4	7.4	8.4	9.5	8.2
x1	1	1	1	0	0.5	1	0	0	0.5	1	1	0	1	1	1	0
x2	1	1	1	1	1	0	1	1	1	1	1	0	1	1	1	1
x3	1	0	0	0	0	1	1	1	1	1	0	0	1	0	1	1
x4	1	1	0	0	0	0	1	0	0	1	1	0	1	0	0	0
x5	1	0	1	1	0	0	1	0	0	1	0	1	1	0	0	0
x6	1	1	0	1	0	0	1	1	1	1	0	0	1	1	0	1
id	17	18	19	20	21	22	23	24	25	26	27	28	29	30	31	32
y1	8.6	9.4	9.1	8.2	8.5	8.8	9.7	8.2	8.5	9.0	8.4	8.5	8.5	8.8	9.0	9.4
x1	0	0	0	0	1	0	1	0	0	0	1	0	1	0	1	1
x2	0	1	1	0	1	1	1	0	1	1	1	1	1	1	1	1
x3	0	0	1	0	1	1	0	0	1	0	1	0	1	1	1	1
x4	0	0	1	0	0	0	0	0	0	0	0	0	0	1	0	1
x5	1	0	0	0	1	0	0	0	1	1	1	0	0	1	0	1
x6	1	0	0	0	1	0	0	1	1	0	1	0	1	1	1	1

Source Own work based on CRM systems rankings and CRM manufactures' websites

$$|s_i| = \left| \sum_{k=2}^{n-1} x_i(k)d2 + \frac{1}{2} x_i(n)d2 \right|$$

$$|s_j| = \left| \sum_{k=2}^{n-1} y_j(k)d2 + \frac{1}{2} y_j(n)d2 \right|$$

$$|s_j - s_i| = \left| \sum_{k=2}^{n-1} [y_j(k)d2 + x_i(k)d2] + \frac{1}{2} [y_j(n)d2 - x_i(n)d2] \right|$$

Table 10 presents a screenshot of the application in which the data collected in Table 9 were processed. The consecutive columns provide the ID of the factor, then the sum in the rows (Sum for y or x), the value of the last row (Last one for y or x), the result of calculating the value of Sj for y (the second row of the column Sj, Si), the result of calculating the value Si for each factor x (rows 3 to 8 in the column Sj, Si)—the first 7 columns are used to improve calculations in the application, and a fragment of the appropriate partial calculations (from the 8th column)—the essence of the logic of the GIA method.

On the basis of data from Table 10, after calculating the values of Sj and Si, the absolute value of Sj–Si was determined, which illustrates the degree of influence of a given factor x on the general characteristics of the CRM system y. The results of the calculation are presented in Table 11.

Table 10 Results of comparison of each company's score to master model

Id	Sum y or x	Last y or x	Sj, Si	Sum Sj, Si	1		1	2	3	…	30	31	32
y1	−36.7	−0.6	s1	36.4	10	y1	0	−2	−1.6	…	−1.2	−1	−0.6
x1	−16.0	0	s2	16.0	1	x1	0	0	0	…	−1	0	0
x2	−5.0	0	s3	5.0	1	x2	0	0	0	…	0	0	0
x3	−14.0	0	s4	14.0	1	x3	0	−1	−1	…	0	0	0
x4	−23.0	0	s5	23.0	1	x4	0	0	−1	…	0	−1	0
x5	−18.0	0	s6	18.0	1	x5	0	−1	0	…	0	−1	0
x6	−13.0	0	s7	13.0	1	x6	0	0	−1	…	0	0	0

Source Own work

Table 11 Sum of x1–x6 values from the table

Sj–Si	y1
x1	20.4
x2	31.4
x3	22.4
x4	13.4
x5	18.4
x6	23.4

Source Own work

The epsilon coefficient indicates how much each of the x factors affects the characteristics of the y system, i.e. customer satisfaction. The formula for calculating the epsilon coefficient (Liu et al. 2017).

$$\varepsilon_{0i} = \frac{1 + |S_0| + |S_i|}{1 + |S_0| + |S_i| + |S_i - S_0|}$$

The final results of the GIA method, after ordering all the factors x by the descending degree of importance, are presented in Table 12.

Table 12 Results of epsilon value in a descending order

0.818428	x4	4.246612	y1
0.750678	x5		
0.723577	x1		
0.696477	x3		
0.682927	x6		
0.574526	x2		

Source Own work

Finally, as part of extending the research process beyond the GIA method, it was decided to determine the ranking of CRM systems based on the weighted mean method. The values for the factors are data from Table 9, the weights for the factors are set out in Table 12. Table 13 presents a weighted ranking of CRM systems.

Table 13 Comparison of weighted mean and user rating

ID	Name	Weighted mean	User rating
9	Hubspot	1.00	9.80
12	Microsoft Dynamics 365 Sales	1.00	7.40
31	Zoho	1.00	9.40
6	EngageBay	0.83	8.90
29	vCita	0.83	8.80
20	Really Simple Systems	0.81	8.50
26	SuiteCRM	0.81	8.40
1	Act!	0.66	8.00
24	Shape Legal	0.64	8.50
28	Upsales Sales and Marketing Platform	0.63	8.50
30	Zengine	0.63	9.00
8	Freshsales	0.55	9.60
10	InfoFlo	0.50	9.10
18	Pipeliner	0.49	9.10
2	Agile CRM	0.48	8.40
3	Avochato	0.47	8.80
14	NetSuite	0.47	9.50
13	NetHunt	0.47	8.40
7	eWay-CRM	0.46	8.50
15	Onpipeline	0.46	8.20
16	Oracle EBS CRM	0.34	8.60
5	Creatio	0.33	9.30
25	Streak	0.31	9.00
22	Salesforce	0.31	9.70
21	Salesflare	0.30	8.80
4	Close	0.22	9.40
11	InStream	0.18	9.40
23	SAP CRM	0.16	8.20
17	Pipedrive	0.14	9.40
27	Sumac	0.14	8.50
19	Prophet CRM	0.00	8.20

Source Own work

One interesting issue is the Prophet CRM system, the result of which is zero (last row in Table 13). This CRM system does not use any customer service channel—it rather serves as a database.

Finally, it was decided to divide the CRM systems listed in Table 13 into classes—the author's view. It was decided to distinguish the following classes and their ranges:

- 1.00—reference CRM systems, support all 5 customer service channels and integrate them with each other; there are 3 reference systems;
- 0.81–0.83—very good CRM systems, not much different from the standards, but with shortcomings—they support 4–5 customer service channels and try to integrate them; for selected industries, they may as well meet the reference systems if the company does not intend to use the service channels missing in these systems; there are 4 very good systems;
- 0.63–0.66—medium class of CRM systems, support 4 different customer service channels, but often do not integrate them; due to the lack of integration and support for only a few customer service channels in the medium class of CRM systems, there are large gaps in customer service, which means that such systems have a low impact on omnichannel; there are 4 medium class systems;
- 0.46–0.55—CRM systems below the medium class, systems in this class support a maximum of 3 channels, very often not integrating them all together; these CRM systems are more suitable for single or multichannel than for omnichannel; there are 9 systems below the medium class;
- <0.34—a class that most often supports a maximum of 2 different customer service channels; channel integration, if present, combines data of only 2 channels; therefore, systems of this class are suitable for enterprises that prefer a single or double channel approach; there are 11 systems in this class.

Referring to Table 13, the results of the authors' research (weighted mean) were confronted with the views of users of CRM systems (user rating). The authors' research was strictly focused on the possibilities of omnichannel functioning in CRM systems for SMEs, while users of CRM systems evaluated these systems from different, broader angles. Therefore, both classifications do not match and cannot be compared with each other.

7 Final Conclusions

Based on the results, it can be seen that in CRM systems evaluated in terms of the omnichannel approach, the social media channel is the customer service channel that has the greatest impact on the value of the system. The use of social media in CRM systems allows to reach a large number of customers in a non-intrusive way. In the second place was chat with an employee, i.e. a form of direct contact with the customer, which is not as official as the letter. Chat can be conducted for a longer period of time, and while doing so one can break away to focus on other activities, which may be the reason for considerable impact of this customer service channel on the CRM system assessment. Only in the third place, somewhat surprisingly, is the integration of the channels. This solution allows customers to smoothly change the service channel used in the rapidly changing reality of everyday life. Website, which is a rather a static source of information, ranks fourth. The fifth position is phone contact—a form that requires the greatest attention from both sides. Email is the customer support channel that has the least impact on CRM rating. The reason for this may be the universality of this channel including the amount of spam customers are flooded with.

To sum up, providers of CRM systems for SMEs focus primarily on modern customer service channels—social media and chat. This corresponds to the profile of today's digitized customer (a cell phone or a laptop in the continuous internet access mode). It is technically feasible today—widespread availability of information technologies (hardware and software). More traditional forms of service—website, telephone contact, e-mail—are beginning to lose their importance.

Omnichannel, identified with the integration aspect of customer service channels (factor x1-3rd position), is already clearly noticed, yet not fully appreciated. Based on the results of the study, it can be concluded that omnichannel is implemented in CRM systems for SMEs only partially (1 for factor x1 in Table 9). Still half of the customer service interactions (0 for factor x1 in Table 9) are implemented through different, non-integrated channels—a multichannel approach.

The market for popular, universal CRM systems is not very saturated—only 31 applications. Among these CRM systems, the authors diagnosed only 7 as pro-channel CRM systems (classes: reference and very good)—it constitutes only 23% of the total number of applications. The integration of customer service channels turned out to be a key selection factor. Most CRM systems for SMEs focus only on the functioning of a certain number of channels (not all possibilities) and they cannot integrate them into one coherent system (fragmentary solutions).

Future research, based on this preliminary study, should focus on changing research subject from SME CRM to large enterprise CRM systems. Another possibility for future research is to focus deeply on SME CRM systems to create a relational network between all objects in terms of using omnichannel.

Acknowledgments This research was funded by the Ministry of Science and Higher Education—project No. 11/140/SBAD/4171.

References

Calderón H, Fayos T, Frasquet M (2019) The transition of small Spanish wineries toward multichannel distribution: the role of ambidexterity. Int J Wine Bus Res 32(1):139–158

Chang C-J, Lin J-Y, Jin P (2017) A grey modeling procedure based on the data smoothing index for short-term manufacturing demand forecast. Comput and Math Organ Theory 23(3):409–422

Chen Y, Wu S, Liu S (2004) Application of fuzzy-grey comprehensive evaluation method in safety management of small and medium enterprises. Proc Int Symp Safety Sci Technology, ISSST 2004:2034–2041

Cheng Q, Wang M (2009) Risk analysis and evaluation of the destructive innovation in small and medium-sized enterprises, Proceedings of international conference on management and service science, MASS 2009, art. no. 5305759

Církovský T, Maryška M (2018) Usage of advanced CRM campaigns in medical therapy. Proceedings of 26th interdisciplinary information management talks, IDIMT 2018, pp 259–265

Critchley S (2018) Dynamics 365 CE essentials: administering and configuring solutions, 1st Edition, Apress, pp 1–613

Domański R, Filipiak P (2019) Omnichannel in CRM systems from the perspective of the software service provider. In: Dujak D (ed.), Proceedings of the 19th international conference business logistics in modern management, Josip Juraj Strossmayer University, Osijek Croatia, pp 21–41

Feng Y, Mei Q (2010) Research on electronic commerce application decision-making in China SMEs. Proceedings of the international conference on e-business and e-government, ICEE 2010, art. no. 5590920, pp 4386–4389

Govindan K, Khodaverdi R, Vafadarnikjoo A (2016) A grey DEMATEL approach to develop third-party logistics provider selection criteria. Indust Manag Data Syst 116(4):690–722

Heidekröger R, Heuchert M, Clever N, Becker J (2018) Towards an omni-channel framework for sme sales and service in the B2B telecommunications industry. Multikonferenz Wirtschaftsinformatik, MKWI 2018:386–397

Hu G-L (2010) Performance evaluation of enterprise CRM system based on grey-fuzzy theory, International Conference on Networking and Digital Society, ICNDS 2010, art. no. 5479195, pp 32–35

Hutchinson K, Donnell LV, Gilmore A, Reid A (2015) Loyalty card adoption in sme retailers: the impact upon marketing management. Eur J Mark 49(3–4):467–490

Ieva M, Ziliani C (2018) The role of customer experience touchpoints in driving loyalty intentions in services. TQM J 30(5):444–457

Khan H, Faisal MN (2015) A Grey-based approach for ERP vendor selection in small and medium enterprises in Qatar. Int J Busin Inform Syst 19(4):465–487

Kim H, Lee D, Ryu MH (2018) An optimal strategic business model for small businesses using online platforms, Sustainability (Switzerland), 10(3), art. no. 579

Kung C-Y (2005) Grey system method to assess enterprise's decision on marketing resource distribution and its performance—the experience of taiwanese medium and small manufacturing industries. Proc Int Confer Sys Man Cybern SMC 2005:1562–1567

Kung C-Y, Yan T-M, Huang C-C (2008) Using grey statistic method to evaluate the optimal distribution for strategic resources in medium and small enterprise. Proceedings of international conference on systems, man and cybernetics, SMC 2008, art. no. 4811732, pp 2866–2871

Li J, Sun D, Xia W, Xue Y (2011) AHP-based grey relational comprehensive evaluation method of the capability of technological SME's financing. Proceedings of international conference on business management and electronic information, BMEI 2011, art. no. 5920924, pp 85–88

Lingyu H, Bingwu L, Juntao L (2009) An ERP system selection model based on fuzzy grey TOPSIS for SMEs. Proceedings of 6th international conference on fuzzy systems and knowledge discovery, FSKD 2009, art. no. 5358953, pp. 244–248

Liu S, Yang Y, Forrest J (2017) Grey data analysis. methods, models and applications. Springer Singapore, pp 76–81

Ma Z-Q, Fan Q (2009) The study on risk evaluation of networked SME cluster based on gray-fuzzy theory. Proceedings of international conference on management and service science, MASS 2009, art. no. 5303826

Orzan G, Ioanas E, Delcea C, Orzan MC (2014) Impact of social media advertising campaigns on buyers' decisions, Proceedings of the 24th international business information management association conference, IBIMA 2014, pp 2038–2051

Park S, Lee D (2017) An empirical study on consumer online shopping channel choice behavior in omni-channel environment. Telematics Inform 34(8):1398–1407

Picot-Coupey K, Huré E, Piveteau L (2016) Channel design to enrich customers' shopping experiences: synchronizing clicks with bricks in an omni-channel perspective—the Direct Optic case. Int J Retail Distribut Manag 44(3):336–368

Prodanova J, Van Looy A (2019) How beneficial is social media for business process management? a systematic literature review. IEEE Access, art. no. 8663288, pp 39583–39599

Sun X-X, Amit M, Cao P-G (2007) Grey correlative degree analysis on factors of small and medium-sized enterprise marketing mix. Wuhan Ligong Daxue Xuebao/Journal of Wuhan University of Technology 29(6):143–146

Tang X, Wang C (2014) GEM listed companies' internal control effectiveness based on grey incidence analysis. J Grey Syst 26(2):38–48

Tang X, Wang C (2013) The research of GEM listed companies' internal control effectiveness based on grey incidence analysis. Proceedings of international conference on grey systems and intelligent services, GSIS 2013, art. no. 6714738, pp 38–43

Taufique Hossain TM, Akter S, Kattiyapornpong U, Wamba SF (2017) The Impact of integration quality on customer equity in data driven Omnichannel services marketing. Procedia Comput Sci 121:784–790

Vasiliev SA, Serov ER (2019) Omnichannel banking economy.Risks 7(4), art. no. 115

Wojciechowski H, Hadaś Ł (2020) Ranking of opportunities for implementing the omnichannel concept. Logforum 16(2):229–237

Won EJS (2018) Pioneering the distribution industry in Korea: dynamic capability at lotte shopping. J Distribut Sci 16(10):5–21

Xiong W, Chen L, Zhan GZ, Qiu Z (2008) RFM value and grey relation based customer segmentation model in the logistics market segmentation. Proceedings of international conference on computer science and software engineering, CSSE 2008, art. no. 4723147, pp 1298–1301

Yong L (2011) Grey incidence analysis on China's e-commerce development. Proceedings of international conference on electronic and mechanical engineering and information technology, EMEIT 2011, art. no. 6023903, pp 3859–3862

Zeng Y-R, Wang L, Xu X-H (2017) An integrated model to select an ERP system for Chinese small- and medium-sized enterprise under uncertainty. Technol Econ Dev Econ 23(1):38–58

Zhang B-X, Hu S-Q, Song J, Cheng S (2012) Analysis about medium-long-term demands forecasting and capacity decision based on the grey GM(1,1) improved models. Proceedings of international conference on management science and engineering, art. no. 6414194, pp 276–281

Zhang Q, Chen R (2014) Financial repression approach to the financing difficulty of the small and medium-sized enterprises: Empirical evidence from China. Proceedings of 6th international conference on business intelligence and financial engineering, BIFE 2013, art. no. 6961129, pp 240–244

Sustainable Supply Chain and Logistics

Sustainable City Mobility—Comparison of Actual State in Selected European Countries

Joanna Oleśków-Szłapka, Irena Pawłyszyn, Francesco Facchini, Agnieszka Stachowiak, and Ana Paula Tanajura Ellefsen

Abstract The topic of green mobility is very interesting and indispensable to consider due to European policy and climate changes. Cities are now actively starting to address these issues by implementing a range of different strategies, such as bringing in congestion charges and parking fees to internalise the negative impacts of transport (E. Commission in How can cities address future mobility challenges? 2019), improving and promoting the use of public transport (PT), encouraging non-motorised transport with dedicated pedestrian-bike paths, and restricting the access of certain vehicles to the city centre. The main objective of hereby chapter is to evaluate actual state in terms of green mobility in selected European countries such as Poland, Ukraine, Italy and Norway. Authors endeavor to assess the availability, comfort, technical condition, safety, environmental friendliness, amenities in the cities from the point of view of existing public transport such as trams, buses, urban scooters, city bikes, electric kick scooters, carsharing.

Keywords Green mobility · Sustainable transport · Urban transportation · Mobility solutions · Sustainability

---

J. Oleśków-Szłapka (✉) · I. Pawłyszyn · A. Stachowiak
Poznan University of Technology, Poznan, Poland
e-mail: joanna.oleskow-szlapka@put.poznan.pl

I. Pawłyszyn
e-mail: irena.pawlyszyn@put.poznan.pl

A. Stachowiak
e-mail: agnieszka.stachowiak@put.poznan.pl

F. Facchini
Polytechnic University of Bari, Bari, Italy
e-mail: francesco.facchini@poliba.it

A. P. Tanajura Ellefsen
Tesla, Norway
e-mail: anapaula_tanajura@hotmail.com

© The Author(s), under exclusive license to Springer Nature Switzerland AG 2020

P. Golinska-Dawson et al. (eds.), *Smart and Sustainable Supply Chain and Logistics – Trends, Challenges, Methods and Best Practices*, EcoProduction,
https://doi.org/10.1007/978-3-030-61947-3_9

1 Introduction

City logistics covers all activities in the urban agglomeration supported by modern information systems, which aim to improve the network of goods and information flow. Activities implemented under its assumptions place great emphasis on both the economic and ecological aspects of the expected effects. The tasks of urban logistics include reducing the number of cars on the streets, reducing exhaust emissions or choosing optimal routes and frequencies for public transport, taking into account various social events.

Transportation is one major source accounting for nearly a quarter of Europe's total greenhouse gas emissions and is the main cause of air pollution in cities (E. Commission 2016). Reducing fossil fuel-based travel is therefore crucial for both the local and global climate and has become a major policy objective for national- and city- governments across the globe (Moran et al. 2018).

The importance of green mobility is underscored in the 2030 Agenda for Sustainable Development, the Paris Agreement on Climate Change, and the New Urban Agenda. Green mobility is a broad concept that aims to simplify a diverse set of environmental impacts caused by the transport sector. Many countries are already taking climate change mitigation seriously through transport system improvement. However, climate action in the transport sector still has a long way to go.

Green and sustainable cities present fundamental opportunities to both apply new technologies (such as public transit, district heating, and green building and design) and bring about major lifestyle changes (such as walking, bicycling, and reductions in consumption).

There are, in fact, many different terms used today in discussing efforts to reduce environmental impacts and to live more lightly on the land. Sustainable development, sustainable communities and sustainable cities are a few of these terms (Beathley 2000).

One increasingly popular shared mobility service is bike sharing and such systems have appeared in cities across the world as a green transportation measure (DeMaio 2009; Shaheen et al. 2013; Fishman 2016; Pucher and Buehler 2017; Hamilton and Wichman 2018).

The main objective of hereby chapter is to evaluate actual state in terms of green mobility in selected European countries such as Poland, Ukraine, Italy and Norway. Authors endeavor to assess the availability, comfort, technical condition, safety, environmental friendliness, amenities in the cities from the point of view of existing public transport such as trams, buses, scooters, city bikes, electric kick scooters etc.

Authors compared city mobility in selected European cities such as Poznan (Poland), Bari (Italy), Oslo (Norway) and Lviv (Ukraine) based on primary and secondary sources of data.

The authors of the article conducted research to answer the following research questions:

RQ1: What means of transport are used the most common in analyzed cities?

RQ2: Do the respondents in the analyzed cities use green transport?

RQ3: What factors influence the choice of a particular means of transport?

Poznan is a city located in central-western Poland in the Greater Poland Lakeland, on the Warta River. It is the capital of the Greater Poland Voivodeship and the Poznan County. It is inhabited by 538,6 thousand people, which makes it the sixth largest city in Poland and the fifth largest in terms of population. Collective public transport in Poznan and the Poznan agglomeration is organized by the City Transport Board (ZTM). The main operator in the city and the surrounding area that provides bus and tram transport services is Miejskie Przedsiębiorstwo Komunikacyjne sp.z o.o. (MPK). The rolling stock belonging to MPK consists of 221 tram sets and 311 buses. Currently, the authorities of the city are implementing the "City of Poznan Cycling Program 2017–2022 with a perspective to 2025", which aims to improve the movement of cyclists and increase the interest of residents in the daily use of bicycles. The plan includes, inter alia, the construction of new bicycle routes, Bike & Ride parking lots, the extension of the Wartostrada, as well as education and promotion for conviction to ride a bicycle.

Lviv is a city located in the west of Ukraine. It is an important industrial centre, an air, rail and road hub. Lviv is the seventh largest city in the country (724.713 people) and is considered as a centre of western Ukraine. There are 4 types of public transport in Lviv: tram, trolley bus, bus and small urban buses (so-called marshrutka).[1] In 2017, the number of public transport vehicles was 631, but the strategy of developing city transport by the end of 2020 assumed an increase in the rolling stock to 730 vehicles, which is expected to result in an increase in transport capacity from 127,680 to 196,840 passengers. The bus transport system in Lviv creates 7 radial bus routes and 45 circular routes. Radial routes connect the outskirts of the city with the central ring and are served by buses with high and very high capacity. Circular routes, in turn, connect the city's outskirts. The capacity of buses serving these routes is 35 people. In total, the city supports 575 buses. In addition to bus lines, the city has 11 tram lines and 9 trolleybus lines. The demand for electric transport has increased in recent years. While building the *Strategy for the development of public transport*, the city authorities plan to implement systems enabling the purchase of an electronic ticket, creating a Park & Ride point on the radial end loops, increasing the share of public transport in the city's road traffic and limiting traffic in the centre of Lviv. A bicycle rental service has also appeared in the city since 2016.

Bari is a big port city on the Adriatic coast; the capital of the Italian region of Puglia. The city itself has a population of 320,257 inhabitants. Within Bari city itself there are urban buses services operated by a municipalized company, founded in 1965, called AMTAB. The company has a fleet of 234 buses, most of them powered by diesel, only a part (less than 20%) is powered by methane (Hosseini et al. 2019). Starting from 2020 will be included 2 electric buses. The average age of the available buses is around 10 years. There are 39 lines allow to serve 303 km, in last year around

[1]Marshrutka—is a small bus, designed to carry 9–20 passengers, not including the driver.

25 million of the persons were transported. According to last 'Urban Sustainable Mobility Plan', around 74% of people prefer private car or motorbike (52% and 22%, respectively) for move around the city, 18% of person prefer the public transport, and 5% of person adopt private bikes (Carli et al. 2015). Currently, only a 3% of persons in Bari adopt the service like bike-sharing, for this scope in last three years (2017–2019) 135 e-bikes have been provided by municipality of Bari (Ranieri et al. 2018; Digiesi et al. 2017). The carsharing service, provided by a national public company (ACI), was started in 2017, when the city received a fleet of 10 electric cars. The town of Bari promotes the use of bicycles, an eco-friendly means of transport that is both fast and efficient, through Bike Sharing (Barinbici), a bike rental service located in different locations around the city. The "Cicloattivi and the University" (biking) is a project promoted by the Puglia region and allows university students to rent folding bikes for a year, as well as promoting sustainable mobility and intermodal transport by both train and bus.

Oslo is an old town, but a relatively young city, and it has developed extensively in twentieth century as Norway has urbanized. In 2018 the city had a population of 673,469. Oslo has made generous use of fossil fuel dependent transportation technologies to access ever larger hinterlands in search of space for populations. Thanks to its green reputation, the city has won two prizes from the European Union for developing as a green city (Luccarelli and Røe 2013). This is a city with ambitious environmental targets aiming at reducing greenhouse gas emissions by 50% within 2030 (Plansamarbeidet 2015). Oslo has had a bikesharing scheme since 2002 (Alvik 2009). City council's proposition suggests several measures for the various municipal sector programs. For the transport sector, the most important measures are to ensure transition from individual car transport to public transport, increased cycling and walking. Public transport in Oslo is coordinated by Ruter AS and constitutes a network of travel possibilities by train, tram, subway, bus and boat almost 24 h per day. By 2020, public transport in Oslo will only use renewable energy. After that, all city buses will also have the Euro VI standard on their engines, which will significantly reduce both noise and local pollution. All public transport in the Oslo metropolitan area must be emissions-free by 2028 (Oslo Komune 2020).

2 Methodology

Authors will utilize following embedded case study methodology and survey research: statistical analysis (sampling scheme), developed survey questionnaire, applied survey and analysis of the results.

Based on obtained results chapter presents a comprehensive analysis and indicate future actions in order to increase green mobility and sustainability of city means of transportation. Authors try to point out how to boost the competitiveness of transport industries in selected European countries and how to achieve a city transport system that is resource-efficient, climate and environmentally friendly, safe and seamless for the benefit of all citizens and society.

Questionnaire developed by authors consists of 64 questions including respondent's particulars.[2] The main section comprises questions with respect to means of transportation, available public transport options in given city and assessment of time, cost, distance of public transport. The other sections are divided dependent on particular means of transportation such as urban bikes, public transport (buses/trams etc.), carsharing, electric kick scooters, small urban buses, urban scooters and other means of transportation. Most of questions use Likert scale – five-point agreement scale used to measure respondents' agreement with a variety of statements by means of unipolar scales.

2.1 Research Results

As a result of the survey, which was conducted at the turn of January and February 2020, over 2000 responses were obtained from all analyzed cities. The respondents to whom the survey was addressed are young people (over 97% of respondents are people under 25 years of age) studying at universities. About 26% of respondents are also employed. In terms of gender, 45% of women and 55% of men took part in the survey. The results of the collected responses of the respondents are presented below.

The first question of survey asked the respondents what means of communication they use most often. Respondents could indicate several means of transport which they most often use on a daily basis. The Fig. 1 shows the distribution of respondents' answers. It can be seen that young people from Poznan, Lviv and Oslo most often choose public transport, i.e. trams, buses and trolleybuses, as the main means of transport (Poznan—almost 95% of respondents; Lviv—91% of respondents, Oslo—73.7% of respondents).

Students from Lviv also often use marshrutkas (small urban buses), which are also part of public transport (38% of respondents). Just over 10% of respondents from Poznan, 3% of respondents from Bari and 1.6% of respondents from Lviv use city bikes. Only 4.5% of respondents use electric kick scooters available in Poznan, currently such means of communication is not available in Lviv, Oslo and Bari. Lviv, Bari and Oslo are also not equipped with urban scooters. On the other hand, In Poznan only 0.5% of respondents use them. Therefore, the conclusion is that the use of city bicycles, electric kick scooters or urban scooters is poorly developed in the analyzed cities, and therefore more attention should be given to them in the context of their development and promotion. On the contrary, data from secondary sources show that the city bike network in Oslo is very well developed, which makes it possible

[2]In this article, the authors will present selected questions from the survey and throw spotlight on the results obtained.

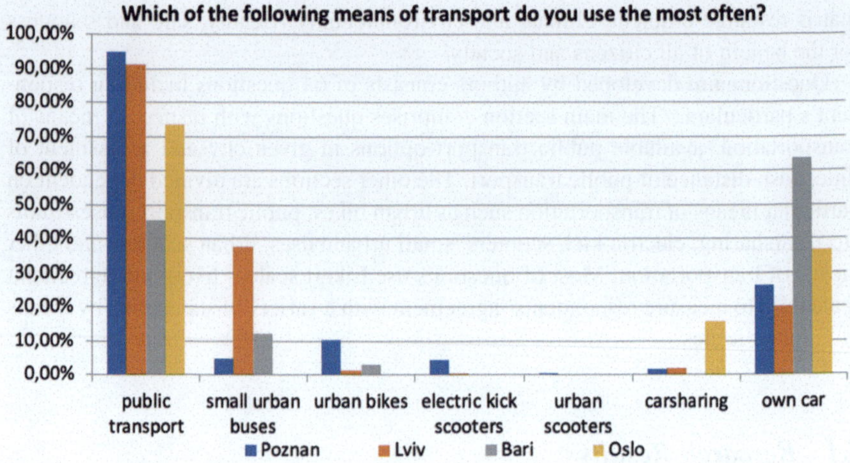

Fig. 1 The distribution of answers on question 1. (*Source* Own elaboration)

to conclude that respondents prefer other means of transport or their own bicycles instead of city bikes.

In Oslo quite big number of respondents (15.3%) use carsharing whereas Poznan and Lviv have a similar level of carsharing (around 2%). In Bari such service is not used by interviewees. It should be noted that in Bari the highest number of respondents (i.e. 63.3%) indicated their own car as the most preferred means of transport.

A fairly large group of respondents, as in Oslo, Poznan and also in Lviv prefer to travel by their own car (36.8%, 26.3% and 20.3% respectively).

The second question concerned the recognition of which of these means of communication are actually available in specific cities (Fig. 2). Respondents from Poznan, Oslo and Bari indicated that all suggested types of communication are available in the city, but there are no marshrutkas in these cities. Most likely, they confused the means of transport with ordinary buses. This means that the question posed has not been formulated precisely enough. A similar situation occurred in the case of Lviv, Oslo and Bari, where there are no urban scooters and electric kick scooters rental points and which, apparently, were treated as privately owned means of transport.

Third and fourth question in the survey verified how many respondents spend time on commuting every day (Fig. 3a), as well as what average distance they travel each day (Fig. 3b).

Figure 3a shows that almost 63% of respondents from Poznan require up to 1 h for travel time every day, and about 32% of those surveyed spend 1–2 h on travel. In Lviv, near to 73% of respondents devote up to 1 h commuting time and about 24%—up to 1–2 h. In Bari and Oslo also most of the respondents spend to transport up to 1 h (respectively near to 52% and 69%).

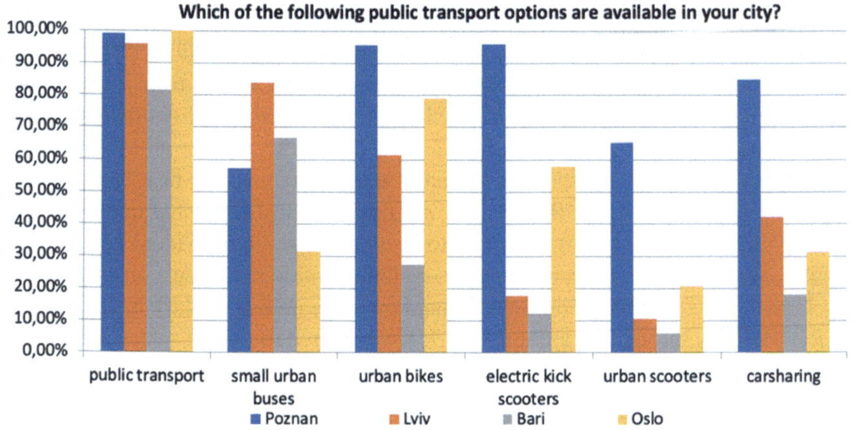

Fig. 2 The distribution of answers on question 2. (*Source* Own elaboration)

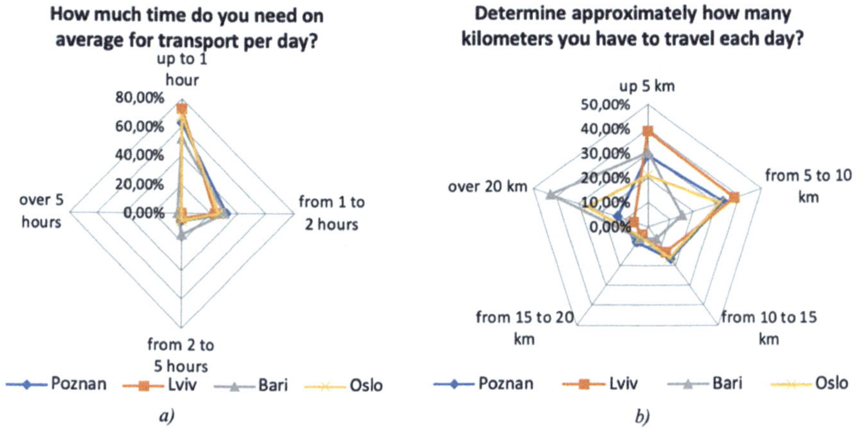

Fig. 3 The distribution of answers on questions 3 and 4. (*Source* Own elaboration)

In turn, Fig. 3b shows that just over 29% of students from Poznan travel daily up to 5 km, 33%—from 5 to 10 km/day and 16.4%—from 10 to 15 km/day. Quite a large group (13.4% of respondents from Poznan) travels over 20 km every day. In Lviv, the situation is slightly different: 39% daily travels up to 5 km, almost 38%—from 5 to 10 km/day and 13%—from 10 to 15 km/day. In Oslo, the results obtained are very similar to the situation in Poznan.

In Oslo, the results obtained are very similar to the situation in Poznan, solely more people travel more than 20 km a day (26.3% of respondents). In Bari, a smaller number of respondents have to pass a distance of 5–10 km (about 15% of respondents) and most surveyed have to travel more than 20 km (almost 43% of respondents).

The above results may be dictated by the fact that Poznan is a more geographically spread city compared to Lviv, and therefore students are forced to travel a greater distance. Whereas In Bari and Oslo students commute to studies from districts and smaller villages outside the city centre, which is confirmed by the results obtained.

In the next—***fifth—question*** authors asked what is the average monthly amount of charges incurred for transport. In Poznan students pay monthly near to 13.5 EUR (56 PLN), in Lviv 10.47 EUR (290.29 UAH), in Bari 40 EURO and in Oslo 82.36 EUR (959 NOK). Comparing the costs to average wages in individual countries, it should be noted that the costs dedicated to communication in Poznan constitute 1.4% in relation to the net remuneration in Poland (average monthly salary in 2019—962 EURO). In Lviv these costs amount to 2.75% compared to the net salary in Ukraine (average monthly salary in 2019—381 EURO). In Bari it is 3.24% respectively (average monthly salary in 2019 in Italy—1233 EURO) and in Oslo—2,72% (average monthly salary in Norway in 2019—3033 EURO). It can be seen that the highest percentage of costs is paid by Bari respondents. Students from Lviv as the second in a comparative analysis pay a fairly large percentage of transport costs. Oslo ranks third in this respect and the smallest percentage of communication costs in relation to average earnings is paid by respondents from Poznan.

The ***sixth question*** refers directly to public transport, and more precisely—the frequency of its use by respondents. Figure 4 shows that 98.25% of people use Poznan trams and buses. Similarly, in Lviv, 96.41% of people use trams, trolley buses and city buses, and almost 81% of respondents use marshrutkas. In Oslo public transport is mainly used by 84.21% of respondents. In the contrary in Bari the percentage share of student commuters using public transport is much lower (48.48% of respondents).

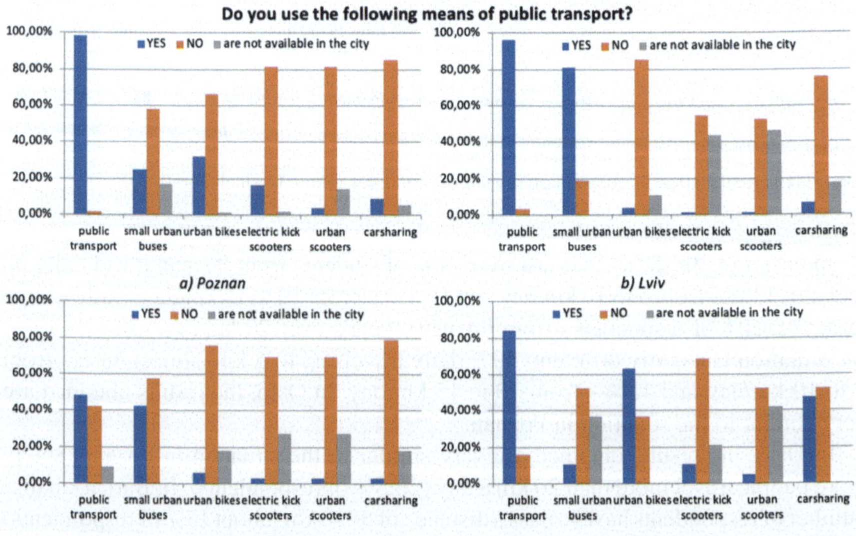

Fig. 4 The distribution of answers on question 6. (*Source* Own elaboration)

A significant difference can be seen in the case of city bikes, which are used by 63.16% of respondents in Oslo, 32.13% of respondents in Poznan and only 6% in Bari and less than 4% in Lviv. 16.62% of students in Poznan use electric kick scooters. The obtained research results draw attention to the need to develop particular types of communication, as it will directly contribute to the development of green mobility in the city. Less than 4% of respondents use urban scooters in Poznan, and 9.15% carsharing. The carsharing has generated small interest among Lviv students and Bari students alike (6.37% and 3.03% respectively). Such mode of transport is not popular because of its cost. On the other hand, 47.37% students from Oslo indicated that they use carsharing. It may also be dictated by the better financial standing of students in Norway.

Figure 5 is the answer to the ***seventh question***—how often do people use public transport? An analysis of the answers obtained indicates that in Poznan 66.41% of respondents use public transport every day, and 22.51% use it at least several times a week. 1.34% and slightly more—2.68%—ride a city bike ride in Poznan every day, several times a week. Almost 95% of respondents from Poznan have never used urban scooters, less than 90% have never used carsharing, almost 77% have never used electric kick scooters and almost 60% have never used city bikes. In Oslo similar to Poznan near to 60% of respondents use public transport every day. In the contrary in Lviv, none of the respondents use city bikes every day, rather young people use a given means of communication several times a month (5.58%) or several times a year (9.56%). In Bari solely 18.18% of respondents use public transport every day and 36,36% surveyed have never used such means of transportation.

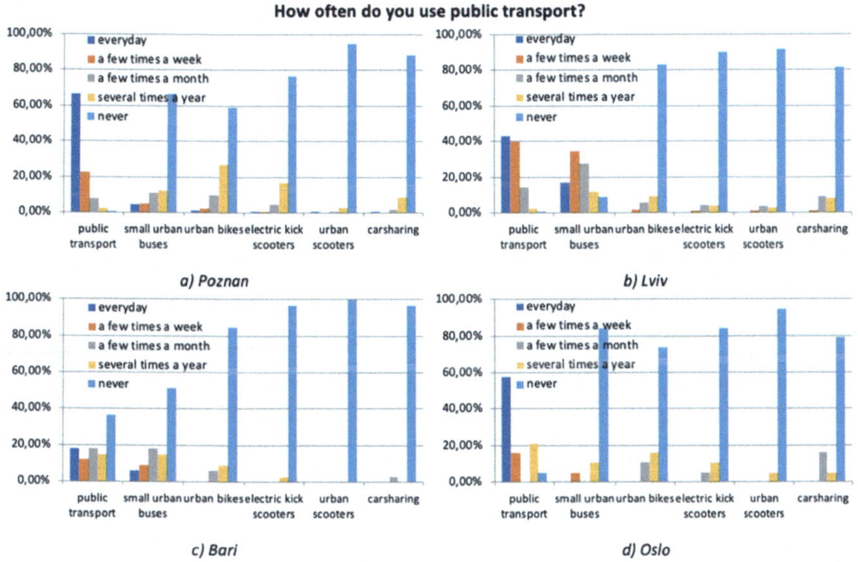

Fig. 5 The distribution of respondents in question 7. (*Source* Own elaboration)

In Lviv, hardly more than 43% of respondents use public transport and 17,13% of respondents use marshrutkas every day. Several times a week, almost 40% of people use public transport, while 34.66% use marshrutkas. Almost 83% surveyed have never used urban bikes in Lviv and 81,27% have never used carsharing.

Respondents were also asked about the reasons for not using certain means of communication. Among the reasons for not using city bikes, respondents in Poznan most often indicated a complicated procedure for renting them (25.8%), lack of comfort (19.9%), too long distance to travel (19.2%) and a high price for the service (17, 9%). In Oslo 66,7% prefer different mode of transport. The other reasons they mentioned: too long distance to travel (40%), bad infrastructure of urban bike paths (26,7%), lack of comfort (26,7%), travel time too long (20%). Lviv respondents pointed out that the lack of use of city bikes is primarily related to poor cycling infrastructure (35.8%), lack of comfort (27.6%), lack of safety (21.1%) and too long distance to travel (20.7%). In Bari the most important obstacles to use city bikes are as follows: prefer different mode of transport (14.8%), too long distance to travel (16.8%), bad technical condition of city bikes (22.2%), complicated procedure for renting them (14.8%). On the basis of above results, it can be seen that students do not prefer city bikes due to lack of comfort and too long distance to travel.

The lack of use of electric kick scooters in Poznan is primarily due to the excessive price (57.3%), lack of comfort (23.8%) and lack of safety (20.8%). Regarding the last reason, the respondents have repeatedly indicated that there are no clear rules regarding how to move the scooter around the city. Exactly the same hierarchy of answers was obtained by the reasons for not using urban scooters in Poznan (25.5%, 20.9% and 16.1% respectively). People who do not use carsharing in Poznan indicated that the price of a given service is too high (50.8%), rental rules are risky (39%) and that the rental procedure is too complicated (19.9%). In Oslo 14,3% of respondents also agreed that rental procedure is too complicated. Almost 43 students in Oslo prefer different mode of transport and 7% of them do not have driving license. In Lviv, respondents also do not use carsharing, because they think the price is too high (58.3%), the rules for renting a car are risky (30.1%), and the renting procedure is too complicated (28%).

It should be mentioned that in all cities respondents clearly see the prospects for the development of such public transport as bicycles, electric kick scooter or urban scooters.

The eighth question was asked to indicate the level of satisfaction of respondents with the use of specific means of transport (Fig. 6). It can be seen that the use of public transport in Poznan is rated. Almost 35% of respondents assess their level of satisfaction as average for the use of trams and buses. In turn, 52.88% of respondents are satisfied with the use of city bikes, and 38.12% are average satisfied with the use of a given means of transport. 55.04% of respondents indicate that their level of satisfaction with the use of electric kick scooters is high or very high, while little above 32% of respondents define this level as average. For urban scooters, the figures are: 52.45% - satisfied and very satisfied, 29.37% - average satisfied. With regard to carsharing, almost 64% of respondents described their level of satisfaction as high/very high, and 26.32% as average. It should be noted here that for any means

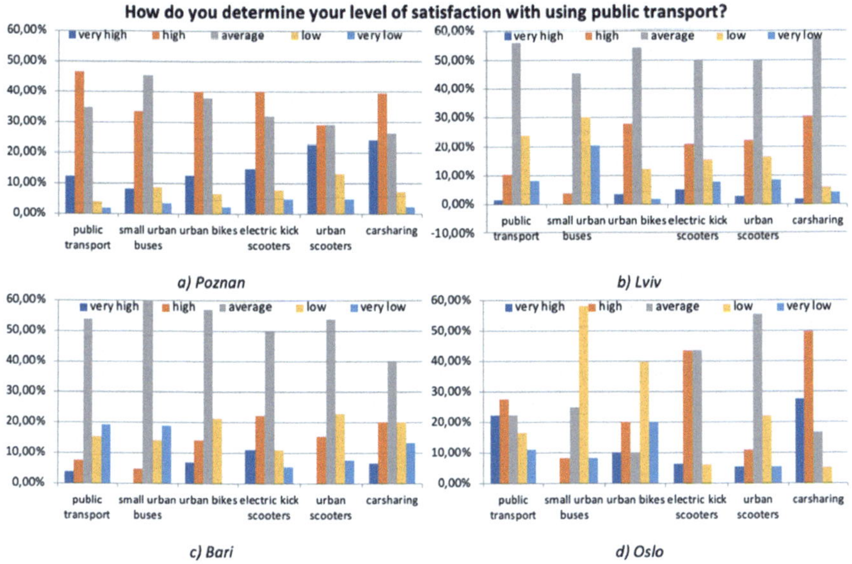

Fig. 6 The distribution of answers on question 8. (*Source* Own elaboration)

of communication the "very low level of satisfaction" option did not exceed even 5%, and the "low level of satisfaction" option—9%, except for city scooters, which received 13.29% of respondents' answers in a given option. This means that travelers are generally satisfied with the available means of transport in the city of Poznan. It is also interesting that public transport received the smallest percentage of negative answers—6.14% (the sum of the answers "low level of satisfaction" and "very low level of satisfaction").

In Lviv, public transport is rated at an average level of satisfaction—55.82% of responses. At a high and very high level, it was rated by slightly over 12%. Even less respondents are satisfied with the use of marshtrutkas—3.9% satisfied and 45.45% average satisfied with traveling on a given means of transport. For city bikes, the satisfaction level reaches 31.58% (the sum of "high" and "very high"). In turn, 54.39% of respondents indicated that they are average satisfied with their use. On the other hand, carsharing shows 32.65% satisfied users and 57.14% average users. Respondents indicate that over 32% are not satisfied with the use of public transport (the sum of "low" and "very low" responses), and more than 50% of respondents are not satisfied with the use of small urban buses (the sum of "low" and "very low"). The results are a signal as well as a direction of development which the city government should focus on.

In Bari the majority of respondents evaluate all means of public transport at average level (public transport—near to 54%, urban bikes—57%, carsharing—40%) Only about 11% of surveyed students are satisfied with using public transport at very high and high level and almost 35% is not satisfied (sum of results "low" and "very low). Similar answers can be observed for carsharing (33% of surveyed students).

From the other hand 6,67 of respondents if very high satisfied and 20% of respondents is high satisfied using carsharing services. The number of answers placed in the middle of the rating scale indicates great opportunities to improve the current level of satisfaction.

In Oslo comparing to other cities the level of satisfaction is the highest. Public transport obtained almost 50% of satisfied answers indicating "very high" and "high" options. 30% of respondents is also satisfied with using city bikes and 77,78% of respondents is satisfied with carsharing services. In both cases the sum of "high" and "very high" answers is taken into account. The number of received responses can on the one hand confirm the quality of transport services offered in Oslo as well as be proof of the level of life satisfaction among students in Oslo.

The goal of the ***ninth question*** was to identify the most important advantages of different modes of public transport (Fig. 7). In Poznan the respondents indicated that public transport's (i.e. trams and buses) biggest advantages are: accessibility (18.17%), price (65.25%) and travel time (49.56%). As the fourth criterion, the respondents distinguished health and ecology (25.66%). Speaking about city bikes, the respondents consider health and ecology (44.43%), accessibility (31.55%) and price (25.95%) as the biggest advantages. In turn, the biggest advantages of electric scooters are accessibility (23.62%), comfort (18.72%), health and ecology (17.90%) and travel time (17.84%). Poznan respondents use scooters because of travel time (12.71%), comfort (11.37%) and accessibility (8.1%). Carsharing is chosen primarily for comfort (23.44%), privacy (19.94%) and travel time (12.01%).

In Lviv, the biggest advantages of public transport are the price (65.34%), accessibility (60.16%), comfort (36.65%) and travel time (27.49%). Marshrutkas are

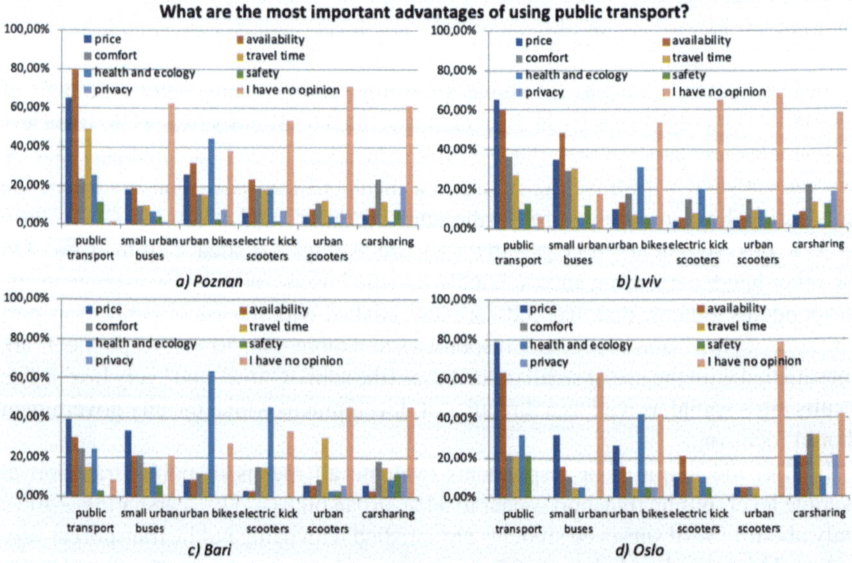

Fig. 7 The distribution of answers on question 9. (*Source* Own elaboration)

chosen primarily for their availability (48.61%), price (35.06%), travel time (30.68%) and comfort (29.08%). Respondents use city bikes because of their health and ecology (31.08%), comfort (17.53%) and accessibility (13.55%). On the other hand, carsharing services in Lviv were used for the same reasons as in Poznan: comfort (21.91%), privacy (18.73%) and travel time (13.55%).

In Bari the respondents pointed out the most important advantages of public transport: price (39.39%), availability (30.3%), comfort (24.24%), health and ecology (24.24%). Whereas urban bikes and electric kick scooters received the most indications on health and ecology (respectively 63.64% and 45.45%). The travel time and comfort were evaluated at lower level comparing to public transport. (12% for city bikes and 15% for electric kick scooters). Carsharing was evaluated at similar level (from 12 to 18% of respondents) regarding travel time, safety, privacy and comfort. Comparing to Poznan and Lviv price for public transport is recognized as smaller advantage probably due to higher prices for public communication.

In Oslo the most respondents selected price as the biggest advantage for public transport (almost 79% of respondents) despite the fact that in this city prices for public transport are the most expensive. The other circumstances for above transport are availability (63.16% of respondents), health and ecology (almost 32% of respondents), travel time and comfort (both about 21% of answers). The health and ecology for city bikes was chosen by a similar number of surveyed as in Bari and Poznan. Another advantage was price (26.31% of surveyed) and availability (15.79% of surveyed). More respondents than in other cities chose the advantage of comfort for carsharing (almost 32%).

In all cities, the most important advantages of public transport are, therefore, price, availability and travel time, bicycles as a means of transport are chosen primarily for health and environmental reasons, and carsharing is enjoyed by people who value privacy and comfort.

Owing to the ***tenth question***, it was possible to obtain information on how respondents assess the technical condition of individual means of communication (Fig. 8). The condition of Poznan trams and buses is rated as good by 33.20% of respondents and rather as good by 51.08% of respondents, which in general constitutes nearly over 84% of positive answers. The condition of city bikes in Poznan is assessed as good by 19.79% of respondents and by 40.28% of respondents as rather good. The distribution of answers regarding the electric kick scooters indicated 27.41% of satisfied users and 42.07% of rather satisfied users, respectively. Subject to urban scooters (it should be mentioned that this is the mode of transport that is least often used by respondents in Poznan) the percentage of the highest indications is about 32.97% and the percentage of "rather good" answers is 34.41%. The technical condition of shared cars is evaluated as good by 46.13% of respondents and rather good by 32.87% of respondents. City bikes received the largest percentage of negative responses (16%). Therefore, one should pay attention to a given means of communication, because poor condition of bicycles may lead to a collision or accident of road users.

In Lviv, the situation is as follows: the technical condition of public transport is assessed as good in 5.65% of cases and in 30.24% of cases as rather good. The sum of positive answers is therefore less than 36%. The technical condition of marshrutkas

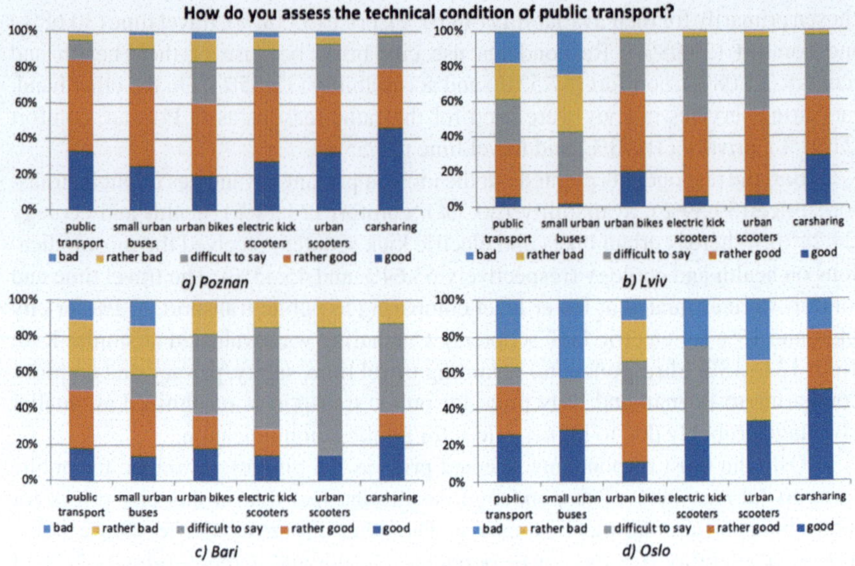

Fig. 8 The distribution of answers on question 10. (*Source* Own elaboration)

is the worst of all these types of communication in Lviv. Only 2.16% of respondents indicated good technical condition of vehicles and less than 15% indicated rather good condition, which in total gives less than 17% of positive opinions. The technical condition of city bikes is definitely better, as 20.31% of respondents indicated that they rated their condition as good, and 45.31%—rather good, which gives 65.63% of positive responses. The condition of rented cars in carsharing is also positive: 29.17% of respondents rated their condition as good and 33.33%—rather good. City authorities should devote considerable attention to the technical condition of public transport, as well as small urban buses, since it is the latter means of transport which has obtained the largest percentage of negative responses (56.71%).

In Bari almost 18% of respondents assess technical condition of public transport as very good and 32.14% as rather good. On the other hand, the similar number of answers select condition rather bad and bad (28.57% and 10.71% respectively). Considering urban bikes 45.45% of respondents cannot decide how to evaluate their technical condition. At the same time 36% of surveyed think that this condition is very good and good. None of respondents suggest that technical condition of urban bikes and carsharing is bad. Almost half of research sample cannot decide how to assess the technical quality of carsharing. This distribution of answers may result from the number of respondents who do not use selected means of transport, preferring their own car.

In the last of the cities surveyed (Oslo) more than 50% of respondents recognized a technical condition of public transport as good and rather good but also nearer to 36% evaluate this condition as bad and rather bad. This may be due to the varied technical condition of different vehicles in public transport. City bikes according to surveyed are in a good condition (25% of respondents) and simultaneously in a bad condition by the same number of respondents. The better situation is for carsharing, 50% of respondents evaluate its technical condition as good and 33.33% as rather good.

In each city, the technical condition of the available urban mobility modes should be analyzed and, if possible, dependent on technical resources further improvements should be done.

The last ***eleventh question*** that the authors will present under this article will be the assessment of the amount of fees for individual means of communication (Fig. 9). 35.2% of Poznan users determine that the prices of city buses and trams are high or very high. Prices are medium for more than half of users (51.23%). As for city bikes, in 17.47% of cases, fixed usage prices are perceived as high or very high, and in 41.45% of cases—as medium. Poznan respondents estimate the prices for renting scooters and scooters as definitely too high (81.53% and 68.38% respectively of the indications "very high price" and "high price"). This confirms the previously presented information on the reasons for not using the means of transport. The fee for car renting is also considered as too high (67.90%).

Lviv public transport is rated as expensive or very expensive in 35.89% of cases. Over half of the respondents (52.82%) determine public transport charges at medium level. In the case of small urban buses, 48.7% of users indicated that the charges for a given means of transport are high or very high. 48.44% think marshrutkas charges are moderate. For bicycles, just over 40% of users' prices are too high or high, and 48.44% see them as medium. Carsharing is perceived as the most expensive in Lviv, where exactly half of the respondents consider car rental too expensive or expensive,

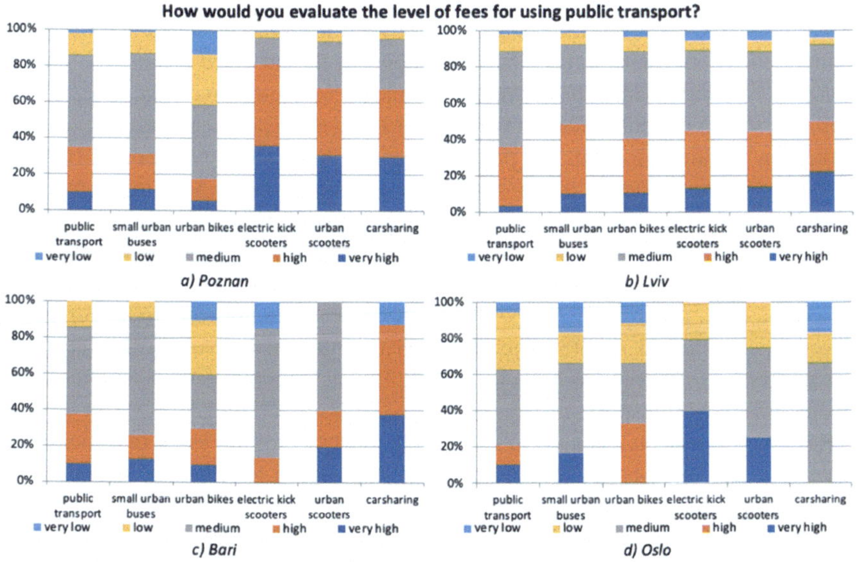

Fig. 9 The distribution of answers on question 11. (*Source* Own elaboration)

and 42.59%—as medium. At this time, information about the reasons for not using carsharing in Lviv due to the high price of the service is also confirmed.

From Bari respondents' point of view the fees for public transport are very high (10.34%) and high (27.59%). At the same time 48.28% of them perceive these fees as medium. The answers for city bikes are divided in very similar proportions: 30% of respondents assess fees as high and very high, medium and low. The fees for carsharing are definitely too high (37.5% of respondents—very high, 50% of respondents—high).

In Oslo almost 20% of respondents perceive fees for public transport as very high and high. Whereas by 42% of respondents their level is medium and by 31.58% of respondents is low. Urban bikes' fee by more than 60% of surveyed are assessed as high and medium (in each group 33% of answers). The level of fees was assessed as low and very low: for carsharing by 33% of respondents, urban scooters by 25% of respondents, city bikes by 33% of respondents.

There are many aspects that can affect the assessment of the price level of various means of public transport in the cities analyzed. Therefore, detailed conclusions depend on the level of earnings, correlation between the frequency of journeys and the distance to overcome. The above results show that in Italy and Norway prices are generally perceived as lower compared to Poland and Ukraine, which may be conditioned by the social status of citizens. The exception is the assessment of carsharing services in Bari.

3 Conclusions

Summing up the results of the conducted analyses, it should be stated that the most common means of transport used by respondents of the surveyed countries is public transport. Similarly, a large proportion of people prefer their own means of transport. Further research directions should include verifying the criteria encouraging owners of their own vehicles to switch to public transport, including bicycles or scooters.

Research also indicates that a significant proportion of respondents never use green transport such as bicycles or electric kick scooters. This is dictated by the lack of services in the city, or a weak marketing campaign encouraging potential users to use the given means of transport. Interestingly, one of the problems with using such services is, according to respondents, the complicated procedure for renting a bicycle or electric kick scooter. The request is a signal to the city authorities to work on simplifying such procedures as well as explaining on a larger scale how to rent a specific means of transport.

Review also reflects the need to modernize transport rolling stock in three of the surveyed cities, as respondents show average satisfaction with its use. Despite this, respondents distinguish the price as the main advantage of public transport in all countries. At this point, further research directions could include queries regarding users' consent to temporarily increase prices for public transport in order to modernize transport rolling stock.

Positive in all four cities is the fact that answerers see opportunities for developing green mobility in their cities. This signals the awareness of the respondents about the impact of transport on the natural environment, so they expect for infrastructural and informational support from the authorities. Public transport showed the most minimal changes, with demand remaining constant over the study period. The main improvements were in technology, such as smart-phone apps to buy tickets and check timetables.

In Italy there are plans underway to develop bike sharing schemes in different cities including Bari, with funding provided by the national government. The environmental benefits of bike sharing are at its best when bike sharing does not substitute walking, cycling or public transport and when it is combined with public transportation in covering the first and last mile of public transportation journeys.

Making cities smarter and more sustainable is a major aim of the European Union. Improving the sustainability of transport is prioritized through measures such as encouraging cycling, improving public transport and providing incentives for low-emission vehicles.

Bicycle sharing programmes serve as an alternative transportation mode in cities and provide public access to pick-up and drop-off bikes at numerous locations (Shaheen 2010; Kuppusamy et al. 2019). The potential environmental benefit of such programmes is however debated as a considerable number of trips are substituting other green transportation modes and the sustainable impact of bike sharing is argued to be limited (Fishman et al. 2013).

The use of a bicycle for everyday travel is particularly advantageous over short distances. According to research carried out in the European Union, the average distance of most displacements carried out in cities does not exceed several kilometres. At such distances—considering the traffic congestion—a bicycle may turn out faster than a car and public transport. Mobility is one of the most difficult topics to face in metropolitan large areas. It involves both environmental and economic aspects and needs both high technologies and virtuous behaviors' people.

Among the solutions that can improve the conditions of cycling in the city include: creating bicycle paths that run in the field regardless of the road system or within the lane, separating bicycle lanes on the road, separating bus and bicycle lanes, two-way bicycle traffic allowed on one-way streets with limited traffic and speed with the possible separation of a counter-cycle for bicycles, introduction of bicycle locks at intersections with traffic lights, introduction of markings for cyclists (organizational and information), allowing the transport of bicycles in public transport, adaptation of interchanges to leave bicycles in the "park and ride" system, introduction of a city bike rental system.

There is a need to continuously monitor mobility initiatives, for example by using the indicator-based assessment to ensure projects are implemented effectively and to understand what aspects of governance help ensure sustainable-transport schemes are realized.

Transport is one of the most important factors determining the country's economic development. A well-developed transport infrastructure strengthens the social, economic and spatial cohesion of the country and contributes to strengthening the competitiveness of the Polish economy. Modern infrastructure and an effective transport system are conducive to the country's economic growth, and the country's location on international transport routes is one of the important competitive advantages. Mobility in urban areas is also an important facilitator for growth and employment and for sustainable development.

The main contribution of this article is the assessment of actual situation in terms of different mode of transport. Authors plan to develop herein research in order to recognize green mobility activities in surveyed cities. Another perspective will be also the evaluation of support given by city authorities and European Union funds. Due to Logistics 4.0 trends it will be valuable to recognize actual trends towards efficient, connected and automated transport systems that are sustainable, safe and accessible for all citizens.

Acknowledgments The authors want to thank everyone who participated in the survey.

References

Alvik A (2009) Bysyklene

Beathley T (2000) Green urbanism, learning from European cities

Carli R, Dotoli M, Pellegrino R, Ranieri L (2015) Using multi-objective optimization for the integrated energy efficiency improvement of a smart city public buildings' portfolio. In: IEEE international conference on automation science and engineering

DeMaio P (2009) Bike-sharing: history, impacts, models of provision, and future. J Public Transp

Digiesi S, Fanti MP, Mummolo G, Silvestri B (2017) Externalities reduction strategies in last mile logistics: a review. In: Proceedings—2017 IEEE international conference on service operations and logistics, and informatics, SOLI 2017

E. Commission (2016) A European strategy for low-emission mobility, 20/07/2016—COM 501. [Online]. Available: https://ec.europa.eu/commission/presscorner/detail/en/MEMO_16_2497.

E. Commission (2019) How can cities address future mobility challenges? [Online]. Available: https://urban.jrc.ec.europa.eu/thefutureofcities/mobility#_edn1.

Fishman E (2016) Bikeshare: a review of recent literature. Transp Rev

Fishman E, Washington S, Haworth N (2013) Bike share: a synthesis of the literature. Transp Rev

Hamilton TL, Wichman CJ (2018) Bicycle infrastructure and traffic congestion: evidence from DC's capital bikeshare. J Environ Econ Manage

Hosseini SM, Carli R, Dotoli M (2019) Robust day-ahead energy scheduling of a smart residential user under uncertainty. In: 2019 18th European control conference, ECC 2019

Kuppusamy GP, Thakkar C, Zheng Y (2019) Bikesharing: analysis and prediction. In: SIGITE 2019—Proceedings on 20th annual conference on information technology education, vol 4, no 2009, p 157 (2019)

Luccarelli M, Røe PG (2013) Green Oslo: visions, planning and discourse

Moran D, Kanemoto K, Jiborn M, Wood R, Többen J, Seto KC (2018) Carbon footprints of 13 000 cities. Environ Res Lett

Oslo Komune (2020) [Online]. Available: https://www.oslo.kommune.no/

Plansamarbeidet (2015) [Online]. Available: https://plansamarbeidet.no/.

Pucher J, Buehler R (2017) Cycling towards a more sustainable transport future. Transp Rev

Ranieri L, Digiesi S, Silvestri B, Roccotelli M (2018) A review of last mile logistics innovations in an externalities cost reduction vision. Sustain

Shaheen SA (2010) Bikesharing in Europe, the Americas, and Asia: past, present, and future. Transp Res Rec J Transp Res Board
Shaheen S, Cohen A, Martin E (2013) Public bikesharing in North America. Transp Res Rec

Sustainable Urban Freight Strategies for Jaipur City, India

Pankaj Kant and Sanjay Gupta

Abstract Urban freight transport is integral to the overall sustainable development of cities in tandem with passenger transport. This research chapter is based on an empirical study carried out in the city of Jaipur in north India. This chapter presents traffic & parking regulation strategies from literature and intends to come up with aggregate weightage and ranking of different strategies for sustainable freight mobility for Jaipur city. Analytical Hierarchical Process (AHP) and Best Worst Method (BWM) are used to assess the weights. Both AHP & BWM are multicriteria decision-making techniques, thus useful in selecting the most relevant freight strategies for sustainable urban goods distribution in case city. Hypothesis for selecting AHP & BWM is that both methods produce similar results. This chapter critically analyses the traffic & parking regulation strategies in the preview of wholesalers, transport operators and local policymakers involved in urban goods distribution in case city. Primary data and response were collected with face to face paper–pencil survey from stakeholders in case city for selection and ranking of freight strategies. Response received from 32 stakeholders of Jaipur is used for analysis. The policy implications based on this study demonstrates the potential utility of AHP & BWM as a decision-making tool in urban freight sector in India.

Keywords Sustainability · Freight strategies · AHP · BWM

P. Kant (✉) · S. Gupta
Transport Planning Department, School of Planning and Architecture, 4-Block-B Indraprastha Estate, New Delhi 110002, India
e-mail: pankaj179phd17@spa.ac.in

S. Gupta
e-mail: s.gupta@spa.ac.in

© The Author(s), under exclusive license to Springer Nature Switzerland AG 2020

P. Golinska-Dawson et al. (eds.), *Smart and Sustainable Supply Chain and Logistics – Trends, Challenges, Methods and Best Practices*, EcoProduction,
https://doi.org/10.1007/978-3-030-61947-3_10

1 Introduction

There are growing concerns for the impacts of urban goods distribution in urban areas for congestion, externalities, infrastructure vulnerability and consistency with land use. Sustainability of urban goods transport is the key policy objective of various countries (OECD 2003). Sustainable transportation is essential for overall sustainable development due to its contribution to externalities, economic and social issues (Gudmundsson et al. 2015). Strategies and key performance indicators (KPI) are used for the decision-making process, travel patterns, physical impacts, environmental effects and economic impacts (Litman 2006). Decision-making, simulation and diagnosis are critical objectives of urban goods studies. Freight strategies are required and helpful for decision-makers to achieve sustainable development objectives (Litman 2019). Local authorities and policymakers have low awareness and knowledge regarding urban freight sector due to complex supply chains of urban goods and multiple stakeholders involvement. It is tough to predict the outcomes of freight policies and strategies (Patier and Routhier 2008). There is a need to raise awareness for freight issues considering all relevant stakeholders (Lindholm 2010).

Heterogeneous nature of urban freight stakeholders and goods movement characteristics in the city poses a key challenge to policymakers to select suitable freight strategies for sustainable goods distribution. Freight strategies for sustainable urban goods distribution are not adequately addressed in India. This research paper evaluates the weights of various traffic and parking regulation strategies for sustainable urban goods distribution for Jaipur city (India) by the Analytical Hierarchy Process (AHP) and Best Worst Method (BWM) method. AHP & BWM are useful tools for dealing with complex decision making. AHP and BWM aid the decision-maker to make the best decisions (Ruescha et al. 2012; Saaty 2008). The next section presents a literature review related to freight strategies used in urban goods distribution. Section 3: case city profile and wholesale commodity market selection. Section 4: research methodology. Section 5: AHP & BWM model formulation. Section 6: AHP & BWM analysis of urban goods strategies and the last part, Sect. 7: presents conclusions and policy implications.

2 Literature Review

Effect of urban freight transport can be quantified under the transport fleet, urban deliveries, economy, environment and safety. Each category required a different kind of freight strategies (Rezaei 2015). Urban freight data of 56 cities from 32 countries were compiled and analysed in-depth for various urban freight strategies implemented based on their final results. A total of 48 freight strategies under seven major group were identified for investigation. A multicriteria scoring function was used to rank & weight urban freight strategies of 56 cities (BESTUFS II 2006). Urban freight strategies under three groups i.e., last-mile, environment and trade node strategies were identified for sustainable urban goods distribution. These three group consists of 21 urban freight strategies already implemented by various cities around the globe. There is more need for research for intracity freight movements and the effectiveness of existing policies and freight measures (BESTUFS 2006; Holguín-Veras et al. 2018). Several types of traffic, building and vehicle regulations freight strategies are identified (Dablanc et al. 2013).

The study concluded Seven groups of sustainable policy measures and company actions to mitigate the negative impact of urban freight for economic vitality and quality of urban life. The study also considers the barriers and difficulties in implementing freight measures for the sustainability of urban freight transport. Policy measures are grouped at the level national, urban scale and also focused for shippers, receivers and transport operators (Transportation Research Board 2013). There is a lack of interaction and understanding between administrators and other stakeholders of urban freight like transport operators and retailers in finalisation and implementation of freight strategies (Ogden 1992).

Preferences and characteristics of stakeholders involved in urban goods distribution are essential for the success of city logistics policies and strategies. City logistics freight strategies and measures should complement the issue related to policymakers, freight demand and supply (Browne and Allen 2011).

International best practice in sustainable urban freight strategies on the basis of their strength and weakness have been extensively compiled for local policymakers and private stakeholders. Aspects of sustainability, i.e. economic, environment and society, has been addressed in organising urban freight strategies (Pronello and Valentinaa 2017). Divergent objectives of shippers and transport act as a barrier in the implementation of urban freight initiatives, and there is a need for dialogue between urban freight stakeholders (Visser et al. 1999). Different freight operation strategies need to be considered for collaborative and systematic examination for management and operations during transportation project design and development (CIVITAS 2015). Urban freight strategies need to be formulated on the basis and issues faced by urban goods movement. Design standards, infrastructure design, land use zoning and truck regulation are some key issues that need to considered for freight strategies (Duin et al. 2018).

Freight strategies and their integration in land use plan enhance the sustainable goods supply and transport in urban areas. No single freight measure but a bundle of freight measures along with close cooperation between the administration and private stakeholders are required for sustainable transport in urban areas (Lindholm 2010). Transport management plan (TMP) strategies are often selected without comprehensive information of its potential benefits and disadvantages (US-DOT 2013). A small improvement in freight efficiency can provide considerable benefits due to large movements of big trucks in goods distribution (Litman 2006). Framework for identification and selection of freight strategies by incorporating stakeholders concerns is essential for sustainable freight transport. Modelling and evaluation of freight strategies are also vital in the range of available freight strategies (NCFRP report 2012).

AHP method was used to evaluate the choice of transportation alternatives in the multimodal freight transport system (DoT 2012). AHP method has a wide variety of applications like Performance type problems, resource management, corporate policy, public policy, political strategy, and planning. AHP method is easy to use, scalable, and it's not data-intensive (Sharma et al. 2017). AHP method can be applied within the context of the stakeholder-driven or institutional approach to transport project evaluation (Kopytov and Abramov 2012). BWM method was used for evaluation and supplier selection in the context of social sustainability (Velasque and Hester 2013). BWM was used to evaluate the different factors affecting the energy efficiency of buildings along with barriers to mitigate these factors. Results show that economic, governmental and technological barriers as the most prominent barriers among all (Brucker et al. 2011). BWM is preferred methods to use when deciding on weights for parameters as it easy to understand (Bai et al. 2019). Various freight strategies are available to improve the sustainability of urban goods distribution. Impact of freight strategies can vary due to different characteristics of stakeholders involved in urban goods distribution. Assessment of freight strategies weights will be helpful for policy intervention in urban freight distribution.

3 Methodology

The objectives of this research study are to assess the weights of freight strategies across and within two wholesale markets in the city of Jaipur. These two wholesale markets are the Electronics market and Building hardware market. Hypothesis for the research study is that freight strategies have the same relevance & importance across two commodity distribution for stakeholders. AHP & BWM methods were used to assess the weights of traffic and parking regulation strategies. Traffic and parking regulation strategies are selected based on the literature review of urban goods distribution for this research study (Holguín-Veras et al. 2018). Parking and traffic regulation strategies are chosen in the purview of the city administrator, as they have an essential role in city logistics for managing freight traffic and providing freight infrastructure. The framework proposed for the assessment of urban goods distribution strategies and its variation for selected wholesale markets the city of Jaipur is presented in Fig. 1. The final weights of freight strategies are geometric means of individual responses with consistency ration less than 10%.

This research study is based on primary survey and data collected by face to face pen-pencil survey of shippers (wholesalers), transport operators (carrier) and local policy planners. The sample size of stakeholder is shown in Table 1.

List of wholesalers was collected from the market association of both markets. Time and date of face to face paper–pencil surveys were fixed via telephonic conversations, similarly in the case of transport operators from the transport association situated in Transport Nagar in Jaipur. Transport operators having more than two commercial vehicles has been selected for the survey. List of traffic police officers having more than two years of experience in and around these two wholesale markets

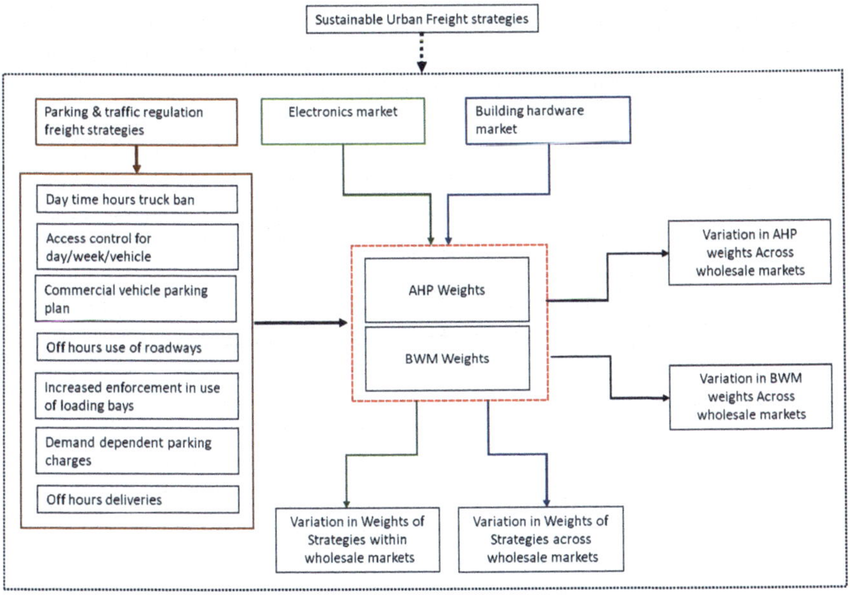

Fig. 1 Framework for evaluating weights for urban freight strategies

Table 1 Data set

S.no	Experts	Sample size AHP	Sample size BWM
1	Shippers (wholesalers)	5	5
2	Transport operators	5	5
3	Traffic management/enforcement officials	5	5
4	Urban planners	1	1

have been obtained from superintendent of police (Traffic) Jaipur. The core urban planning team of Jaipur development authority and Jaipur municipal corporation have been consulted for a detailed interview as per their available time. A team of four trained enumerators were assigned to collect eight samples per day. Consistency of results from both methods was obtained each day for the further need of interviews on the next day until the required sample size is collected as per Table 1.

4 Case City Profile

Jaipur is the capital city for the state of Rajasthan in India. The city is a major attractor of tourists and major hub for handicrafts, gems, jewellery, die printing (textile) and stone crafting. Jaipur city is located in the north-eastern part of Rajasthan state,

280 km from Delhi. Jaipur extends to an area of 2939 km^2 in which walled city (old area) constitute 17 km^2, 281 km^2 city area under Jaipur municipal corporation and rest is 2650 km^2 under Jaipur development authority. The total population of Jaipur city is 30.5 lacs (yr. 2011). Residential land use of city constitutes 44.8% (13,825 ha), commercial 6.7% (2064 ha), industrial 6% (1862 ha), governmental 2% (602 ha), mixed land use 3.3% (1034 ha), public & semi-public 10.5% (3241 ha), recreational 11.3% (3461 ha) and circulation 15.4% (4741 ha). In Jaipur city, there are 11 major wholesale markets. Electronics, handicrafts and pharmaceutical markets are situated in the walled city area. Fruits and the vegetable wholesale market at *Muhana* and *Lal Kothi* area. Food grain markets at *Surajpole* and *Kukasheda* area. Dairy and meat products at *Malviya Nagar* and NH11 bypass. Construction material market at *Aatish Nagar* and *Chandpole* area. Furniture market at *Sitapura* area. Chemical and fertilisers market in *Durgapaura* area. Industrial products at *Viskarma* industrial area.

Building Hardware market and Electronic market were selected for the assessment of parking and regulation strategies. The building Hardware market is planned market recently developed by Jaipur development authority, whereas the electronics market is situated in the old city (walled city) area. Goods distribution in building hardware market is weigh based, whereas, it is number and item-based in the Electronics market.

5 AHP & BWM Methods

5.1 AHP Method

Analytical Hierarchical Process (AHP) is a multicriteria decision-making technique that can help to express the general decision operations. It decomposes a complicated problem into a multilevel hierarchical structure of objective, criteria and alternatives. AHP can combine quantitative and qualitative factors to handle different groups of stakeholders. In the AHP method, the opinions of many experts can be mixed (Serrai et al. 2017). The relative importance of criteria and alternatives are calculated by the pairwise comparison in each level of the hierarchy. AHP method can also be used to assess the relative importance among criteria and sub-criteria. AHP is a commonly used mathematical tool, especially where subjectivity may affect the overall result of the decision-making process (JDA 2011). The relative importance of criteria or alternatives is determined on an absolute scale during a paired comparison of the hierarchy of structures.

The nine-point scale is used for paired comparison. Each point of this scale is assigned according to the intensity of importance among alternatives. Point 1 on the scale reflects that both alternatives (A&B) during comparison are equally important. Point 3 when later alternative (B) is slightly favourable than the former alternative. Point 5 when later alternative (B) is strongly favourable than the former alternative.

Point 7 when the later choice (B) is strongly favoured due to its demonstrated dominance. Point 9 on the scale when later option (B) affirmed on the highest possible order. Even scales (2, 4, 6 8) are used when there is a compromise between successive odd scales (1, 3, 5, 7, 9). If the former alternative (A) is more important than the latter alternative (B), then a reverse scale to be used (1/3, 1/5, 1/7, 1/9) (Sharma et al. 2008).

Results obtained from AHP methods for each criterion is in ratio scale. Weightage of each alternative is proportional to the total score of all criteria's in the hierarchy (Semih and Seyhan 2011). The result of the pairwise comparison on n criteria can be summarised in an (n x n) evaluation matrix A. This pairwise comparison can be shown by a square matrix (1).

$$A = (a_{ij})_{n \times m} \begin{vmatrix} a_{11} & a_{12} & * & a_{1m} \\ a_{21} & a_{22} & * & a_{2m} \\ * & * & * & * \\ a_{n1} & a_{n2} & * & a_{nm} \end{vmatrix} \quad (1)$$

In the last step of AHP, the comparison matrix is normalised to get relative weights. The right eigenvector gives the relative weights of criteria (w) corresponding to the largest eigenvalue (λ max) as Eq. (2),

$$Aw = \lambda_{max}\, w \quad (2)$$

AHP results from the output are related to the consistency of the pairwise comparison judgments. The Consistency Index (CI) of judgments is calculated by Eq. (3) (Saaty 1987).

$$CI = \frac{\lambda_{max} - n}{n - 1} \quad (3)$$

Consistency ratio (C.R.) is the final check of results in AHP to evaluate the consistency of results. The number 0.1 (10%) is the accepted upper limit for C.R. check. The ratio of the CI and the random index (R.I.) is C.R., as shown in Eq. (4).

$$CR = \frac{CI}{RI} \quad (4)$$

Value of random index when there are 3 alternatives is 0.58, 4 alternatives then 0.90, 5 alternatives then 1.12, 6 alternatives then 1.24, 7 alternatives than 1.32, 8 alternatives then 1.41, 9 alternatives then 1.45 and when there are 10 alternatives, then the value is 1.49 (Schoner and Wedley 1989).

5.2 *Best Worst Method (BWM)*

The Best–worst method (BWM) is a vector-based, multicriteria decision-making method. BWM also characterised by pairwise comparison of alternatives like AHP. For BWM analysis, a set of decision criteria needs to be determined for comparison, secondly the selection of best and worst criteria without any comparison by selection maker. Then the comparison of other criteria with the best criteria. Similarly, comparison of other criteria with the worst criteria. In the last step, the optimal weights of criteria are computed. Comparison of alternatives is conducted by using a scale between 1 and 9, where $1 =$ equal preference and a $9 = 9 \times$ times more important. Higher consistency ratio results in less reliable comparisons. BWM requires fewer comparison data (alternatives) and leads to a more consistent comparison and reliable results. BWM can be combined with other MCDM methods. BWM is a more comfortable alternative of multicriteria decision-making methods (Duin et al. 2018; DoT 2012).

6 Results

Seven traffic and parking regulation strategies have been evaluated for the city of Jaipur by AHP and BWM for two wholesale markets. Table 2 represents the results obtained by these two methods. All results confirm the consistency ratio of less than 10%, and final weights are the geometric mean of all individual stakeholders results.

Table 2 Traffic and parking regulation strategies for urban goods distribution

Strategies	Electronics market		Building hardware market	
	AHP weights (%)	BWM weights (%)	AHP weights (%)	BWM weights (%)
Day time hours truck ban	17	13	21	16
Access control for day/week/vehicle	14	19	32	25
The commercial vehicle parking plan	9	4	5	7
Off-hours use of roadways	12	17	11	10
Increased enforcement in use of loading bays	6	6	4	4
Demand dependent parking charges	8	9	6	12
Off-hours deliveries	19	28	20	22

Results confirm that there is variation in weights of strategies across and within markets by both methods (Table 2).

In the electronics market, both methods have given the highest weight to off-hours deliveries and the same weights to increased enforcement of loading–unloading bays. Weights for demand dependent parking charges are almost similar. For the rest of the freight strategies, there is variation in weights. In the case of building hardware market, increased enforcement of loading–unloading bays has the same weights. Off-hours deliveries, off-hour use of roadways and commercial vehicle parking plan strategies have minimal variation in their weights by both methods. Off-hours delivery is most preferred, followed by day time truck ban for the electronics market. Access control for commercial vehicle is most preferred, followed by off-hours deliveries in building hardware market. Increased enforcement in the use of loading bays strategy is least preferred, after commercial vehicle demand dependent parking charge strategies in both markets by both methods. Maximum variation of weights by both methods across markets are observed for day time/hours truck ban and Access control for day/week/vehicle strategies. Access control for day/week/vehicle strategy is not common in India, especially in the preview of the wholesale markets in India. Weights for this strategy can be superseded with weights of day time/hours truck ban strategies instead of omitting from the data set. In the electronics market variation in weights of off-hour deliveries has maximum (10%) by both methods, but still, it is the most preferred freight strategy. Electronics market in Jaipur is not a planned market and situated in the walled city where parking and logistics facilities are minimal with truck ban already in place. Wholesalers in this market receive their good at very early morning at the periphery of the walled city and use the small vehicle to place their goods in their respective warehouses. Similarly, in case of building hardware market, most interurban deliveries take place in night as big trucks can enter in the night only, this confirms the preferring off-hour deliveries and day time truck ban strategies. Low weights for increased in enforcement in the use of loading bays strategy is due to no demarcation and designated place for commercial vehicle in case of electronics market and availability of parking place for commercial vehicle in the planned building hardware market in Jaipur.

7 Conclusions

Sustainable urban goods distribution is vital for the overall sustainable development of cities. This research paper presents how weights of freight strategies vary for different goods distribution due to their supply chains and the inherent nature of the commodity, which needs to be incorporated in sustainable urban freight transport policy. This paper demonstrates the potential utility of AHP & BWM as a decision-making tool for policymakers in the urban freight sector.

The study results confirm that there are variation and similarity in weights of freight strategies across markets by both AHP and BWM methods. Off-hours deliveries in the electronics market and day time truck ban strategies in building hardware market are most preferred. Parking related strategies like dedicated commercial vehicle parking plan, parking charges and use of loading–unloading bays are least preferred in both markets. All these strategies have an impact on the end-users cost. Study results suggest that there is a need for the combination of freight strategies at city level according to goods characteristics and its supply chain for sustainable urban goods distribution.

BMW method is more comfortable then AHP method in the context of surveying as it has the simple and lesser comparison of alternatives than AHP method, especially when the number of paired comparison is large. Both methods are very tedious and too time-consuming for data collection. Assigning of the rating scale to alternatives is hard to explain to transport operators and wholesalers due to their low education level, especially with the AHP method.

More urban commodities and industrial products need to be explored to assess the overall weights of freight strategies within and across cities of varying sizes with easy multicriteria decision-making method in the purview of sustainable city development in India.

References

BESTUFS II (2006) Best urban freight solutions II. https://www.bestufs.net/download/BESTUFS_II/key_issuesII/BESTUFS_II_results_datacollection/BESTUFS_II_data_collection_synthesis_report.pdf. Accessed 21 Mar 2018

BESTUFS II (2006) Quantification of urban freight transport effects I. NET, BESTUFS

Bai C, Kusi-Sarpong S, Sarkis HB (2019) Social sustainable supplier evaluation and selection: a group decision-support approach. Int J Product Res 7046–7067

Browne M, Allen J (2011) Enhancing the sustainability of urban freight transport and logistics. Transp Commun Bull Asia Pac no 80(2011):80

Brucker KD, Macharis C, Verbeke A (2011) Multicriteria analysis in transport project evaluation: an institutional approach. Eur Transp 47:3–24

CIVITAS (2015) Civitas. Eu. from 'Making urban freight logistics more sustainable'. Civitas policy note, pp 1–63. https://www.eltis.org/resources/tools/civitas-policy-note-making-urban-freight-logistics-more-sustainable. Accessed 12 June 2019

Dablanc L, Giuliano G, Holliday K, Obrien T (2013) Best practices in urban freight management: lessons from an international survey. TRB, Transp Res Rec (TRR), pp 29–38

Duin RV, Slabbekoorn M, Lori Tavasszy HQ (2018) Identifying dominant stakeholder perspectives on urban freight policies: a q-analysis on urban consolidation centres in the Netherlands. Transport 33(4):867–880

Gudmundsson H, Hall RP, Marsden G, Zietsman J (2015) Sustainable transportation—indicators, frameworks, and performance management. Samfundslitteratur, Frederiksberg

Gupta P, Anand S, Gupta H (2017) Developing a roadmap to overcome barriers to energy efficiency in buildings using best-worst multicriteria decision-making methodology. Sustain Cities Soc 244–259

Holguín-Veras J, Leal JA, Sánchez-Diaz I, Wojtowicz MB (2018) State of the art and practice of urban freight management. Transp Res Part A Polic Pract

JDA (2011) Master development plan—Jaipur region. Jaipur development authority. https://www.jda.urban.rajasthan.gov.in/content/dam/raj/udh/development.../jda-jaipur/.../Part1.pdf. Accessed 13 May 2019
Kopytov E, Abramov D (2012) Multiple-Criteria analysis and choice of transportation alternatives in multimodal freight transport system. Transp Telecommun 13(2):148–148
Lindholm M (2010) A sustainable perspective on urban freight transport: factors affecting local authorities in the planning procedures. Procedia Soc Behav Sci 2:6205–6216
Litman T (2006) Mobility management innovative management strategies to transport problems'. In: SATC : 25th annual Southern African transport conference
Litman T (2019, Mar 18) Developing indicators for sustainable and livable transport planning. www.vtpi.org/wellmeas.pdf. Accessed 9 May 2019
NCFRP 23 (2013) Synthesis of freight research in urban transportation planning. Transportation Research Board, Washington, DC
NCFRP report 14 (2012) Guidebook for understanding urban goods movement. Transportation Research Board, Washington, DC, p 2011
OECD (2003) Delivering the goods: 21St century challenges to urban goods transport. https://www.itf-oecd.org/sites/default/files/docs/03deliveringgoods.pdf. Accessed 12 Apr 2019
Ogden KW (1992) Urban goods movement: a guide to policy and planning. Ashgate, London
Patier D, Routhier J-L (2008) Hoe to improve the capacity of urban goods movement data. In: Annecy: 8th international conference on survey methods in transport
Pronello C, Valentinaa CC (2017) Last-mile freight distribution and transport operators' needs: which targets and challenges. Transp Res Procedia 25:888–899
Rezaei J (2016) Best-worst multicriteria decision-making method: some properties and a linear model. Omega 126–130
Rezaei J (2015) Best-Worst multi-criteria decision-making method. Omega 49–57
Ruescha M, Hegi P, Haefeli U, Matti D, Schultz B, Rütsche P (2012) Sustainable goods supply and transport in conurbations: freight strategies and guidelines. Procedia—Soc Behav Sci 39:116–133
Saaty R (1987) The analytic hierarchy process—what it is and how it is used. Math Model 9(3–5):161–176
Saaty TL (1980) The analytic hierarchy process: Planning, priority setting, resource allocation. McGraw-Hill, New York
Saaty TL (2008) Decision making with the analytic hierarchy process. Int J Serv Sci 1(1)
Saaty TL, Vargas LG (1991) Prediction, projection and forecasting. Kluwer Academic, Boston
Schoner B, Wedley WC (1989) Ambiguous criteria weights in AHP: consequences and solutions. Decis Sci 20:462–475
Semih T, Seyhan S (2011) A multi-criteria factor evaluation model for gas station site selection. J Glob Manag 2(1):12–21
Serrai W, Abdelli A, Hammal LM (2017) Towards an efficient and more accurate web service selection using MCDM methods. J Comput Sci 22
Sharma MJ, Moon I, Bae H (2008) Analytic hierarchy process to assess and optimise distribution network. Appl Math Comput 202:256–265
Sharma S, Shelton J, Warner GV (2017) Methodologies and models for selection of optimal truck freight management strategies. Transp Res Board (TRB)
US-DOT (2013) Designing for transportation management and operations a primer. Washington, DC
US DoT (2012) Assessing the effectiveness of transportation management plan (TMP) strategies. Federal Highway Administration, Washington, DC 20590
Velasque M, Hester PT (2013) An analysis of multi-criteria decision-making methods. Int J Oper Res 10(2):56–66
Visser J, Binsbergen AV, Nemoto T (1999) Urban freight transport policy and planning. City Logist 39–70

Approaches, Challenges and Impacts of Asian Regional Multimodal Logistics for Supply Chain Integration and Interdependencies

Michael Smith, Adam Voak, and Don Gunasekera

Abstract Asia's successful economic growth over the past few decades is well known and is receiving increasing attention from current and potential trading partners. The region is home to some of the world's largest emerging and dynamic economies, and this includes the 10 countries of the Association of Southeast Asian Nations (ASEAN). Along with this growth, there is a recognition that trade costs, along domestic and international supply chains in these Asian economies, can be significantly reduced by improving the logistics performance in each mode of transport involved in various logistics and supply chain transactions. These improvements may be conveniently facilitated by the optimisation of the transitioning strategies from unimodal to multimodal (or combined) transport services. In this context, we examine here the status of current multimodal logistics in the provision of supply chain integration and interdependencies practiced in the key Asian economies. For this chapter we have particularly focused on the current approaches, challenges and potential impacts of transforming and enhancing the levels of logistics and supply chain multimodal integration and interdependencies on the emerging economies of this region.

Keywords Multimodal logistics · Supply chain management · Integration · Supply chain interdependencies · ASEAN region

M. Smith (✉) · A. Voak · D. Gunasekera
Deakin University, Melbourne, Australia
e-mail: michael.smith1@deakin.edu.au

A. Voak
e-mail: adam.voak@deakin.edu.au

D. Gunasekera
e-mail: don.gunasekera@deakin.edu.au

© The Author(s), under exclusive license to Springer Nature Switzerland AG 2020

P. Golinska-Dawson et al. (eds.), *Smart and Sustainable Supply Chain and Logistics – Trends, Challenges, Methods and Best Practices*, EcoProduction,
https://doi.org/10.1007/978-3-030-61947-3_11

1 Introduction

Within the current growth of the Asian economic sphere, there has arisen an ever-increasing demand for end-to-end handling of multimodal cargo which has driven the growth of multimodal logistics and supply chain management services. A well-integrated and sustainable multimodal process and handling system is an imperative for these Asian economies, particularly those within the ASEAN block of nations. For ASEAN to take full advantage of the emerging regional economic development opportunities and for capturing further regional trade growth, it needs to become cognisant of the conceptual nature of supply chain integration and global competitiveness. Such a realisation will enable it to leverage its location at the cross roads of Northeast Asia, South Asia and Oceania and to further bolster its growing strategic significance in the region.

It is important to comment here that, from the point of view of logistics and supply chain management, accessibility and efficiency become key decision-making considerations when utilising multimodal logistics and supply chain services. In this respect, it is the reliability and security of the service provision, together with an acute understanding of transit times and cost, that play an essential factor in informing the level of multimodal efficiency, the extent of seamless integration achievable and the completeness of the unified interdependencies of the service offerings.

We assert that improving and maintaining the quality of relevant, reliable and efficient multimodal logistics services throughout the supply chain, is extremely important for multimodal integration and improved supply chain performance. This is especially so for economies transitioning from mainly unimodal and intermodal operations towards integrated, interdependent and more sophisticated multimodal transport contracts. The current absence of consolidated regulatory frameworks within the ASEAN region, along with the somewhat incongruent accreditation systems for Multimodal Transport Operators (MTOs), presents a clear problem. This barrier, together with a shortage of the required human resource skills and capabilities, could combine to hinder the successful transition to multimodal transport integration and greater supply chain connectivity.

Our overall research intention is to contribute to the more efficient and effective utilisation of Asian regional multimodal logistics and supply chain integration, and to inform key players and stakeholders. This could be done through better understanding of those interdependencies along and within supply chains that are often neglected in broader systems-thinking discussions. To this end, this chapter focuses on a review of current approaches, challenges and impacts associated with evaluating the role played by regional multimodal transport, logistics and supply chain integration and their resultant interdependencies within ASEAN. Analysis was based on the research findings and investigation of a series of in-country missions to each of the 10 ASEAN contributing nations to assess their progress and development in the implementation of multimodal transport and logistics. The outcome resulted in the development and full adoption by these countries of an agreed implementation framework for the realisation of the ASEAN Framework Agreement on Multimodal

Transport (AFAMT) 2020–2025. It is noted that this implementation framework details a comprehensive suite of strategies and deployment guidelines, including a description of the regional action plans, which chart out, in detail, the key activities, implementation mechanisms, recommended measures, timeline and actions required for strengthening the regional approach for the realisation and implementation of the AFAMT.

2 Other Studies on Multimodal Integration and Connectivity

The concept of integration and connectivity between trading economies have often been studied through the lens of single modality (Arvis and Shepherd 2016), as well as between modes within a single economy (Shepherd et al. 2011). While a number of multimodal connectivity assessments have been undertaken in relation to passenger transport (Litman 2017; OECD 2016; Krul et al. 2010), similar freight-based studies are not common. The Asia–Pacific Economic Cooperation (APEC) completed a detailed study on the economic impact of multimodal connectivity in the APEC region in 2010 which revealed, for each of the 19 countries analysed, the changes from 2005 to 2010. In addition, the relative position with regard to maritime, air and land transport, logistic competence and an aggregated multimodal transport indicator was presented. In addition, the correlation between exports and multimodal transport performance has been estimated using the Gravity Model of Bilateral Trade (Shepherd et al. 2011).

It is appreciated that the importance of engagement with intermediaries has long been recognised in both empirical (Grubel and Lloyd 1975) and theoretical (Batra and Casas 1973; Woodland 1977) studies. More recently, whilst these interrelationships have increased in importance, they are, however, being represented and becoming manifested in new phrases, terms and descriptive terminology (Hummels et al. 2001; Kohler 2004; Antras et al. 2006; Grossman and Rossi-Hansberg 2008; Koopman et al. 2010; Johnson and Noguera 2012). Moreover, it is becoming glaringly obvious that the supply-chain trade concerns goods that will be inputs into production processes in other nations. In this respect, missing information in the chain regarding the 'next step' in each part or segment of the supply chain is becoming a central problem to production. Further, it is the lack of visibility within the 'next step' along supply chains that leads to uncertainty and ambiguity around global patterns of cross-border multimodal transport freight data (in terms of volume and value), as data collection becomes increasingly more complex.

Zhao et al. (2019), postulate that many firms in a traditional supply chain setting constitute self-organizing networks, and, as a result, these supply chains can be innately adaptive or resilient in the face of external perturbation. However, due to the complexity, uncertainty of supply and interdependence in more globalised supply chains, there is an increased risk of loss of continuity in these more finely balanced networks due to an interceding disruption event (Bode et al. 2011; Bode and Wagner 2015; Kamalahmadi and Parast 2016; Scheibe and Blackhurst (2018) contend that emerging new supply chain trends built around stringent efficiency demands, such as Just-In-Time (JIT) operations, can significantly increase the need for coupling and interdependence between entities within the structures. De Sá et al. (2019), also conclude that, in this environment, there are distinct managerial and social implications, especially around the interdependence of a firm's robustness and overall supply chain resilience. Nye (2020), commenting on power relationships in supply chains, also believes a situation in which interdependence exists without asymmetry of control generates little power imbalance, but when a discrepancy of control exists, interdependence within the chain creates situations that can be used to one party's advantage in strategic competitive systems. In this respect, Michalski et al. (2019) observed that the level of trust and innovation developed in symmetric relationships differs markedly from that in asymmetric situations with respect to organizational performance. It is suggested that the growth of 'deep dependence' comes from a Principal-and-Agent type of relationship based upon asymmetric knowledge, whilst 'deep interdependence' suggests a wider-ranging relationship between equal partners, which is manifested in a symmetrical unity and greater mutual reliance (Deif and Mohib 2019).

However, what is generally poorly understood, is the extent of the lack of general understanding regarding the level of complexity and attendant interdependencies within supply chains. These arise from an array of multifaceted issues which are essential to integrated multimodal transport and logistics services across borders. These intricacies and characteristics of modern supply chains are particularly evident in the interdependence and organisation of supply-and-demand processes in a global market. This complexity is manifested through a multitude of factors including raw material forecasting, asset investment, risk management and response to change. Successful supply chain enterprises must place a greater focus on a systems-thinking approach in order to better understand the increasingly complex nature of global supply chains. For example, supply chain professionals need to understand and appreciate the interdependent and interconnected system of labour, energy sources, transport and logistics operations, finance, business processes, service expectations, information technology, sourcing and procurement, commercial power, legal and regulatory frameworks and strategic asset investment. Added to this impressive list of requirements, we note that with cross-border supply chains, there are subtle barriers related to cross-cultural issues and competing political stances. As we have intimated earlier, the term 'global supply chain' is not a simple extension of a national supply chain.

3 Challenges Facing ASEAN Supply Chain Development

Given the dynamic environment in which ASEAN connectivity is now taking place, it is crucial to consider the forecasted emerging trends that are likely to influence the successful implementation of the 'Master Plan on ASEAN Connectivity 2025' ('MPAC 2025'). However, since the formal adoption of MPAC, it is clear that many challenges still face the 10 member nations, with much more needing to be done to realise the MPAC vision in achieving cross-border seamless logistics connectivity. Particular areas that will need to be addressed include the introduction of hard and soft infrastructure, the development of the various logistics service sectors, relevant education and skills programs, together with the challenges relating to the mobility of skilled labour, cross-border regulatory standards and multimodal transport and logistics connectivity. In this regard, there are particular ASEAN issues which we see as needing to be addressed before more systematic conjoint connectivity initiatives might be introduced. As a consequence, we note that there are words and concepts, which are essential to an understanding of the complexities of this area that are being continually introduced and, even more importantly, which are being modified as conditions and situations change. It is these to notions, and their introduction, to which we now turn our research attention.

4 Methodology

The key purposes of this study were to (i) identify core ASEAN values, objectives and plans with regard to national and international transportation, logistics and connectivity strategies, (ii) assess the strengths, opportunities, weaknesses and related threats apparent within those logistical networks that potentially affect progress toward broader ASEAN goals, (iii) identify infrastructure deficiencies in the logistics environment of the ASEAN region, (iv) determine policy deficiencies which may impede economic growth and development, and (v) offer policy recommendations that may contribute toward achieving MPAC 2025's objectives by improving, expanding, or otherwise enhancing logistics infrastructure within the ASEAN region.

In order to meet these research aims, it was necessary to determine the current status of understanding and practice of multimodal logistics strategies related to the provision of supply chain integration and organisation of interdependencies common to the ASEAN economies. In order to achieve this, we needed to focus our research on establishing the perceived efficacy of current approaches, challenges and potential impacts of transforming and enhancing the levels of logistics and supply chain multimodal integration and interdependencies amongst key functionaries in the area. This level of understanding is clearly time and place dependent, thus we employed a qualitative data collection and analysis methodology. Primary data sources consisted of a collection of statements and perceptions provided by a purposively selected range of supply chain professionals and government officials. To facilitate this data collection, a series of meetings in each ASEAN country was organised under the auspices of the ASEAN Regional Integration Support, using an opportunity which arose from the EU ARISE Plus program 'Supporting the Implementation of the ASEAN Framework Agreement on the Facilitation of Multimodal Transport'. These meetings, which were for invited delegates, included open and frank discussions with concerned government agencies including National Ministries of Transport and other relevant groups related to membership of the ASEAN National Transit Transport Coordinating Committee (NTTCC). The investigation was conducted with the support and guidance of the ASEAN Secretariat based in Jakarta (Indonesia), the ASEAN Transit Transport Coordinating Board (TTCB), the ASEAN Transport Facilitation Working Group (TFWG) and ARISE Plus. This primary data was collected directly by the researchers, with permission from the organising committee, and all comments were de-identified before analysis to protect the positions of the delegates.

To provide a sound background for the investigation, secondary qualitative and descriptive quantitative data were retrieved electronically from relevant governmental, intergovernmental, academic, trade and industry sources. In addition, a detailed literature review was carried out in order to compare ASEAN objectives and plans with existing conditions, to gain a sense of how successful existing plans have been. It is acknowledged here that our research findings have been advised by the 2019 technical assistance project related to the ASEAN Framework Agreement on Multimodal Transport (AFAMT).

5 Results

It was found that the most urgent matter that needed to be addressed by ASEAN member nations is the matter of intra-regional connectivity. It was noted that transportation and communications networks which serve as a conduit for the robust free trade that the union hopes to generate are currently below par. In this respect, Blyde and Molina (2015) have reported that the quality of logistics infrastructure impacts on the level of foreign direct investment, and it is also apparent that currently, infrastructure quality in eight of the 10 ASEAN states (excluding Singapore and Malaysia) lag behind the global average quality, which is seen to drag down regional economic productivity. Further, poorly constructed and managed transportation networks are found to act as trade barriers. Transportation safety and security are seen as major concerns that require immediate attention, and disparate customs policies across the region were found to increase time and cost in the import/export process. As part of the new open-border agenda, ASEAN members have been urged to harmonize their customs procedures, reduce documentation requirements, and allow for electronic processing of shipments. Our research has indicated that increased intergovernmental cooperation is desperately required to provide adequate security and safety compliance in shipping industries around the region. Infrastructure investment opportunities were found to be in abundance, but investment policies need to be amended to attract further foreign investors. Another strong recommendation in this regard includes the establishment of a regional consumer credit rating bureau.

6 Discussion

There is little disagreement that infrastructure clearly continues to play a big role in assuring basic connectivity and access to gateways for most developing countries. In all income groups, survey respondents reported that infrastructure was improving, but it was indicated that in all ASEAN countries except Singapore and Malaysia, the scores for infrastructure were lower than the overall scores. The introduction of better logistics connectivity, not only within each ASEAN member nations but between countries in ASEAN, will boost the region's value chain and economy. However, there are, in fact, a number of deep-seated problems that we see on the horizon for the logistics and supply chain industry that could hinder the future growth of the ASEAN Economic Community (AEC).

The largest economy in the AEC (Indonesia) spends 26% of its GDP on logistics, one of the highest rates of spending on logistics per capita in the world. In the meantime, there are major commodity price gaps between provinces in Indonesia a well as major logistics inefficiencies that hamper economic development and connectivity. In a similar way, Thailand, despite offering easy access to Cambodia, Vietnam and the Lao PDR, still spends almost 20% of its GDP on logistics.

Limited facilities at ASEAN ports make congestion problems worse, but the greatest challenge is the reducing the cost of getting products from manufacturers to consumers, and addressing the problem of why land transport through ASEAN areas actually takes up more time than processing in the ports. Inadequate roads because long transport times since rural roads are often closed for maintenance, and poorly maintained open roads can only be used by small vehicles, a restriction which implies high operating costs and which urgently needs to be redressed. Rail and air cargo networks are also inadequate in many ASEAN nations, with more on-dock rail facilities needed throughout Vietnam and the Philippines. If these problems are not addressed, it is likely that China will step in to fill the infrastructure void, thus influencing and accelerating its own transport development across Asia to the detriment of ASEAN.

In the logistics and supply chain sector, transport is the central issue determining performance, but there are still many companies that do not think seriously about integrated multimodal logistics and supply chain management. For example, many small manufacturers still use containers only for loading and unloading in ports, rather than at the origin and destination of their cargo. Highly fragmented supply and demand scenarios for road transport services means that truckers frequently return home empty, which incurs costs and erodes margins. In Indonesia, for example, trucking accounts for 72% of transport costs, and yet trucks are only half full most of the time. At a time of great change in the logistics sector, ASEAN governments have indeed started to focus on policy development and have ramped up investment in an effort to revamp long-neglected infrastructure. However, by itself, this action will not help the ASEAN countries to improve their performance and boost trade growth and competitiveness. In Indonesia, Thailand and several other AMS, there is an urgent need to reform the management of human resources and the adoption of new technologies to support more efficient and effective multimodal logistics and supply chain systems.

7 Conclusion

Emerging from this work is support for the notion that successful modern multimodal logistics development and supply chain integration can be summarised as 'one network, one contract, and one set of standards. 'One network' stresses the importance of an integrated transport infrastructure network conducive to convenient, reliable, low-cost logistics services. 'One contract' is the result of an open, collaborative and orderly freight market. 'One set of standards' provides unambiguous operating rules and performance appraisal across all national jurisdictions.

However, at present, it is perceived that the ASEAN multimodal market is largely fragmented and disorganised, with different government agencies remaining divided in their approaches. Currently, the lack of modern unified and, mutually agreed specifications and standards has led to high logistics costs, holding back ASEAN's multimodal logistics development and integration. These are key issues in the development of ASEAN's regional logistics and supply chain connectivity agenda and this hiatus should be the main focus of future policy reforms.

It is widely recognised that efficient multimodal transport is an essential part of the ASEAN connectivity agenda. In this respect, the AFAMT framework provides a suitable policy tool for a systems approach to integrate the different transport modes into one coherent transport system which caters for the needs of ASEAN Member State governments and industry. However, this research has identified that whilst the desire to reach uniformity of the laws governing multimodal transport is recognised to be viable, much work remains to make their introduction a reality. In this light, the challenge now, for ASEAN and Member States, is to adopt the action plan presented by the AFAMT framework to meet the milestone aspirations of the Kuala Lumpur Transport Strategic Plan 2016–2025 and to realise the AFAMT aims in the spirit of the agreement signed in 2005. This will require the willingness and co-operation of Member State governments, industry peak bodies and transport service providers and operators.

Finally, given the obvious growth of the logistics industry and the significant changes it is undergoing, there is a need to review the methodology for assessing the economic value of multimodal logistics and supply chain integration to the ASEAN regional economy, and we believe any detected shortcomings would be largely addressed by the development of a customised version of an economy-wide

Computable General Equilibrium (CGE) model, emphasising supply chain and logistics activities in ASEAN. Our aim is to stimulate further research into the way multimodal logistics for supply chain integration and interdependencies has transformed the nature and impacts of ASEAN regional supply chains. The importance of this topic has, to date, been overlooked.

Acknowledgments This study was funded by the ASEAN Regional Integration Support from the European Union ARISE Plus program, which provides technical support for ASEAN economic integration. We wish to thank ARISE Plus for giving permission to use the work funded by them to be used in this paper. We also thank the ASEAN participants who were involved in the in conduct of the research, and ensure them of their anonymity.

References

Antras PL, Garicano E, Rossi-Hansberg E (2006) Offshoring in a knowledge economy. Quart J Econ 121(1):31–77

Arvis J, Shepherd B (2016) Measuring connectivity in a globally networked industry: the case of air transport. the World Economy 39(11):369–385

Batra R, Casas F (1973) Intermediate products and the pure theory of international trade: a Neo-Hecksher-Ohlin framework. Am Econ Rev 63(3):279–311

Blyde J, Molina D (2015) Logistic infrastructure and the international location of fragmented production. J Int Econ 95(2):319–332

Bode C, Wagner S (2015) Structural drivers of upstream supply chain complexity and the frequency of supply chain disruptions. J Oper Manag 36:215–228

Bode C, Wagner S, Petersen K, Ellram J (2011) Understanding responses to supply chain disruptions: insights from information processing and resource dependence perspectives. Acad Manag J 54(4):833–856

De Sá MM, Miguel P, Brito R, Pereira S (2019) Supply chain resilience: the whole is not the sum of the parts. Int J Oper Prod Manag 40(1):92–115

Deif A, Mohib A (2019) A Typology to understand some dynamics of supply chain innovation location. J Supply Chain Oper Manag 17(1):47–55

Grossman G, Rossie-Hansberg E (2008) Trading tasks: a simple theory of offshoring. Am Econ Rev 98(5):1978–1997

Grubel H, Loyd P (1975) Intra-industry trade: the theory and measurement of international trade in differentiated products. Econ J 85(339):646–648

Hummels D, Ishii J, Yi K (2001) The nature and growth of vertical specialization in world trade. J Int Econ 54(1):75–96

Johnson R, Noguera G (2012) Proximity and production fragmentation. Am Econ Rev 102(3):407–411

Kamalahmadi M, Parast M (2016) A review of the literature on the principles of enterprise and supply chain resilience: major findings and directions for future research. Int J Prod Res 171:116–133

Kohler W (2004) Aspects of international fragmentation. Rev Int Econ 12(5):793–816

Koopman R, Powers W, Wang Z, Wei S-J (2010) Give credit where credit is due: tracing value added in global production chains. National Bureau of Economic Research. In working paper 16426, Cambridge, MA

Krul B, Kaikai S, Sheran R (2010) Hill uptown Oakland multimodal connectivity assessment. In: Report submitted to Oakland Planning and Development Corporation (OPDC), Pittsburgh, PA

Litman T (2017) Introduction to multi-modal transportation planning: principles and practices. Victoria Transport Policy Institute, Victoria, BC, V8V 3R7, Canada

Michalski M, Montes J, Naramsimhan R (2019) Relational asymmetry, trust, and innovation in supply chain management: a non-linear approach. Int J Logist Manag 30(1):303–328
Nye J (2020) Power and interdependence with China. Washington Quart 43(1):7–21
OECD (2016) Intermodal connectivity for destinations. A policy paper prepared by the OECD Centre for Entrepreneurship, SMEs and Local Development, as part of the Tourism Committee's Programme of Work for 2015–2016, Paris
Scheibe K, Blackhurst J (2018) Supply chain disruption propagation: a systemic risk and normal accident theory perspective. Int J Prod Res 56(1–2):43–59
Shepherd B, Serafica R, Bayhaqi A, Jing H (2011) The trade impact of enhanced multimodal connectivity in the Asia-Pacific region. J Econ Integr 26(4):624–650
Woodland A (1977) Joint Outputs, intermediate inputs and international trade theory. Int Econ Rev 18(3):517–533
Zhao K, Zuo Z, Blackhurst J (2019) Modelling supply chain adaptation for disruptions: an empirically grounded complex adaptive systems approach. J Oper Manag 65(2):190–212

Electric Vehicle Routing Problem with Time Windows and Cargo Weight

Sina Rastani, Tuğçe Yüksel, and Bülent Çatay

Abstract There is growing interest in the utilization of electric vehicles (EVs) in logistics operations as they can cut dependency on fossil fuels, hence, significantly contribute to the efforts on reducing carbon emissions and air pollution. However, their limited driving range still remains as a major barrier in their adoption despite the advancements in battery technology. In this study, we extend the well-known Electric Vehicle Routing Problem with Time Windows by taking into account the cargo weight, which may play a crucial role in the operational efficiency of the EVs since it can affect the energy consumption significantly. We present the mixed-integer linear programming formulation of the problem and perform an extensive experimental study to investigate the influence of load on the routing decisions. We solve small-size instances using a commercial solver, and for the large-size instances, we develop a Large Neighbourhood Search algorithm. The results show that cargo weight may create substantial changes in the route plans and fleet size.

Keywords Vehicle routing · Electric vehicles · Time windows · Load · Energy consumption

S. Rastani (✉) · T. Yüksel · B. Çatay
Faculty of Engineering and Natural Sciences, Sabanci University, Istanbul, Turkey
e-mail: srastani@sabanciuniv.edu

T. Yüksel
e-mail: tugce.yuksel@sabanciuniv.edu

B. Çatay
e-mail: bulent.catay@sabanciuniv.edu

Smart Mobility and Logistics Lab, Sabanci University, Istanbul, Turkey

© The Author(s), under exclusive license to Springer Nature Switzerland AG 2020

P. Golinska-Dawson et al. (eds.), *Smart and Sustainable Supply Chain and Logistics – Trends, Challenges, Methods and Best Practices*, EcoProduction,
https://doi.org/10.1007/978-3-030-61947-3_12

1 Introduction

Electric vehicles (EVs) have recently attracted great attention in transportation and logistics sector as they considerably reduce dependency to oil and consequently air pollution. The EVs are more efficient than conventional vehicles due to their electric motor and transmission system which transfers mechanical power to the wheels (Wu et al. 2015). Nevertheless, there are some technical drawbacks in case of using EVs. The most significant drawback is their battery capacity, which is low, and users should charge their batteries frequently in order to reach their destinations. Due to this restriction, routing an EV fleet has appeared as a challenging combinatorial optimization problem in the Vehicle Routing Problem (VRP) literature.

Electric Vehicle Routing Problem (EVRP) is an extension for VRP, where EVs are used in the fleet instead of fossil fuel vehicles. EVs reduce tailpipe emission and enhance green logistics. It tries to handle distribution tasks of logistics companies by minimizing the total energy consumption cost of serving customers and satisfying their demands. EVRP with Time Window (EVRPTW) is introduced by Schneider et al. (2014) where a full-recharge strategy was adopted. The authors developed the mathematical programming formulation of the problem and proposed a hybrid Variable Neighbourhood Search (VNS) and Tabu Search (TS) algorithm to solve it. Different variants of EVRP and EVRPTW were addressed in several studies including the cases of partial recharge (Bruglieri et al. 2015; Keskin and Çatay 2016), mixed fleet (Goeke and Schneider 2015; Hiermann et al. 2016), location routing (Schiffer and Walther 2017), fast charging (Felipe et al. 2014; Çatay and Keskin 2017; Keskin and Çatay 2018), non-linear charging function (Montoya et al. 2017; Froger et al. 2019), battery swapping (Yang and Sun 2015; Hof et al. 2017; Paz et al. 2018). Desaulniers et al. (2016) also studied EVRPTW and proposed a branch-price-and-cut algorithm to solve four different recharging strategies. Some recent studies have dealt with the availability of recharging stations and queueing for recharging service (Froger et al. 2017; Kullman et al. 2018; Keskin et al. 2019). A comprehensive review of the EV technology and survey of the EVRP variants may be found in (Pelletier et al. 2016; Pelletier et al. 2017; Erdelić and Carić 2019).

Energy consumption on the road does not only depend on the distance travelled but many other factors including the vehicle's weight, velocity, auxiliary equipment (internal factors) as well as ambient temperature and road gradient (external factors). These factors have been often neglected in the VRP literature either because they make the problem too complex to solve or the driving range is not an issue as the vehicles can easily refuel at a nearby gas station. However, they may play a critical for in the operational efficiency of the EVs since they can increase their energy consumption significantly (Rastani et al. 2019). Among them, the weight of the transported cargo may play a crucial role in route planning. The logistics operations of hypermarkets, hardware stores and other companies that deal with heavy loads are examples for which a load-dependent model produces more efficient transportation plans in comparison with basic routing models (Zachariadis et al. 2015), which constitutes the main motivation of this study.

Load Dependent Vehicle Routing Problem (LDVRP) was introduced in (Kara et al. 2007). They used the weighted distance objective and relate it with the energy requirements of vehicles. They proposed mathematical formulations for collection and distribution cases. Xiao et al. (2012) attacked the same problem by emphasizing the relation of the weighted distance with the fuel consumption of the vehicles within the context of Fuel Capacitated VRP. Zachariadis et al. (2015) extended LDVRP by considering simultaneous pick-ups and deliveries and proposed a local-search algorithm to solve large-scale instances.

In this study, we address the load-dependent variant of EVRPTW with partial recharges by taking into account the energy consumption associated with the cargo carried on the vehicle. We adopt a hierarchical objective function where the primary objective is to minimize the fleet size whereas the secondary objective is to minimize total energy consumption. We solve small-size instances using a commercial solver, and for the large-size instances, we develop a Large Neighbourhood Search (LNS) algorithm. The remainder of the chapter is organized as follows: Sect. 2 introduces the problem and formulates its mathematical programming model. Section 3 describes the proposed LNS method. Section 4 presents the experimental study and discusses the results. Finally, concluding remarks are provided in Sect. 5.

2 Problem Description and Mathematical Model

We tackle EVRPTW where a homogeneous fleet of EVs serve a set of customers with known demands, time windows, and service times. As opposed to previous studies in the literature which assume that the energy on the battery is consumed proportional to the distance traveled, we take into account the additional energy consumption related to freight load. Carrying more load by an EV causes more energy consumption. Furthermore, we allow partial recharging and its duration depends on the amount of energy transferred. Since it is a common practice in the real world to operate within the first phase of recharging where the energy transferred is a linear function of the recharge duration in order to prolong the battery life (Pelletier et al. 2017), we also assume a linear charging function. In addition, we assume that the EV can be recharged at most once between two consecutive customers, which is practical in last-mile logistics. We consider a pick-up problem where the load of the EV increases along its tour as it visits the customers. Each EV departs from the depot with full battery since it can be recharged overnight.

2.1 Mathematical Formulation

In line with the mathematical notation and modelling convention in the literature (Schneider et al. 2014; Keskin and Çatay 2016; Rastani et al. 2019) we define $V = \{1, \ldots, n\}$ as the set of customers and F as the set of recharging stations. Vertices 0 and $n + 1$ denote the depot where each vehicle departs from 0 (departure depot) and returns to $n + 1$ (arrival depot) at the end of its tour. We define $V_0 = V \cup \{0\}$, $V_{n+1} = V \cup \{n + 1\}$ and $V_{0,n+1} = V \cup \{0, n + 1\}$. Then, the problem can be represented on a complete directed graph $G = (N, A)$ with the set of arcs, $A = \{(i, j) \mid i, j \in N, i \neq j\}$ where $N = V_{0,n+1} \cup F$ is the total set of nodes on the network.

The energy consumption depends on the distance traveled and the total weight of the EV, which is affected by the cargo load carried on the EV. Each customer $i \in V$ has a positive demand q_i, service time s_i, and time window $[e_i, l_i]$. All EVs have a cargo capacity of C and a battery capacity of Q. At each recharging station, one unit of energy is transferred in g time units. The direct distance from node i to j is represented by d_{ij}.

Travel time from customer i to customer j is denoted by t_{ij} if the journey is direct and $\hat{t}_{ijs} = t_{is} + t_{sj} - t_{ij}$ is the additional travel time if it is via station s. Note that $\hat{t}_{ijs}$ does not include the recharging time at station s. The amount of extra energy needed in order to move one unit of cargo is represented by w. The total energy consumption starting from customer i to customer j is calculated as $(h + wu_i)d_{ij}$, where u_i is the weight of the load on the vehicle upon departure from customer i.

The decision variables ($y^k{}_i$, $y^k{}_{ijs}$, and $Y^k{}_{ijs}$), keep track of battery SoC of vehicle k at arrival at customer/depot i, at arrival at station s on route (i, s, j), and at departure from station s on route (i, s, j), respectively. τ_i denotes the time when the loading starts at customer i. The binary decision variable x^k_{ij} takes value 1 if vehicle k travels from node i to node j, and 0 otherwise whereas the binary decision variable z^k_{ijs} takes value 1 if vehicle k traverses arc (i, j), through station s.

$$\text{Minimize} \sum_{k \in K} (y_0^k - y_{n+1}^k) + \sum_{i \in V_0} \sum_{j \in V_{n+1}} \sum_{k \in K} \sum_{s \in F} (Y_{ijs}^k - y_{ijs}^k) \tag{1}$$

subject to

$$y_0^k = Q \ \ \forall k \in K \tag{2}$$

$$\sum_{\substack{j \in V_{n+1} \\ i \neq j}} \sum_{k \in K} x_{ij}^k = 1 \ \ \forall i \in V \tag{3}$$

$$\sum_{\substack{i \in V_0 \\ i \neq j}} x_{ij}^k - \sum_{\substack{i \in V_{n+1} \\ i \neq j}} x_{ji}^k = 0 \ \ \forall j \in V,\ k \in K \tag{4}$$

$$\sum_{s \in F} z_{ijs}^k \leq x_{ij}^k \quad \forall i \in V_0,\ j \in V_{n+1},\ k \in K,\ i \neq j \tag{5}$$

$$\tau_i + (t_{ij} + r_i)x_{ij}^k + \sum_{s \in F} (\hat{t}_{ijs} z_{ijs}^k + g(Y_{ijs}^k - y_{ijs}^k)) - l_0(1 - x_{ij}^k) \leq \tau_j \quad \forall i \in V_0,\ j \in V_{n+1}, k \in K, i \neq j \tag{6}$$

$$e_j \leq \tau_j \leq l_j \quad \forall j \in N \tag{7}$$

$$0 \leq y_j^k \leq y_i^k - (h + wu_i)d_{ij} + M(1 - x_{ij}^k + \sum_{s \in F} z_{ijs}^k) \quad \forall i \in V_0, j \in V_{n+1}, k \in K, i \neq j \tag{8}$$

$$y_j^k \leq Y_{ijs}^k - (h + wu_i)d_{sj} + M(1 - z_{ijs}^k) \quad \forall i \in V_0, j \in V_{n+1}, s \in F, k \in K, i \neq j \tag{9}$$

$$0 \leq y_{ijs}^k \leq y_i^k - (h + wu_i)d_{is} + M(1 - z_{ijs}^k) \quad \forall i \in V_0, j \in V_{n+1}, s \in F, k \in K, i \neq j \tag{10}$$

$$y_{ijs}^k \leq Y_{ijs}^k \leq Q z_{ijs}^k \quad \forall i \in V_0, j \in V_{n+1}, s \in F, k \in K, i \neq j \tag{11}$$

$$y_j^k \leq Q \sum_{i \in V_0} x_{ij}^k \quad \forall j \in V_{n+1}, k \in K \tag{12}$$

$$u_j \geq u_i + q_j \sum_{k \in K} x_{ij}^k - C(1 - \sum_{k \in K} x_{ij}^k) \quad \forall i \in V_0, j \in V_{n+1}, i \neq j \tag{13}$$

$$0 \leq u_i \leq c \quad \forall i \in V_{0,n+1} \tag{14}$$

$$x_{ij}^k \in \{0, 1\} \quad \forall i \in V_0, j \in V_{n+1}, k \in K \tag{15}$$

$$x_{ijs}^k \in \{0, 1\} \quad \forall i \in V_0, j \in V_{n+1}, s \in F, k \in K \tag{16}$$

The objective function (1) minimizes the total energy consumption. Constraints (2) set the initial battery SoC of EVs at departure to full. The connectivity of customer visits is imposed by constraints (3) whereas the flow conservation at each vertex is ensured by constraints (4). Constraints (5) make sure that vehicle k serves customer j after customer i if it travels from i to j by recharging its battery en-route. Constraints (6) guarantee the time feasibility of arcs emanating from the customers (the depot). Constraints (7) establish the service time windows restriction. Constraints (6) and (7) also eliminate the formation of sub-tours. Constraints (8)–(11) keep track of the battery SoC at each node and make sure that it never falls below zero where $M = Q + (h + w.\sum_{i \in V} q_i).max\{d_{ij}\}$. Constraints (8) establish the battery SoC consistency if the vehicle travels from customer i to customer j without recharging en-route. Constraints (9) determine battery SoC at the arrival at customer j if the vehicle visits a recharging station after it has departed from customer i whereas constraints (10) check battery SoC at the arrival at a station if the battery is recharged en-route. Constraints (11) set the limits for battery SoC when the vehicle departs from a station. Constraints (12) allow positive battery SoC at the arrival of an EV at customer j only if that EV serves customer j. Constraints (13) keep track of the load of the vehicle throughout its journey. Constraints (14) ensure the non-negativity of the load on the vehicle and guarantee that the cargo capacity is not exceeded. Finally, constraints (15) and (16) define the binary decision variables.

2.2 Energy Consumption Function

The energy consumption of an EV that travels from one node to another depends on various factors such as its mass, shape, road gradient, acceleration, etc. By using tractive power requirements placed on the vehicle at the wheels, the power demand of a vehicle can be obtained using function (18) (Demir et al. 2012):

$$F = Ma + Mg\sin\theta + 0.5C_a\rho AV^2 + MgC_r\cos\theta \tag{17}$$

$$P_{tract}(kW) = Fv/1000 \tag{18}$$

where F shows the force function as calculated in (17), M is the total weight of the vehicle that consist of its curb weight and the cargo load (kg), a is the acceleration (m/s^2), g is the gravitational constant, θ is road gradient, C_d is the coefficient of aerodynamic drag, ρ is the air density in (kg/m^3), A is the frontal area, v is the speed (m/s), and C_r the coefficient of rolling resistance. The tractive power requirement can be converted to second-by-second engine power output (kW) as follows:

$$P = P_{tract}/\mu_{tf} + P_{acc} \tag{19}$$

where the vehicle's drive train efficiency is shown by μ_{tf} and P_{acc} is the power demand associated with the accessory equipment such as air conditioning, audio system and cabin lights, which is neglected in this study. Then, the energy consumption in (kWh/km) can be calculated as follows:

$$E = P/v. \tag{20}$$

3 Solution Methodology

We attempt to solve small-size instances using a commercial solver. To solve the large-size instances, we develop an LNS method. LNS was introduced by Shaw (1998) and aims at improving an initial solution by using several destroy and repair mechanisms iteratively. In each iteration, some customers are removed from the solution and reinserted into the routes to create a new feasible solution. This procedure is repeated for a predetermined number of iterations. LNS and Adaptive LNS (ALNS) have been successfully applied to many VRP variants including EVRPs and EVRPTWs (Keskin and Çatay 2016, 2018; Goeke and Schneider 2015; Hiermann et al. 2016; Schiffer and Walther 2017; Keskin et al. 2019; Wen et al. 2016; Schiffer et al. 2018).

We create the initial solution using the insertion heuristic in (Keskin and Çatay 2016) where the cost of inserting a customer into a route is calculated as $(h + wu_i)d_{ik}$ + $(h + wu_k)d_{kj}$− $(h + wu_i)d_{ij}$. This insertion cost is calculated for all unvisited customers and the minimum cost insertion is performed by ensuring that the related constraints are not violated. If an EV runs out energy, a station may be inserted to make its tour energy feasible. We use First-Feasible Station Insertion algorithm which will be described in Sect. 3.3. If no customer can be feasibly inserted in the route, a new route is initialized, and the procedure is repeated until all customers are served.

Our LNS consist of customer removal and insertion mechanisms. In each iteration, a customer removal algorithm is applied on a feasible solution to remove a subset of customers from the routes. If any station is no longer needed in the partial solution, they are removed as well. Next, we apply a customer insertion algorithm that inserts all the customers removed to repair the solution in an attempt to obtain a new improved solution. Stations may be inserted to maintain the energy feasibility along the route. This procedure continues until the stopping criterion is satisfied, which is a limit on the number of iterations in our implementation. Note that the set of stations that can be visited between any two customers is reduced by using the dominance rules presented in (Bruglieri et al. 2016).

3.1 Customer Removal Operators

The current feasible solution is destroyed by removing γ customers. We use Worst-Consumption, Random Worst-Consumption, Shaw, Random Worst-Time, Random, Random Route Removal and Greedy Route Removal procedures of Keskin and Çatay (2016) by modifying them for the load dependent problem. The destroy operators are selected randomly.

- *Worst-Consumption* algorithm selects the customers with high energy consumption imposed to the route by visiting that customer, which is calculated as $(h +$

$wu_i)d_{ik} + (h + wu_k)d_{kj} - (h + wu_i)d_{ij}$ that considers distance and cargo load effect in energy consumption.

- *Random Worst-Consumption* sorts the customers with respect to the associated energy consumptions, considers a subset of $\sigma \times \gamma$ customers with highest costs to select γ customers randomly and remove them.
- *Shaw Removal* removes similar customers with respect to their energy consumption, earliest service time, being in the same route, and their demand. It randomly selects customer i and calculates the relatedness measure as $R_{ij} = \phi_1 h_i d_{ij} + \phi_2 |e_i - e_j| + \phi_3 l_{ij} + \phi_4 |D_i - D_j|$ to find similar customers j. $\phi_1 - \phi_4$ are the Shaw parameters, $l_{ij} = -1$ if i and j are in the same route, 1 otherwise. Small R_{ij} shows high similarity. So, using the non-decreasing order of the relatedness value with customer i, γ customers are removed from the solution.
- *Random Worst-Time* algorithm is a version of Shaw Removal where ϕ_1, ϕ_3, ϕ_4 are set equal to 0. The customers are sorted in the non-decreasing order of their relatedness values and γ customers are randomly removed from the subset of $\sigma \times \gamma$ customers with lowest relatedness values.
- *Random Removal* mechanism randomly removes γ customers from the solution.
- *Random Route Removal* algorithm randomly removes ω routes from the solution.
- *Greedy Route Removal* mechanism sorts the routes in the non-decreasing order of the number of customers visited and removes ω routes which serve the least number of customers.

Note that the Route Removal algorithms attempt to reduce the fleet size.

3.2 Customer Insertion Operators

We adapt Random Greedy, Regret-2, Random Time-Based, Random Greedy with Noise Function, and Regret-2 with Noise Function repair algorithms in (Keskin and Çatay 2016; Demir et al. 2012) for our load-dependent case. In addition, we propose Exhaustive Greedy, Exhaustive Time-Based, Exhaustive Time-Based with Noise Function, and Random Time-Based with Noise Function mechanisms. The repair operators are selected randomly.

- *Random Greedy Insertion* selects a customer and inserts it in the best position which leads to least increase of energy consumption.
- *Regret-2 Insertion* try to avoid the higher costs in the subsequent iteration. It calculates the difference between the cost of the best insertion and the second-best insertion for all customers and selects the customer with the highest difference.
- *Random Time-Based Insertion* calculates insertion costs similar to the Exhaustive Time-Based algorithm, however, at first an unassigned customer is selected randomly, and the algorithm inserts it in its best position.
- *Random Greedy Insertion with Noise Function* is an extension of the Random Greedy Insertion mechanism with a degree of freedom. We use the same noise

function presented in (Demir et al. 2012). The cost of insertion using the freedom degree is calculated as $NewCost = ActualCost + \bar{d}\mu\epsilon$, where $\bar{d}$ represents the maximum distance in the network, the noise parameter used for diversification is shown by μ, and ϵ is a random number between [−1, 1].
- *Exhaustive Greedy Insertion* considers all possible insertion positions for all not-inserted customers and selects the customer-position matching which leads to least increase of energy consumption.
- *Exhaustive Time-Based Insertion* calculates the difference between the route duration after and before inserting a customer as the insertion cost. For all customers, the insertion costs in all possible positions are calculated and the customer with least insertion cost is selected.

Note that *Regret-2 with Noise Function, Exhaustive Time-Based with Noise Function, Random Time-Based with Noise Function* are extensions of Regret-2, Exhaustive Time-Based and Random Time-Based insertion mechanisms, respectively, using a similar noise function.

3.3 Station Removal and Insertion Operators

As we mentioned earlier, the unnecessary stations are removed from the partial solution obtained using the destroy operator. During the repair procedure, the insertion of a customer may not be feasible with respect to battery SoC. In that case, we first attempt to increase the recharge quantity if a station is visited prior to arriving to that customer. If the energy recharged at the station cannot be increased or no station is visited en-route we apply a station insertion operator to make the insertion feasible. We modified Best Station Insertion and Multiple Station Insertion operators from the literature (Keskin and Çatay 2016; Rastani et al. 2019) and applied them for the load dependent problem. Also, we develop First Feasible Station Insertion operator for this problem. Note that at most one station can be inserted between two consecutive customers in a route.

- *First-Feasible Station Insertion* considers the first customer (or depot) where the vehicle arrives at with negative SoC and checks the insertion of a station in the preceding arcs backwards. The first station which makes the problem feasible is inserted.
- *Best-Station Insertion* algorithm checks all possible stations in all possible arcs before the first customer (or depot) with negative SoC and inserts the best station in its best position.
- *Multiple-Station Insertion* algorithm inserts multiple stations into a route when the insertion of a single station cannot make the route feasible. A station is inserted on the arc traversed immediately before arriving at the customer (or depot) with a negative SoC where the vehicle is recharged up to the maximum level allowed by the battery capacity and time windows restrictions of the succeeding customers. If the SoC is still negative at that customer or if the vehicle runs out of energy before

reaching the inserted station, we attempt to insert another station prior to the last customer visited before traveling to the recently inserted station. This procedure is repeated until the route becomes energy feasible.

One of the First-Feasible Station Insertion and Best-Station Insertion algorithms is selected randomly. If it does not make the route feasible, we resort to Multiple-Station Insertion algorithm. Note that, we remove all stations in the solution after every β iterations and use Best-Station Insertion algorithm to insert stations to obtain an improved feasible solution.

4 Computational Study

We performed our computational tests using the dataset of Schneider et al. (2014) and Desaulniers et al. (2016) for the small-size and large-size instances, respectively. The small-size dataset consists of 36 instances involving 5, 10, and 15 customers, and the large-size dataset includes 56 instances generated based on the VRPTW instances of Solomon (1987). The instances are classified according to the geographic distribution of the customers: clustered (c-type), random (r-type), and half clustered half random (rc-type). Furthermore, in type-1 problems (i.e., subsets r1, c1, rc1) the planning horizon is shorter, and customers' time windows are narrower compared to type-2 problems (i.e., r2, c2, rc2). In our study, we only consider type-1 problems from the large-size dataset as they better exhibit the influence of recharging decisions on route planning (Keskin and Çatay 2016, 2018; Rastani et al. 2019).

In order to deal with realistic vehicle cargo capacity and customer demands, we assumed an electric truck based on the specifications provided in (Demir et al. 2012). Since capacity of this vehicle is 3650 kg, we converted the demand quantities to reasonable weights by multiplying each by (3650/original capacity) in order to observe the effect of cargo weight on energy consumption. We assumed a drive train efficiency of 0.9 as EVs are more efficient than internal combustion engine vehicles. Furthermore, since the EVs in the original data are assumed to consume one unit of energy per unit distance/time travelled, we used Eq. (20) to calculate the actual energy consumption of an empty vehicle (i.e., 6350 kg) per unit distance and scaled it to $h = 1$. We used the same approach to determine the energy consumption w associated with unit load carried. We consider a flat network where road gradients are zero and we neglected vehicle acceleration.

The small-size instances were solved using Gurobi 9.0 with a 2-hour time limit. LNS was employed to solve both small- and large-size instances. LNS was coded in Python 3.7.1 and all runs were performed on an Intel Core (TM) i7-8700 processor with 3.20 GHz speed and 32 GB RAM. We performed five runs for each instance. The number of LNS iterations is set to 15,000 for the small-size instances and 25,000 for the large-size.

The results for small-size instances are provided in Table 1. Column "Gurobi" shows the results using Gurobi and "LNS" provides the results obtained by the

Table 1 Results of small-size instances obtained using Gurobi and LNS

Instance	Gurobi						LNS			
	Load independent			Load dependent			Load dependent			
	#Veh	*EC*	*t* (s)	*#Veh*	*EC*	*t* (s)	*#Veh*	*EC*	*t* (s)	*Δ* (%)
r104c5-s3	2	137	<1	2	142	<1	2	142	11	0.00
r105c5-s3	2	156	<1	2	159	<1	2	159	8	0.00
r202c5-s3	1	129	<1	1	144	<1	1	144	16	0.00
r203c5-s4	1	179	<1	1	181	<1	1	181	9	0.00
c101c5-s3	2	258	<1	2	266	<1	2	266	9	0.00
c103c5-s2	1	175	<1	1	187	<1	1	187	10	0.00
c206c5-s4	1	243	<1	1	251	<1	1	251	10	0.00
c208c5-s3	1	164	<1	1	169	<1	1	169	9	0.00
rc105c5-s4	2	233	<1	2	257	<1	2	257	8	0.00
rc108c5-s4	2	254	<1	2	264	<1	2	264	10	0.00
rc204c5-s4	1	185	<1	1	189	<1	1	189	16	0.00
rc208c5-s3	1	168	<1	1	171	<1	1	171	14	0.00
r102c10-s4	3	249	<1	3	336	37	3	336	27	0.00
r103c10-s3	2	206	8	2	220	1507	2	220	29	0.00
r201c10-s4	1	242	<1	1	262	6	1	270	14	2.97
r203c10-s5	1	223	<1	1	227	7200	1	227	4	0.00
c101c10-s5	3	388	<1	3	410	117	3	410	23	0.00
c104c10-s4	2	274	<1	2	309	7200	2	308	48	– 0.26
c202c10-s5	1	304	<1	1	319	8	1	319	11	0.00
c205c10-s3	2	228	<1	2	234	148	2	234	26	0.00
rc102c10-s4	4	424	<1	**5**	475	435	5	475	17	0.00
rc108c10-s4	3	348	<1	3	365	4472	3	365	23	0.00
rc201c10-s4	1	413	<1	1	424	<1	2	327	28	–
rc205c10-s4	2	326	<1	2	335	432	2	335	25	0.00
r102c15-s8	5	413	3	5	431	7200	5	431	28	0.00
r105c15-s6	4	336	2	4	350	7200	4	350	32	0.00
r202c15-s6	1	507	594	**2**	365	7200	2	365	30	0.00
r209c15-s5	1	313	11	1	362	7200	1	360	25	– 0.60
c103c15-s5	3	348	33	**4**	393	7200	**3**	402	73	–
c106c15-s3	3	275	2	3	371	7200	3	352	52	– 5.27
c202c15-s5	2	384	11	2	408	7200	2	393	44	– 3.80
c208c15-s4	2	301	1	2	310	7200	2	310	47	0.00
rc103c15-s5	4	398	117	**5**	416	7200	**4**	416	45	–
rc108c15-s5	3	370	2002	**5**	453	7200	**3**	418	40	–
rc202c15-s5	2	394	1	2	403	7200	2	403	48	0.00
rc204c15-s7	1	382	7200	1	444	7200	1	446	9	0.45

proposed LNS algorithm. Column "Load Independent" reports the results for the case that does not consider the increased energy consumption associated with the cargo carried whereas column "Load Dependent" show the results for the case that considers the load on the vehicle. The comparison of these two columns exhibits the influence of the load on routing decisions. "#*Veh*", "*EC*", and "*t*" refer to the fleet size, energy consumption, and run time (in seconds), respectively. The results show that #*Veh* increases by one in four instances and by two in one instance (shown in bold). Notice that these increases are very significant considering the size of the fleet. Furthermore, we observe that *EC* values obtained in the load-independent case are far from the actual energy consumption found by taking into account the cargo load. Finally, we see that LNS finds (near-) optimal solutions in most of the instances while improving the solutions given by Gurobi in three instances with respect to #*Veh* and in four instances with respect to *EC*.

We solved the large-size instances for both load-independent and load-dependent cases using LNS. The results are provided in Table 2. We observe that #*Veh* increases by one in fourteen instances (shown in bold) in the load-dependent case compared to the load-independent. Furthermore, in the remaining 15 instances, *EC* increases by 14.3% on the average. These results show the importance of considering cargo weight in route optimization.

5 Conclusions and Future Research

In this study, we addressed EVRPTW with partial recharge by taking into account the energy consumption associated with the cargo carried on the vehicle. We formulated its 0–1 mixed integer linear programming model and used it to solve small-size instances. For solving the large-size instances, we proposed an LNS method. Our computational tests showed how the fleet size and/or energy consumption increase in comparison to the case where the load factor is neglected and revealed the importance of considering the weight of the vehicles for more accurate route planning. Future research on this topic may consider the road gradient as well. A loaded vehicle going uphill will consume significantly more energy. On the other hand, when it travels downhill it can recharge its battery with energy recuperation.

Table 2 Results of large-size instances obtained using LNS

Instance	Load independent			Load dependent		
	#Veh	*EC*	*t* (s)	*#Veh*	*EC*	*t* (s)
r101	19	1666	1397	19	1833	1323
r102	16	1491	1459	**17**	1746	1397
r103	14	1223	2365	14	1375	1910
r104	12	1115	3954	**13**	1253	2959
r105	15	1425	1460	**16**	1481	1690
r106	14	1298	2019	**15**	1469	2138
r107	12	1232	2482	**13**	1357	2211
r108	12	1082	3567	12	1190	3317
r109	14	1245	1952	14	1457	2188
r110	12	1151	3767	**13**	1256	3355
r111	12	1151	2749	**13**	1271	2955
r112	12	1091	4689	12	1204	3646
c101	12	1050	1740	12	1198	1460
c102	11	1206	2132	**12**	1210	1820
c103	11	1295	3231	11	1433	2418
c104	11	1049	5564	11	1456	3392
c105	11	1210	1696	**12**	1167	1829
c106	12	1034	2800	12	1171	3107
c107	12	1032	2860	12	1183	3116
c108	12	1087	3108	12	1215	3876
c109	11	1141	3905	**12**	1256	3987
rc101	17	1735	1785	17	1947	1612
rc102	15	1679	1911	**16**	1812	1653
rc103	14	1462	2363	14	1653	1985
rc104	12	1373	3547	**13**	1490	3340
rc105	15	1499	2151	15	1734	1712
rc106	14	1443	2331	**15**	1630	2136
rc107	13	1320	2924	13	1523	2339
rc108	12	1322	3730	**13**	1527	2634

Acknowledgments This study was partially supported by The Scientific and Technical Research Council of Turkey through Grant #118M412.

Appendix Parameters

The parameters used in LNS algorithm are displayed in Table 3.

Table 3 Parameter values

Param	Description	Value
γ	Number of customers removed	Random between [20%, 55%] of all customers
σ	Parameter used in Random Worst-Consumption and Random Worst-Time algorithm	1.5
ω	Number of routes removed	Random between [10%, 40%] of all routes
ϕ_1	First Shaw parameter	0.5
ϕ_2	Second Shaw parameter	0.25
ϕ_3	Third Shaw parameter	0.15
ϕ_4	Fourth Shaw parameter	0.25
μ	Noise parameter	0.1
ϵ	Random number for noise function	Random between [–1, 1]
α	First-Feasible Station Insertion selection probability	0.7
δ_1	Worst-Consumption selection probability	0.077
δ_2	Random Worst-Consumption selection probability	0.308
δ_3	Shaw selection probability	0.231
δ_4	Random Worst-Time selection probability	0.077
δ_5	Random selection probability	0.154
δ_6	Random Route Removal selection probability	0.077
δ_7	Greedy Route Removal selection probability	0.077
λ_1	Exhaustive Greedy Insertion selection probability	0.077
λ_2	Random Greedy Insertion selection probability	0.308
λ_3	Regret-2 Insertion selection probability	0.154
λ_4	Exhaustive Time-Based Insertion selection probability	0.077
λ_5	Random Time-Based Insertion selection probability	0.077

(continued)

Table 3 (continued)

Param	Description	Value
λ_6	Random Greedy with Noise Function Insertion selection probability	0.077
λ_7	Regret-2 with Noise Function insertion selection probability	0.077
λ_8	Exhaustive Time-Based with Noise Function insertion selection probability	0.077
λ_9	Random Time-Based with Noise Function insertion selection probability	0.077
β	Number of iterations to remove and reinsert stations	50

References

Bruglieri M, Mancini S, Pezzella F, Pisacane O (2016) A new mathematical programming model for the green vehicle routing problem. Electron Notes Discret Math 55:89–92. https://doi.org/10.1016/j.endm.2016.10.023

Bruglieri M, Pezzella F, Pisacane O, Suraci S (2015) A variable neighborhood search branching for the electric vehicle routing problem with time windows. Electron. Notes Discret. Math. 47:221–228. https://doi.org/10.1016/j.endm.2014.11.029

Çatay B, Keskin M 2017 The impact of quick charging stations on the route planning of electric vehicles. In: IEEE symposium on computers and communications (ISCC), 2017IEEE, vol 2017, pp 152–157. https://doi.org/10.1109/ISCC.2017.8024521.

Demir E, Bektaş T, Laporte G (2012) An adaptive large neighborhood search heuristic for the Pollution-Routing Problem. Eur J Oper Res 232(2):346–359. https://doi.org/10.1016/j.ejor.2012.06.044

Desaulniers G, Errico F, Irnich S, Schneider M (2016) Exact algorithms for electric vehicle-routing problems with time windows. Oper Res 64:1388–1405. https://doi.org/10.1287/opre.2016.1535

Erdelić T, Carić T (2019) A survey on the electric vehicle routing problem: variants and solution approaches. J Adv Transp. https://doi.org/10.1155/2019/5075671

Felipe Á, Ortuño MT, Righini G, Tirado G (2014) A heuristic approach for the green vehicle routing problem with multiple technologies and partial recharges. Transp Res Part E Logist Transp Rev 71:111–128. https://doi.org/10.1016/j.tre.2014.09.003

Froger A, Mendoza JE, Jabali O, Laporte G (2019) Improved formulations and algorithmic components for the electric vehicle routing problem with nonlinear charging functions. Comput Oper Res 104:256–294. https://doi.org/10.1016/j.cor.2018.12.013

Froger A, Mendoza JE, Jabali O, Laporte G (2017) A matheuristics for the electric vehicle routing problem with capacitated charging stations. CIRRELT.

Goeke D, Schneider M (2015) Routing a mixed fleet of electric and conventional vehicles. Eur J Oper Res 245:81–99. https://doi.org/10.1016/j.ejor.2015.01.049

Hiermann G, Puchinger J, Ropke S, Hartl RF (2016) The electric fleet size and mix vehicle routing problem with time windows and recharging stations. Eur J Oper Res 252:995–1018. https://doi.org/10.1016/j.ejor.2016.01.038

Hof J, Schneider M, Goeke D (2017) Solving the battery swap station location-routing problem with capacitated electric vehicles using an AVNS algorithm for vehicle-routing problems with intermediate stops. Transp Res Part B Methodol 97:102–112. https://doi.org/10.1016/j.trb.2016.11.009

Kara I, Kara BY, Yetis MK (2007) Energy minimizing vehicle routing problem. Combinatorial optimization and applications. Springer, Berlin Heidelberg, pp 62–71

Keskin M, Laporte G, Çatay B (2019) Electric vehicle routing problem with time-dependent waiting times at recharging stations. Comput Oper Res 107:77–94. https://doi.org/10.1016/j.cor.2019.02.014

Keskin M, Çatay B (2016) Partial recharge strategies for the electric vehicle routing problem with time windows. Transp Res Part C 65:111–127. https://doi.org/10.1016/j.trc.2016.01.013

Keskin M, Çatay B (2018) A matheuristic method for the electric vehicle routing problem with time windows and fast chargers. Comput Oper Res 100:172–188. https://doi.org/10.1016/j.cor.2018.06.019

Kullman N, Goodson J, Mendoza JE (2018) 2018, June. Dynamic electric vehicle routing with mid-route recharging and uncertain availability, Seventh international workshop on freight transportation and logistics

Montoya A, Gueret C, Mendoza JE, Villegas JG (2017) The electric vehicle routing problem with nonlinear charging function. Transp Res Part B Methodol. 103:87–110. https://doi.org/10.1016/j.trb.2017.02.004

Paz JC, Granada-Echeverri M, Escobar JW (2018) The multi-depot electric vehicle location routing problem with time windows. Int J Ind Eng Comput 9:123–136. https://doi.org/10.5267/j.ijiec.2017.4.001

Pelletier S, Jabali O, Laporte G (2016) 50th anniversary invited article—goods distribution with electric vehicles: review and research perspectives. Transp Sci 50(1):3–22

Pelletier S, Jabali O, Laporte G, Veneroni M (2017) Battery degradation and behaviour for electric vehicles: Review and numerical analyses of several models. Transp Res Part B Methodol 103:158–187. https://doi.org/10.1016/j.trb.2017.01.020

Rastani S, Yüksel T, Çatay B (2019) Effects of ambient temperature on the route planning of electric freight vehicles. Transp Res Part D: Transp Environ 74:124–141

Schiffer M, Schneider M, Laporte G (2018) Designing sustainable mid-haul logistics networks with intra-route multi-resource facilities. Eur J Oper Res 265:517–532. https://doi.org/10.1016/j.ejor.2017.07.067

Schiffer M, Walther G (2017) The electric location routing problem with time windows and partial recharging. Eur J Oper Res 260(3):995–1013. https://doi.org/10.1016/j.ejor.2017.01.011

Schneider M, Stenger A, Goeke D (2014) The electric vehicle routing problem with time windows and recharging stations. Transp Sci 48:500–520. https://doi.org/10.1287/trsc.2013.0490

Shaw P (1998) Using constraint programming and local search methods to solve vehicle routing problems. In International conference on principles and practice of constraint programming. pp 417–431. Springer, Berlin, Heidelberg. https://doi.org/10.1007/3-540-49481-2_30

Solomon MM (1987) Algorithms for the vehicle routing and scheduling problems with time window constraints. Oper res 35(2):254–265. https://doi.org/10.1287/opre.35.2.254

Wen M, Linde E, Ropke S, Mirchandani P, Larsen A (2016) An adaptive large neighborhood search heuristic for the electric vehicle scheduling problem. Comput Oper Res 76:73–83. https://doi.org/10.1016/j.cor.2016.06.013

Wu X, Freese D, Cabrera A, Kitch WA (2015) Electric vehicles' energy consumption measurement and estimation. Transp Res Part D 34:52–67. https://doi.org/10.1016/j.trd.2014.10.007

Xiao Y, Zhao Q, Kaku I, Xu Y (2012) Development of a fuel consumption optimization model for the capacitated vehicle routing problem. Comput Oper Res 39(7):1419–1431. https://doi.org/10.1016/j.cor.2011.08.013

Yang J, Sun H (2015) Battery swap station location-routing problem with capacitated electric vehicles. Comput Oper Res 55:217–232. https://doi.org/10.1016/j.cor.2014.07.003

Zachariadis EE, Tarantilis CD, Kiranoudis CT, (2015) The load-dependent vehicle routing problem and its pick-up and delivery extension. Transp Res Part B 71, 158–181. https://doi.org/10.1016/j.trb.2014.11.004

Applicability of Daoism, Confucianism, and Mencian Thought to Modern Corporate Governance in the Maritime Shipping Industry

Shiang-Min Meng

Abstract This study explores the applicability of ancient Chinese philosophy—namely, Daoism, Confucianism, and Mencian thought—to current considerations regarding corporate governance in the maritime shipping industry. It adopts the ethical thought in Laozi's *Daodejing*, considerations on humaneness and soul in Confucius' *Analects*, and Mencius' account of moral spirit and humane governance. These ideas are then applied to the principles of corporate governance put forward by the Organization for Economic Development. The results show that Daoism has positive effects on Confucianism and on Mencian thought, but only indirect effects on corporate governance. However, Confucianism has a positive influence on the thought of Mencius, and Confucianism and Mencian thought play a significant role in corporate governance. This study considers ways by which Chinese corporate leaders can establish ethical criteria based on Confucianism and Mencian thought. An empirical study was conducted among 118 stakeholders in the maritime industry. To test the relationship, structural equation modeling was used to uncover the relationships among Daoism, Confucianism, Mencian thought, and corporate governance.

Keywords Daoism · Confucianism · Mencian · Corporate governance

1 Introduction

As demonstrated by the Enron scandal, executives can create a negative culture that sets the stage for self-interested action and unethical accounting that influences employees and organizational behavior (Treviño et al. 2003) In 2008, the Lehman Brothers scandal played a major role in the financial crisis that subsequently destroyed worldwide economic development and political stability. Such events provide lessons on the importance of corporate governance. Leadership is influenced by a set of processes, customs, policies, laws, and institutions that affect

S.-M. Meng (✉)
Department of International Logistics and Transportation Management, Kainan University, Taoyuan City, Taiwan
e-mail: eddymeng@mail.knu.edu.tw

© The Author(s), under exclusive license to Springer Nature Switzerland AG 2020

P. Golinska-Dawson et al. (eds.), *Smart and Sustainable Supply Chain and Logistics – Trends, Challenges, Methods and Best Practices*, EcoProduction,
https://doi.org/10.1007/978-3-030-61947-3_13

how a corporation is directed, administered, and controlled. Shareholders influence the behavior of managers to ensure that a company is run according to their interests (Hemailin and Weisbach 2003) and according to Western models of governance whereby shareholders bear the ultimate responsibility for avoiding moral hazards (Ward et al. 2009). Indeed, corporations act in their own interest and can change the rules of the game to ensure that their interests are protected (Bragues 2008).

Inadequate corporate governance is also a key factor in failures of Asian corporations (Meng and Liang 2011) and these problems can be attributed to the ethics and moral character of individuals. According to the Organization for Economic Development (OECD), companies are encouraged to disclose policies relating to business ethics, the environment, and other public policy commitments (OECD 2004), and government has an important responsibility and must be sufficiently flexible to allow markets to function effectively and to respond to the expectations of corporate governance.

Kelly (1990) argued that ethical leaders are valuable and that prosperity-seeking organizations should obtain the social services of such leaders, whose ethical behavior is the result of a socialization process (Jones 1995). Thus, this is empirical evidence that suggests preventing future ethical breakdowns. Leaders have received much attention as a means of protecting corporate governance and providing structures and processes by which to direct and manage organizations. They deliver on management accountability and thus protect corporate governance. An ethical leader will increase ethical awareness within organizations linked to corporate objectives as an element of management discourse (Husted and Allen 2000). Fulmer and Barry (2009) demonstrated that a positive mood can reduce critical thinking and concern regarding dubious unethical activities. Treviño et al. (2003) argued that leaders should engage with ethical figures who are concerned with the community and society. Ethical leaders are pressured to behave in a socially responsible manner (Campbell 2007). Effective corporate governance relies on both the internal moral behavior of leaders and the external ethical considerations that control the systems that cover corporate managerial boards when disclosing unethical or unlawful behavior in corporations (Ward et al. 2009). Leadership structures encourage and facilitate the occurrence of occupational crime (Zahra et al. 2005).

This study focuses on the applicability of ancient Chinese philosophy—specifically, Daoism, Confucianism, and the thought of Mencius—to modern Chinese corporations, and compares ancient ethical considerations to corporate governance guidelines set forth by the OECD. Ethical behavior is sometimes defined merely as acts and behavior that exceed moral minimums (Treviño et al. 2006). OECD efforts on corporate governance will help to develop a culture of values that pertain to professional and ethical behavior on which well-functioning markets depend. Corporate governance focuses on legislative policy to deter fraudulent activity and promote transparency—although, as such, it merely treats the symptoms and not the cause of the problem (Gabrielle 2003). Indeed, structural reforms of the type promoted by the corporate governance movement will do little to prevent the recurrence of widespread wrongdoing (Bragues 2008). Thus, the aim of this study is to investigate corporate governance from the perspective of Daoism, Confucianism, and Mencian thought, with particular focus on the maritime shipping industry.

2 Criteria of Daoism, Confucianism, and Mencian Thought

China is one of the cradles of world civilization, ancient China venerated Daoist wisdom as the harmony between humans and nature, and the need for calmness during the vigorous interaction of *yin* and *yang*. The Bible in the West and Laozi's writings in East have been translated frequently, and they have spread widely around the world. Laozi's *Daodejing* has been venerated, and Laozi himself has been worshipped as a god. The ideal of humaneness and the concept of harmony is central to Confucianism, and its founder, Confucius, was venerated by later generations as a sage and a holy teacher. Mencius was posthumously given designated the Second Sage of Confucianism during the Yuan Dynasty, insofar as he reformed Confucian doctrine and served as its interpreter. As an academic value system, Mencian thought continued to be so used during the Ming and Qing Dynasties, forming the core of the new classical studies. Confucian indicated the governance paths that initially step is making personal will sincere to depends on setting one's soul right to regulate the home, govern the nation, and, ultimately, to restore peace in the world. Meng (2013a) argued that ethical leaders often adopt Mencian ideas as a key element in the long-term development of enterprises, whereas leaders with poor ethics will negatively influence the operation of organizations. Thus, Mencian thought can be used by ethical leaders to solve governance problems. High ethical standards are in the long-term interests of a company as a means of credibility and trustworthiness, not only in daily operations, but also with respect to longer-term commitments (OECD 2004). Zheng (2004) showed that the existence of corporate governance can restrain the morally hazardous behavior of entrepreneurs in China. It is thus important that Chinese leaders follow Daoism and Confucianism to establish suitable corporate governance criteria. Although corporate governance is confronted with business laws and mechanisms, the ethical and moral virtues of leaders are more useful to governance and less at risk of being mistreated. A survey of the literature on Daoism, Confucianism, and Mencian principles for current corporate governance revealed evidence suggesting that regulatory and legal structures should be altered to improve the ethical and moral codes of corporate governance. Modern laws and regulations hold little promise, but the prevention of corporate misconduct can be facilitated by ancient Chinese philosophical codes (Meng and Liang 2011). This means that workable Confucian and Mencian ideas can be used by ethical leaders to alleviate negative selfish actions by providing internal control systems with appropriate audits, and by communicating with boards of directors and treating corporate governance fairly.

This study first established the parameters for four constructs: Daoism, Confucianism, Mencian thought, and corporate governance. Thus, we obtained an overall definition of an ethical leader. First, Laozi's *Daodejing* (DDJ) and the work of Meng and Liang (2011) were used to generate the ethical criteria for ten survey items based on Daoism. After reviewing the *Analects* related to humaneness and adulthood in Meng (2019), we added ten survey items based on Confucianism. Considerations regarding moral spirit and humane governance, as described in Meng (2013a), provided ten additional items based on Mencian thought. Finally, the OECD principles of corporate governance in Meng (2013b) comprised the last ten survey items based on modern corporate governance.

3 Methodology and Model

The aim of this study was to determine the main factors that influence the ethics and moral behavior of corporate management in terms of ancient Chinese philosophy. The popular seven-point Likert scale was used (Gallarza et al. 2002; Meng and Liang 2011). Based on ancient Chinese philosophy and OECD guidelines, a list of 40 items (see Table 1) was included in the questionnaire Windsor (2006). argued that corporate self-restraint and altruistic activities can serve to expand public policy and benefit stakeholders. Using discriminant analysis of our four constructs of corporate governance, we developed and validated a concise model that establishes ethical mechanisms to prevent internal control failures. The viewpoint of various stakeholders can provide an acceptable overview of the different theoretical perspectives of corporate governance (Dignam and Lowry 2006). Thus, these questionnaire items were pre-tested on 30 stakeholders from the shipping industry. After validating all 40 items, the survey was distributed to 150 members of the shipping industry from August to October, 2019. Of these, 124 questionnaires were returned, and six questionnaires were discarded because they contained incomplete information. Therefore, there were 118 usable responses in total, for an overall response rate of 78.6%. Statistical analyses and factor analysis were conducted using SPSS. Factor analysis was performed to investigate any separate underlying factors, and to reduce redundancy. To test the relationship, we used structural equation modeling on the four structures (viz., Daoism, Confucian, Mencian thought, and corporate governance) using AMOS.

All of the candidate fitness assessments indicated a good fit, with $\chi^2 = 3002.480$, df = 681, a comparative fit index of 0.713, a goodness-of-fit index of 0.677, a Tucker–Lewis index of 0.803, and a root mean squared error of approximation of 0.150. The results shown in Table 1 indicate that the measurement model met the discriminant validity criterion. The structural model used to test the relationships consisted of all four latent constructs.

The test results for the proposed model are shown in Table 2 and Fig. 1.

For all structural path estimates, Daoism was positively linked to Confucianism ($\beta = 0.877$, $P < 0.001$) and Mencian thought ($\beta = 0.548$, $P < 0.001$), but was not significantly linked to corporate governance ($\beta = 0.031$, $P > 0.05$). Confucianism was positively linked to Mencian thought ($\beta = 0.381$, $P < 0.001$). Confucianism ($\beta = 0.335$, $P < 0.001$) and Mencian thought ($\beta = 0.463$, $P < 0.001$) were both positively

Table 1 Standardized loading (SL), Cronbach's α, CR, and AVE for the model

Construct and items	SL	*t*-value
Daoism (α = 0.812; CR = 0.871; AVE = 0.78)		
Leader behaves with civility and humility	0.716	14.190
Leader is caring and virtuous	0.734	15.216
Leader is committed to advocacy of moral concepts	0.784	14.261
Leader has industrious and frugal habits	0.612	11.127
Leader has an ethical mind from the perspective of people	0.788	21.486
Leader is selfless in relation to the company and society	0.729	17.418
Leader has great moral faith	0.789	19.114
Leader exhibits goodness and integrity	0.746	23.482
Leader sacrifices his own interests to share with others	0.812	16.657
Leader shows patience and peace of mind	0.788	17.472
Confucianism (α = 0.768; CR = 0.801; AVE = 0.78)		
Humaneness characteristics behave with ritual heart	0.641	18.719
Humaneness characteristics is moral to be worthy of appreciation	0.785	17.647
Humaneness characteristics are committed to within the class of charity	0.785	17.420
Humaneness characteristics stay alive at the expense of faithfulness	0.818	19.427
Humaneness characteristics are determined intellectual man	0.765	17.567
Manhood soul is being trustworthy treats employees	0.689	16.541
Manhood soul purses the goal being a bravery man	0.785	17.850
Manhood soul is responsibility without forming cliques	0.798	16.912
Manhood soul cares for working environment, protect employee benefits	0.801	18.401
Manhood soul takes care of humane love	0.831	17.454
Mencian thought (α = 0.871; CR = 0.867; AVE = 0.83)		
Moral spirit behaves with care and humility	0.712	14.962
Moral spirit is loving and righteous	0.856	19.142
Moral spirit is committed to the resolution of righteousness	0.805	19.084
Moral spirit behaves with care and humility	0.799	17.121
Moral spirit is selfless to society	0.451	8.610
Humane governance implements taking care of a humane government	0.767	19.452
Humane governance implements the mind to produce an enduring nature	0.901	21.481
Humane governance implements producing a sudden seizure of righteousness used to control the mind	0.785	19.149
Humane governance implements taking care of a humane government	0.804	16.422
Humane governance implements treating employees with respect for their culture and civil rights	0.912	24.312

(continued)

Table 1 (continued)

Construct and items	SL	t-value
Corporate governance (α = 0.864; CR = 0.844; AVE = 0.86)		
No approval of material transactions without a majority vote at shareholder meetings	0.811	23.478
Remuneration for directors and supervisors determined at shareholder meetings	0.772	18.412
Details on the fairness and legitimacy of management remuneration and bonuses are disclosed	0.798	18.118
Details on employee salary, benefits, on-the-job training, pension plans, and company support of employee rights are disclosed	0.713	15.542
Insider trading and abusive self-dealing are prohibited	0.687	11.442
Disqualified directors should be removed from office by a resolution adopted at a shareholder meeting	0.612	9.412
Implementation of major resolutions approved by management	0.785	15.954
Annual dividends fairly distributed among shareholders	0.745	16.412
Relevant and material information on the corporation disclosed in the annual report to shareholders	0.779	15.897
Leaders secure the methods of ownership registration for corporate governance	0.912	19.987

Table 2 Test results

Relationship		Standardized parameter estimate	t-value	Result
Independent variable	Dependent variable			
Daoism	Confucianism	0.877	11.14	Linked
Daoism	Mencian thought	0.548	7.64	Linked
Daoism	Corporate governance	0.031	0.42	Unlinked
Confucianism	Mencian thought	0.381	5.65	Linked
Confucianism	Corporate governance	0.335	7.03	Linked
Mencian thought	Corporate governance	0.463	9.21	Linked

linked to corporate governance. Overall, five of the six proposed relationships tested using the structural model were linked.

4 Results and Conclusion

According to the research results, Daoism positively affects Confucianism and Mencian thought, but has only an indirect effect on corporate governance. Confucianism positively influences Mencian thought; in addition, Confucianism and

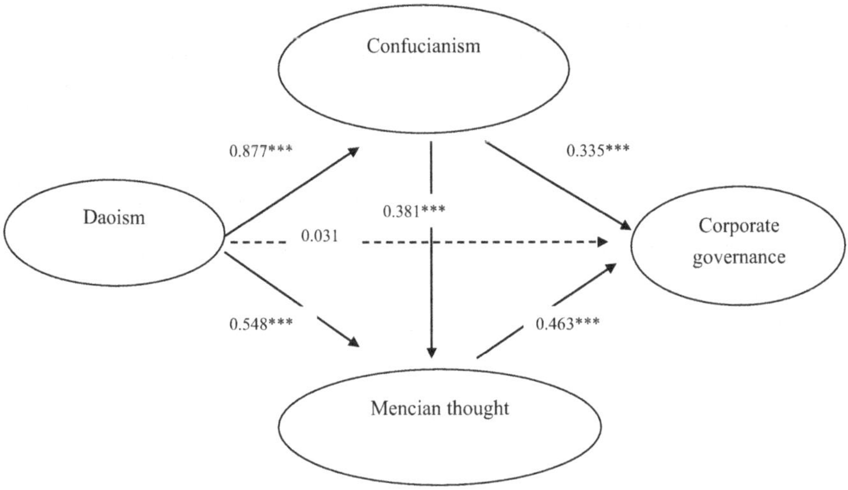

Fig. 1 Internal consistency of the model. *** P < 0.001

Mencian thought play a significant role in corporate governance. The overall framework for Daoism goes beyond Confucianism and Mencian thought. This process should be underpinned by legal protections for corporate leaders in the maritime industry. Company-specific characteristics have been suggested as important contributors to understanding the differences in corporate governance mechanisms in a given system (Weir et al. 2002). Effective corporate governance can disperse exploitative action by unethical leaders. Confucianism and Mencian thought provide mechanisms that can ensure that modern corporate leaders operate ethically. La Porta et al. (2000) concluded that the role of corporate governance is to enforce ethical codes. **The type of corporate governance favored by various ethical and moral codes reflects differences in corporate governance.** Daoism, Confucianism, and Mencian thought can be implemented by today's boards of directors, audit committees, management, and personnel, to provide reasonable assurances of reliable financial reporting, operating efficiency, and compliance with laws and regulations. Arjoon (2005) argued that there is a tendency to overemphasize legal compliance mechanisms. Philosophical practices based on the principle of equity should be understood in terms of the governance environment. China has a more relation-based governance system, as it lacks a comprehensive rule of law (Li and Filer 2007), and corporate governance has traditionally sought to use this as the main parameter when evaluating leaders. Hence, this research suggests that property rights—and, by implication, an increasing amount of corporate governance-based practices—are gaining awareness regarding the role of a firm's Confucian policies. Board members are also playing an increasingly important role in monitoring the actions of corporations and their efforts to conduct business in environmentally and socially responsible ways (Armour et al. 2003). The study showed that for the common aim of facilitating Confucianism and Mencian thought, the choice and ultimate design depends on the overall research framework. Business ethics and corporate awareness of the environmental and societal interests of the communities in which companies operate also have an impact on their reputation and long-term success (OECD 2004). Both Confucianism and Mencian thought should interact with corporate governance in a transparent manner to monitor the conduct of management.

According to the OECD (2004), corporate governance objectives are also formulated in voluntary codes and standards that do not have legal status, leaving management and other stakeholders with uncertainty concerning their status and implementation. Organizations must address procedures and structures to repair moral legitimacy and illegitimate organizational structures, which have a negative effect on the moral legitimacy of Chinese firms (Wang 2010). Hence, this study identified 40 governance criteria for leaders to take into account for effective enforcement, including the ability of leaders to deter dishonest behavior and to impose effective sanctions for violations. An ethical leader is a key person, and evaluations of ethical leadership are likely to depend on subjective perceptions of a leader's character and motives (Treviño et al. 2003). This study conceptualized the viewpoint of stakeholders regarding empirical applications of new evaluations of the ethical and moral criteria. This study demonstrated that Daoist ideas of leadership are negatively related to corporate governance, when leaders cannot address corporate governance through OECD principles. It also found that Confucian and Mencian programs and policies are successful. Garriga and Mele (2004) argued that leaders can cement the relationship between business and society. Doing so would provide a system for structuring, operating, and controlling companies with a view to achieving long-term strategic goals that satisfy corporate governance, creditors, employees, customers, and suppliers, in compliance with legal and regulatory requirements and environmental and local community needs. Meng (2019) emphasized Confucianism in the characteristics of leaders to instill a sense of virtue in terms of the methods and power a leader uses for managerial activities, as opposed to directing leaders with a mixture of laws, regulations, and checks and balances (Meng 2013b). Ancient Chinese thought has modern merit, particularly in terms of corporate governance and the integration of core ethical and moral insights. This study thus proposes that Chinese leaders should establish virtues based on Confucianism and Mencian thought to improve modern commercial societies.

References

Arjoon S (2005) Corporate governance: an ethical perspective. J Bus Ethics 61:343–352

Armour J, Deakin S, Konzelmann SJ (2003) Shareholder primacy and the trajectory of UK corporate governance. Br J Ind Relat 41:531–555

Bragues G (2008) The ancients against the moderns: focusing on the character of corporate leaders. J Bus Ethics 78:373–387

Campbell JL (2007) Why would corporations behave in socially responsible ways? An institutional theory of corporate social responsibility. Acad Manag Rev 32:946–967

Dignam A, Lowry J (2006) Company law. Oxford University

Fulmer IS, Barry B (2009) Managed hearts and wallets: ethical issues in emotional influence by and within organizations. Bus Ethics Q 19:155–191
Gabrielle OD (2003) Change management-a board culture of corporate governance. Mondaq Business Briefing.
Gallarza MG, Saura IG, Garcia HC (2002) Destination image towards a conceptual framework. Ann Tour Res 29:56–78
Garriga E, Mele D (2004) Corporate social responsibility theories: mapping the territory. J Bus Ethics 53:51–71
Hemailin BE, Weisbach M (2003) Boards of directors as an endogenously determined institutions: A survey of the economic literature. Econ Policy Rev 9:7–26
Husted WH, Allen DB (2000) Is it ethical to use ethics as strategy? J Bus Ethics 27:21–31
Jones HB (1995) The ethical leader: an ascetic construct. J Bus Ethics 14:867–874
Kelly DJ (1990) Ethics: the tone at the top. Manag Account 70:18–19
La Porta R, Lopez-De-Silanles F, Shleifer A, Vishny R (2000) Investor protection and corporate governance. J Financ Econ 58:3–27
Li S, Filer L (2007) The effects of the governance environment on the choice of investment mode and the strategic implications. J World Bus 42:80–98
Meng SM (2013a) The application of Mengzi to today's ethical criteria for maritime leader in Taiwan. Int J Asian Soc Sci 3:1227–1235
Meng SM (2013b) Application of Laozi's DAODEJING to current corporate governance. Int J Asian Soc Sci 3:2114–2133
Meng SM (2019) The ancient Confucian of Analects to today's moral criteria of maritime leader in Taiwan. Int J Asian Soc Sci 9:66–73
Meng SM, Liang GS (2011) The application of Laozi's DAODEJING to today's maritime leaders: an empirical study from stakeholders' viewpoints in Taiwan. Afr J Bus Manag 5:11955–11967
OECD (2004) OECD principles of corporate Governance. OECD Publications Service
Treviño LK, Brown M, Hartman LP (2003) A qualitative investigation of perceived executive ethical leadership: perceptions from inside and outside the executive suite. Hum Relat 56:5–37
Treviño LK, Weaver GR, Reynolds SJ (2006) Behavioral ethics in organizations, a review. J Manag 32:951–990
Wang P (2010) Restructuring to repair legitimacy-a contingency perspective. Corp Gov Int Rev 18:64–82
Ward AJ, Brown JA, Rodriguez D (2009) Governance bundles, firm performance, and the substitutability and complementarity of governance mechanisms. Corp Gov Int Rev 17:646–660
Weir C, Laing D, McKnight PJ (2002) Internal and external governance mechanisms: their impact on the performance of large UK public companies. J Bus Financ Account 29:579–611
Windsor D (2006) Corporate social responsibility: three key approaches. J Manage Stud 43:93–114
Zahra SA, Priem RL, Rasheed AA (2005) The antecedents and consequences of top management fraud. J Manag 31:803–828
Zheng Z (2004) The conflicts of interest among investors and integration of corporate governance mechanisms. Econ Res J 2:115–125

Human Factor in Logistics and Supply Chain Management

Case Study Research in Humanitarian Logistics: Challenges and Recommendations for Action

Christian Hein, Nora Eichhorn, and Rainer Lasch

Abstract Providing help to people in need is the responsibility of society and does not exclude logistical issues. Especially in this area of logistics research, a strong practical orientation is of great importance. Case study research is often seen as a suitable instrument to bridge the gap between research and practice and is therefore particularly recommended. However, since the contextual influences in humanitarian operations are very diverse, this research method is exposed to specific challenges. There is a lack of a systematic consideration of these challenges and a comprehensive guideline for researchers. Therefore, this chapter aims at systematically identifying and categorizing the challenges in conducting case study research in the context of humanitarian logistics. Practical recommendations for the application of this research method will be developed to overcome these obstacles. To achieve the research objectives, a mixed-method approach was chosen. First, a systematic literature review was carried out. In addition, six interviews with experts were conducted, all of whom have extensive experience in the field of case study research in the context of humanitarian logistics. The recommendations for action are based on a qualitative and deductive methodology. The results of the study show that four characteristics of humanitarian logistics represent unique challenges for case study research and must therefore be considered particularly: the dynamic environment, the political as well as the international context, and the general complexity.

Keywords Humanitarian logistics · Logistics ethics · Research method · Case study

1 Introduction

Natural disasters cause devastating damage worldwide. In 2018, losses resulting from 315 natural catastrophes worldwide totaled approximately US$ 132bn. Despite the alarming consequences of these catastrophes for the global economy, the impact on

C. Hein (✉) · N. Eichhorn · R. Lasch
TU Dresden, Chair of Business Management, esp. Logistics, 01062 Dresden, Germany
e-mail: christian.hein@tu-dresden.de

© The Author(s), under exclusive license to Springer Nature Switzerland AG 2020

P. Golinska-Dawson et al. (eds.), *Smart and Sustainable Supply Chain and Logistics – Trends, Challenges, Methods and Best Practices*, EcoProduction,
https://doi.org/10.1007/978-3-030-61947-3_14

human life remains the most significant challenge. Although the year 2018 was rather mild compared to previous years, approximately 11,800 people died as a direct result of the natural disasters, and more than 68 million were affected (CRED 2019). But it is not only natural disasters that cause tremendous suffering. Man-made disasters are also a significant challenge for humanitarian actors nowadays. According to UNHCR (2020a), more people are currently fleeing than ever before. The crisis in Syria has forced more than 5.6 million people to leave the country, and a further 6.6 million Syrians are registered as Internally Displaced Persons (UNHCR 2020b).

In the event of a major disaster, the difficulty is to provide sufficient emergency supplies exactly when and where they are needed. Logistics is, therefore, an essential element of any disaster relief operation and may determine the level of success of the operation. Financial expenditure on logistics can cause high costs, especially in the acute phases following the outbreak of a disaster (Hein et al. 2020). Humanitarian organizations are, therefore, under pressure to make logistics activities efficient and transparent (van Wassenhove 2006). For this reason, so-called humanitarian logistics is in the focus of an increasingly comprehensive research base that aims to address the challenges in disaster management (Kovács and Spens 2007). The scientific interest has manifested itself in the publication of articles in several journals, in special editions, as well as in a series of conferences on this topic (Kovács and Spens 2009). However, there is criticism that humanitarian logistics research lacks the necessary practical relevance (Kunz and Reiner 2012). In response to this criticism, qualitative research, and in particular case study research, has emerged in the field of humanitarian logistics as the most commonly used method (Chiappetta Jabbour et al. 2019). According to Vega (2018), transparently documented, trustworthy, and rigorously conducted case studies have the potential to contribute to advancements in humanitarian logistics research. However, his findings also show that the rigor with which such scientific papers are produced is not necessarily provided. Especially among the numerous challenges in humanitarian logistics, there is a lack of specific guidance on how to conduct case studies. This paper is thus the first to focus on the systematic analysis of the challenges in case study research in humanitarian logistics. The paper aims to identify and categorize the associated challenges and to develop practical recommendations for action. In particular, the following research questions are addressed:

RQ1: What are the challenges in conducting a case study, especially in the context of humanitarian logistics?

RQ2: How can these challenges be categorized?

RQ3: How can these challenges be addressed when conducting a case study in the context of humanitarian logistics?

To answer these explorative research questions, two methods are applied: a systematic literature search as well as expert interviews. In summary, the work provides an opportunity for researchers, humanitarian organizations, and other actors in the humanitarian sector to develop a better understanding of case study research requirements. This work is structured as follows: In the following Sect. 2, the theoretical foundations are outlined, including a brief overview of humanitarian logistics and an introduction to the method of case study research. The methodology of this work is described in Sect. 3. The evaluation and presentation of the results will take place in Sect. 4. This paper concludes in Sect. 5 with a summary, its limitations, and an outlook on further research.

2 Theoretical Background

2.1 Humanitarian Logistics

Humanitarian logistics aims to support disaster-affected people by delivering the right material, in the right quantity, with the right quality, at the right time, and to the right place. This objective is very similar to that of commercial logistics. It is therefore not astonishing that Thomas and Kopczak (2005) define humanitarian logistics as "*the process of planning, implementing and controlling the efficient, cost-effective flow and storage of goods and materials, as well as related information […] for the purpose of alleviating the suffering of vulnerable people*". Apte (2009) specifies that humanitarian logistics is a specific branch of logistics that manages the response chain of critical supplies and services with challenges such as surges in demand, uncertain deliveries, critical time windows, and a wide operational scope. Differentiation is often made based on the phases of disaster management to systematize the tasks and requirements of humanitarian logistics. The following three phases are distinguished, among others: preparedness, immediate response, and reconstruction (Kovács and Spens 2009).

Although commercial and humanitarian logistics are closely related, there are significant differences which characterize the humanitarian environment and pose specific challenges. A considerable feature affecting the operations of humanitarian organizations is that the ultimate ambition of humanitarian organizations is to save lives and to reduce suffering while the strategic objective in the commercial context is to generate financial returns; for the former, finances are more seen as constraints rather than objectives (Kovács and Spens 2007). This difference has significant influences on logistical activities. Holguín-Veras et al. (2012) note that, in humanitarian logistics, so-called social costs ought to be reduced. In contrast to commercial logistics, where externalities are neglected, and only the logistics (private) costs are considered, these social costs must include externalities. After all, not-performed logistics service in a relief operation can cause losses in human life, which could be treated as an externality in the comprehension of commercial logistics. However, in humanitarian logistics, this aspect cannot be neglected, and therefore, externalities must be considered. Moreover, humanitarian organizations often establish strict principles which determine their actions; they attempt to act with humanity, neutrality, impartiality, and independence.

Relief operations involve a multitude of stakeholders with heterogeneous needs and aims, such as the recipients of service, donors, staff and volunteers, aid agencies and NGOs, governments, and the military (Beamon and Balcik 2008). During the Indian Ocean tsunami in 2004, over 40 countries and 700 NGOs were involved (Chia 2007) and as most of them are operating in independent supply chains and have their funding and organizational systems (Hein 2019), coordination indeed poses a major challenge (Jahre and Jensen 2010). The lack of information sharing and cooperation creates a highly dynamic, informal, and often improvised decision-making process, in contrast to the commercial sector which is featured by formal structures, standardized procedures, and clearly defined roles and responsibilities (Holguín-Veras et al. 2012). Nevertheless, Schulz and Blecken (2010) figured out that, in the humanitarian sector, the same potential benefits from cooperation can be realized as in the private sector, such as cost reduction and concentration on core competencies. Balcik et al. (2010) note that some cooperation efforts have already been undertaken, especially those mechanisms with the greatest potential for implementation (e.g. collaborative procurement and third-party warehousing). Also, joint logistics activities and partnerships between private sector companies and relief organizations are increasing.

Humanitarian logistics is shaped by an environment which is highly volatile and dynamic. As the time, the location, the type, and the size of a disaster are often unpredictable and the lead times can be very shortened or even nonexistent, humanitarian logistics has to cope with these uncertainties on demand (Beamon and Balcik 2008). The relatively steady flow of goods, characterized by repeated activities in the commercial sector, is not realizable for a once-in-a-lifetime disaster with a surge of demand (Holguín-Veras et al. 2012). There are further unknowns in capabilities, personnel, the process stability and even the other activists on-site, especially at the beginning of a relief operation (Blecken 2009). Besides, humanitarian logisticians have to work under enormous time pressure; while in the commercial context delays might be accepted, the timely response in a humanitarian operation truly can be a matter of life and death (van Wassenhove 2006). In addition, disaster situations are highly complex and chaotic. Damaged infrastructure may block access to regions and people in need. The infrastructure can be further susceptible to seasonal conditions. Transportation and storage capabilities might be damaged, and an overall lack of equipment and resources is possible. In this context, the skills of humanitarian staff should also be highlighted, as there is often a deficiency of logistics expertise and career paths (Sandwell 2011). Another considerable problem concerns the communication and information infrastructure, which is notably lacking in developing countries. However, supply chains have become more and more dependent on information (Oloruntoba and Gray 2006), which makes the corresponding infrastructure crucial for disaster management. All these uncertainties and challenges, as well as the deficits in infrastructure and capacities, must be dealt with within the context of humanitarian logistics (c.f. Table 1).

Table 1 Comparison of commercial and humanitarian logistics (Beamon and Balcik 2008; Holguín-Veras et al. 2012; Kovács and Spens 2007)

	Commercial logistics	Humanitarian logistics
Main objective	Generate financial return (reduction of private costs)	Save lives, reduce suffering (reduction of social costs)
Time effects	Delays might be accepted	Delays might result in lost lives
Decision-making/procedures	Formal structures, standardized procedures, clearly defined roles and responsibilities	Dynamic, informal, and often improvised (decision-making) processes
Logistical activities	Relatively steady flow of goods, characterized by repeated activities	Unpredictable, once in a lifetime
Infrastructure and resources	Stable and functional	Restricted/lack of equipment, resources, and skilled staff possible
Actors	Predetermined and homogeneous actors	Multitude of diverse and heterogeneous actors with complex relations
Environment	Relatively constant	Highly volatile and dynamic

2.2 *Case Study Research*

Case studies in the scientific sense are empirical investigations that examine a phenomenon in its real context. This qualitative research method aims to develop an understanding of a complex topic and is particularly suitable when the boundaries between phenomenon and context are not apparent. A case study is based on several sources of evidence (Yin 2009). Merriam (1998) considers qualitative case studies as an intensive, holistic description and analysis of a limited phenomenon. The phenomenon can be a program, an institution, a person, or a process. To distinguish the case study method from other research methods, Merriam (1998) emphasizes the specific characteristics of a case study: particularistic, descriptive, heuristic. Stake (1995) also deals with characteristic properties for case studies, namely that they be holistic, empirical, interpretive, and emphatic. A case has a delimiting character to its environment (Stake 1995; Merriam 1998; Yin 2009). Ragin (1992) considers this demarcation in spatial and temporal terms. Merriam (1998) describes a case as a kind of phenomenon that occurs and is influenced in a specifically defined context. Yin (2003) describes a case as a contemporary phenomenon in its real context. The researcher has little control over the phenomenon and the context. The procedure of case study research can be divided into four phases: planning, data collection, analysis, and reporting (c.f. Fig. 1).

The planning phase aims to develop a guideline for further procedure, by concretizing the design used within the case study. Yin (2009) proposes four types of designs. These types differ in the selection of the research object (case). Regarding the

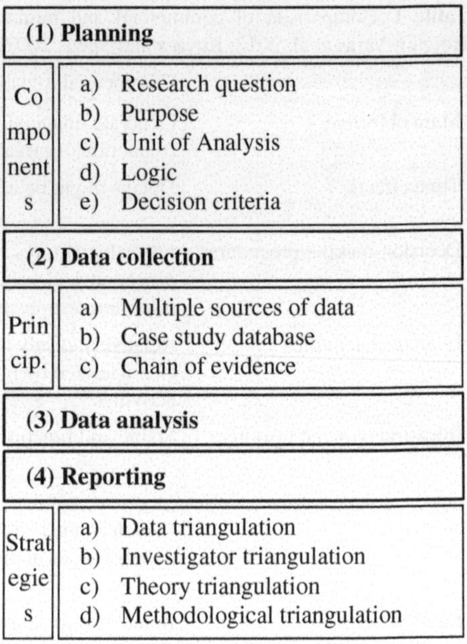

Fig. 1 Process of case study research (Modified from Yin 2003)

case selection, a decision must be made on the number of cases. Single case studies consider only one single case. In general, it is recommended to examine several cases (multiple case). A further distinction is the scope of the case. Possible design types are holistic or embedded case studies. Yin (2009) recommends choosing the design that offers the highest possible instrumentality to answer the research questions. Therefore, the strengths and weaknesses of each design should be considered. The selected design of the case study research consists of five components: research questions, purpose, unit of analysis, the logic linking the data to the purpose, and the criteria for interpreting the results. When designing the research, the researcher should ensure that these components are coherent and consistent. The first component includes the problem definition and the research objectives. To be able to develop hypotheses for case studies, a literature review should be conducted in advance. Regarding the second component, the research must have a distinct purpose. The third component is related to the fundamental problem of defining the case. The fourth and fifth components relate to planning the steps of data analysis in the case study method. The fourth component includes time series analysis and data analysis, while the fifth component requires the researcher to identify and address rival explanations (Yin 2003).

Yin (2003) points out that after the start of the second phase of data collection, minor changes can be made to the research design. For significant changes, it is recommended to return to the first step of conceptualization to redesign the case study. To capture the complexity and completeness of the case under investigation, the task of the researcher in data collection is to obtain data from multiple sources of evidence (Yazan 2015). Yin (2003) suggests using six sources of evidence: Documents, archive data, interviews, direct observations, participant observation, and physical artefacts. While interviews, observations, and participant observation are time-consuming, they provide data which originates closer to reality. Documents, artefacts, and archive data may be partially limited in availability. Their advantage is that these secondary data sources provide supporting material. Furthermore, Yin (2009) lists principles that apply to the entire data collection process. The first principle states the use of multiple sources of evidence. Evidence must refer to the same facts for the purpose of triangulation. The second principle involves the creation of a case study database. The formal collection of evidence helps in the handling and management of data and guides the formulation of the final case study report. The third principle is the creation of chains of evidence, i.e. explicit links between the research questions, the data collected, and their conclusions (Yin 2009).

In the data analysis phase, both quantitative and qualitative evidence is examined, categorized, and tabulated (Yin 2003). Four criteria can measure the quality of the analysis. These include construct validity, internal validity, external validity, and reliability. Construct validity is achieved by triangulating multiple sources of evidence, chains of evidence, and checks by other team members. By using analytical techniques such as pattern matching, internal validity can be achieved. External validity describes the analytical generalization of findings. Reliability can be ensured through case study protocols and databases. As an overriding criterion, researchers must always respect the principle of objectivity. Compliance with the principles of proper case study research leads to a certain degree of transparency and traceability. In the final phase of report preparation, relevant target groups are identified, and the results are made available through appropriate presentations. This phase is essential for the exchange of findings and conclusions for confirmation by the case study participants.

3 Methodology

3.1 Literature Review

A systematic literature review is an essential part of a research project to analyze a specific scientific field. Using this method, a solid basis is created by identifying and subsequently evaluating all relevant literature sources (vom Brocke et al. 2009). A literature analysis should provide an overview of the research topic, develop new approaches for the research area, and uncover gaps in research (Webster and Watson 2002). Fink (2014) stresses that attention should be paid to ensuring transparent implementation. The approach of the literature review in this work combines the two approaches of vom Brocke et al. (2009) and Fink (2014) and is briefly outlined below. The steps are divided into problem formulation, concept development, implementation, and presentation of results.

Table 2 Combination of search terms and selected criteria for the literature search

Challenges in case study research	Humanitarian	Logistics
Challenge, boundar, problem, obstacle, difficult*, limitation,* OR *barrier* AND *case stud**	*Humanitarian, emergency, catastrophe,* OR *disaster*	*Logistic*, fleet management,* OR *supply chain management*
	-Language: English -Publication period: no restriction -Publication medium: no restriction -Search field: title, abstract	

To guarantee a large number of matches, the five academic databases *Academic Search Complete, Business Source Complete, ScienceDirect, Emerald Insight*, and *IEEE Xplore* were used for the search. The selected search terms were derived from the theoretical foundations of this work and according to the research questions. They are divided into three sub-aspects, which were combined in the search. For the first aspect, "challenges in case study research", the search terms *challenge, boundary, problem, obstacle, difficulty, limitation,* or *barrier* were combined with the term *case study*. The second sub-aspect, "humanitarian", includes the search terms *humanitarian, emergency, catastrophe*, and *disaster*, while the third sub-aspect, "logistics", included *logistics, fleet management*, and *supply chain management*. Following, the different search terms were combined to one search phrase. The formulation of the search terms was supported by the online tool *LitSonar*. The search was conducted in English, without any limitation of the publication period, and carried out in the abstract and title. Table 2 summarizes the procedure of the database research process.

The search results were analyzed using a two-stage screening process. The relevance of the articles was first checked based on the title and abstracts (rough selection) and finally based on the full text (final selection). The execution of the database search, in which all three sub-aspects of the search term were used, resulted in only one relevant match. To find more literature sources, the previous phase of concept development was revised concerning the search terms. For this purpose, the theory of the degree of subject linkage could be used. In general, the degree of subject linkage indicates the extent to which the content of the respective source is related to the topic or the exact objective of the analysis (Karmasin and Ribing 2012). In the literature search, the search terms were systematically generalized in order to obtain more results. When using the subject linkage level I, the search terms as described above were used. The second level of topic linkage (subject linkage level II) only includes search terms on the challenges in case study preparation in a humanitarian or a logistical context. Subject linkage level III only includes search terms that refer to general difficulties in case study research (without reference to a specific research area). Table 3 shows the number of results per database for the search with subject linkage level III. The final results of the research process, according to the three levels of subject linkage, are presented in Table 4. A total of eight scientific articles

Table 3 Search results by database for subject linkage level III

Number of results (all search fields)	Number of results (only title)	Potential results (screening)	Final results
Academic search complete (08.12.2019)			
18,417	73	4	4
Business source complete (08.12.2019)			
8,417	35	1	1
Science direct (10.12.2019)			
58,929	86	1	0[a]
Emerald insight (10.12.2019)			
1	0	0	0
IEEE Xplore (10.12.2019)			
0	0	0	0

[a]The potential result was already included in final results of the EBCSO databases

Table 4 Results of literature review according to the level of subject linkage

		Explanation	Final results	Number
Subject linkage level	I	Sources examining case study research challenges in humanitarian logistics	Vega (2008)	1
	II	Sources in which case study research challenges in related fields (logistics without humanitarian reference or humanitarian sector without reference to logistics) are examined	Duboisa and Araujo (2007), Wood (2006)	2
	III	Sources that examine case study research challenges in other areas and without reference to a specific area of logistics	Eisenhardt and Graebner (2007), Ishak and Bakar (2014), Lloyd-Jones (2003), Malterud (2001), McCharthy et al. (2003)	5

were identified and considered in this study. The last step, the presentation of results, takes place in Sect. 4, where the relevant literature is analyzed and interpreted.

3.2 Expert Interviews

Qualitative research is generally suitable for exploratory studies in which a phenomenon has not been fully researched. It is, therefore, suitable for investigating the challenges of case study research in the context of humanitarian logistics. The researchers are supported in their development of a comprehensive and detailed picture of the complex phenomenon. The qualitative research interview attempts to describe the meanings of central issues in the subjects' environments. The main task of the interview is to understand the meaning of the respondents' statements (Kvale 1996). With regard to standardization, interviews are divided into unstructured, semi-structured, and structured interviews, whereas in qualitative research, only the latter forms of standardization are common. For semi-structured interviews, an interview guide is prepared in advance. Corbetta (2003) states that the order in which the different topics are dealt with is left to the interviewer. Within the topic complexes, the questions can also be formulated at the interviewer's discretion. It is possible to ask additional questions that are not foreseeable at the beginning of the interview. It is not the primary goal of semi-structured interviews to test hypotheses. Rather, the aim is to explore the opinions of the interviewee (Gray 2004). In this paper, qualitative expert interviews were applied, since this enables access to the knowledge of researchers who have experience in conducting case studies in the context of humanitarian logistics. The semi-structured form was chosen because its flexibility is suitable for application in this relatively unestablished research field.

The guidelines used for the interviews followed the procedure proposed by Bell (2014). A total of five sets of questions were formed. The first set of questions deals with the context of humanitarian logistics, with the following set addressing challenges that the interviewee experienced in the case study research. In the third set of questions, the experts are asked to explain possible recommendations for action to overcome the challenges mentioned. The fourth set of questions seeks to structure the diversity of the identified problems. Questions regarding the categorization of challenges are formulated. The last set of questions summarizes the identified recommendations. The primary research question, as well as the interview course, was communicated to the experts in advance. This allows the interviewees the opportunity to prepare for the interview.

With regard to the selection of experts, intensive research was carried out using publicly available information on the experts in advance. A precondition for contacting the expert was that the researchers had expertise in the field of humanitarian logistics and case study research. The number of publications and citations on this topic were used as references. Six expert interviews were conducted in total; with this number of experts a theoretical saturation could be achieved. All interviewees work at different universities in four European countries. They all have scientific expertise on the topic. The number of publications ranges from seven to several hundred, with the number of citations ranging from twenty to several thousand. Four of the six experts have additional practical experience. All interviews took place at the end of 2019 either via Skype or WhatsApp. On average, the interviews, which were conducted in English, lasted around fifty minutes, with the lengthiest interview lasting ninety and the shortest around thirty minutes (c.f. Table 5). The interviews were then transcribed.

A qualitative content analysis was conducted to evaluate the interview data. The basic concept of qualitative content analysis, according to Mayring (2002) is to assign predefined or emerging codes to text material systematically. Coding allows

Table 5 Overview of experts interviewed

	Research area	Practical experience	Number of publications	Interview date	Medium	Duration
Expert 1	SCM, social responsibility, humanitarian logistics	Yes	10–24	04.11.2019	Skype	90 min.
Expert 2	Business ethics, marketing management, humanitarian logistics	Yes	10–24	12.11.2019	Skype	36 min.
Expert 3	SCM, disaster management, humanitarian logistics	Yes	<10	11.11.2019	WhatsApp	44 min.
Expert 4	SCM, humanitarian logistics, procurement	–	25–49	21.11.2019	Skype	32 min.
Expert 5	Humanitarian logistics, warehousing, information sharing	Yes	25–49	25.11.2019	Skype	62 min.
Expert 6	Closed-loop logistics, sustainable logistics, humanitarian logistics	–	>100	25.11.2019	Skype	28 min.

for the possibility to categorize data as it organizes the data by concepts, key ideas, or topics. This implies that coding is not only the analysis but also the interpretation of the data (Recker 20132014). Corbin and Strauss (1990) distinguish between three different coding techniques: open, axial, and selective. Selective coding, which was used in this current work, is used to put identified categories into a superordinate context. The goal is to develop a coherent category scheme. The coding and evaluation were performed using the software tool MAXQDA 2020. The results of both the analysis of the literature and the interviews are presented in the following section. The challenges of case study research will be illustrated by means of the distinctive characteristics of humanitarian logistics. The recommendations for action are outlined and clearly presented according to the individual phases of the research process (stated in Sect. 2.2).

4 Results

4.1 It's (Still) Case Study Research

The method of case study research in the context of humanitarian logistics certainly comprises particular challenges. However, the experts surveyed also agreed that the challenges in applying this case study research cannot be attributed solely to the humanitarian context but can be described as inherent difficulties of this method. Expert 6 remarks that "*it might, in general, be not entirely different to other case study research, but it would certainly have to deal with* [complex] *characteristics*". Expert 5 agrees with this thesis. Therefore, many of the challenges mentioned in the interviews and the literature can be classified as general challenges of case study research. The main issue pointed out was that in the application of the case study methodology, it is mainly the diversity of definitions that poses a problem. A lack of differentiation from other methods, such as interview studies, for example, can be seen here ("*You cannot just pick up the phone and ask a few short interviews*", Expert 5). Furthermore, a lack of rigor is identified as a key problem, which has given case study research a bad reputation. Expert 6 criticized that "*ninety percent of case study research, in general, is* [...] *bad; that is just not good research*". In particular, the lack of theoretical foundation, as well as the arbitrary adaptation of the researchers' own methodologies and the lack of information on the research process (e.g. in terms of the case selection), were mentioned (Vega 2018). Expert 4 states, in this context, that "*case study research is taken too lightly*". Another obstacle is that certain circumstances make it challenging to publish case studies in scientific journals (Experts 1, 3, and 4). The limitation of the word count can be mentioned here ("'*Shrink it just* [to] *two lines and then focus on the results!' I do not think that this is the best way*", Expert 1).

General recommendations for action, which in their nature are certainly not unique to the context of humanitarian logistics, were given by the interviewed experts. The conscious application of case studies as research methodology and the theoretical foundation, including a definition, should precede the entire research process. In particular, a clear distinction from other related research methods such as interview studies is claimed. "*Sometimes people just do like a case and they think it is case study research without really taking into account what kind of question they want to answer, and what kind of case they're looking at, and what kind of conclusions they can make* (Expert 4). Besides, adherence to methodological rigor throughout the entire research process is elementary, as "*there is enough literature on how you should do things properly, in terms of methodology*" (Expert 6). Standard literature on the topic should be consulted and supplemented by further literature (Expert 1), such as the framework of Vega (2018). To meet the challenges of publishing, a mixed-methods approach (e.g. combining qualitative interviews with quantitative surveys) was recommended (Expert 5). All these points influence the researcher across all phases (c.f. Sect. 2.2). In the following, it will be outlined which particularities of the humanitarian context have an influence on the conduction of case studies and how to address them.

4.2 It's Dynamic

A distinctive feature of humanitarian logistics is the dynamic and volatile environment (Beamon and Balcik 2008; van Wassenhove 2006). This applies both to the occurrence of a disaster itself and to the humanitarian actors. This dynamic poses several challenges that influence case study research. Humanitarian operations often have a project character; they can be initiated ad hoc and take place under time pressure and with limited financial resources. Dealing with these dynamics in the context of case study research is sometimes difficult, as long-term planning is not always possible. On the other hand, time windows for data collection can be indefinite and small. All this leads to the fact that the effort of case study research in this environment should not be underestimated: "*It will take more work to get to the level when you can actually start seeing things, so we understand that phenomenon that we are looking at*" (Expert 5). In contrast, the researcher's time, and also the funds available, are often limited (Experts 1 and 6). Besides, there is the problem of the right time for data collection. Collecting the data on-site and in close temporal contact with the actual event can be useful from a methodological point of view. Still, unfortunately, it is not always practicable (Expert 1). The dynamics in humanitarian aid, however, do not only relate to the event of the disaster itself; the humanitarian actors are also quite dynamic. A high staff turnover poses particular challenges (Sandwell 2011), which also have an impact on the implementation of a case study. On the one hand, there is the challenge of assigning individual items from interviews to a particular humanitarian organization. "*It is very difficult because each person that you interview, they will have worked in so many different organizations and so many different positions and places.* […] *How do you make sure what they say is really for this organization?*" (Expert 5). On the other hand, it can be challenging if the interviewee is no longer working in the same organization throughout the duration of the study. In addition, aid organizations often lack explicit logistics knowledge on the part of their staff, and there is little imagination regarding the application of theoretical models in practice ("[…] *people that take decisions, that take operational decision, are not necessarily logisticians and they might not have been trained in logistics* […]", Expert 4). Table 6 provides an overview of the challenges mentioned.

The challenges posed by the dynamic environment of humanitarian operations influence different phases of conducting case studies. The interviewed experts have identified some approaches to dealing with them. When planning a case study on humanitarian logistics, the inherent dynamics of the environment should be taken into account. This means that more flexibility on the part of the researcher's time should be included from the beginning (Experts 1 and 5). There are further reasons for this (see Sects. 4.4 and 4.5). For certain temporary research projects, e.g. within the framework of a master's program, the use of the case study as a research methodology

Table 6 Challenges caused by dynamics and recommendations for action

	Challenges	Phase	Recommendations for action
Dynamic	Long-term research processes in the humanitarian context Limited time and funds for research Small and indefinite time windows for data collection Choosing the right time for interviews High fluctuation of staff in the humanitarian sector Lack of explicit logistics knowledge in humanitarian organizations	(1)	Long-term time planning for case study process Apply for funds at an early stage Build networks within the humanitarian sector
		(2)	Use appropriate language in interviews
		(3)	Abstract the statements made in the interviews Link answers in interviews with specific organization/experience

might not be appropriate. Expert 1 considers it a more suitable methodology for advanced research programmers, such as PhDs. The high turnover of staff also has an impact on the planning process of a case study. It is recommended that a network of potential interview partners be established as early as possible. "*If you have those networks, I think that is very valuable*" (Expert 5). Regarding the lower level of logistics knowledge in humanitarian organizations, an appropriate language level should be applied. It is ultimately the task of the researcher to abstract the statements made in an interview to a scientific level (Expert 1). The recommended actions are summarized in Table 6.

4.3 *It's Political*

Humanitarian aid operations always take place in a certain political context. This is particularly true of conflict-based disasters. This area of tension has to be dealt with within the context of case study research (Wood 2006). Access to the objects of study might be made more difficult for political reasons and due to strong regulations. "*Governmental organizations are very restricted and very regulated. It is hard to get into its data if you do not have access to contacts*" (Expert 2). In some instances, security aspects even make this access impossible. Consideration should also be given to the question of who can collect data in such areas. "*I would not want someone, when he just has research and academic experience, to go to Iraq and start asking questions to the different actors that are typically in the field. That has some constraints and implications that can be very difficult to manage*" (Expert 3). As explained in the previous section, this can also lead to an extension of the research project. Furthermore, especially in security-related contexts, a high level of sensitivity with regard to data protection is required (Experts 1 and 2). This leads to challenges in the process of data collection as well as in reporting. Furthermore, the political dimension should not be underestimated within interviews, as political circumstances

Table 7 Challenges caused by political issues and recommendations for action

	Challenges	Phase	Recommendations for Action
Political	Difficult physical access due to political restrictions and regulations Security aspects of conflict-based disasters High-level data security aspects Influence of political circumstances on interview responses Intra- and inter-organizational power structure	(1)	Take political circumstances into account at an early stage of planning
		(2)	Compliance with data protection Involvement of different organizational entities Critical attitude towards political issues in interviews
		(3)	Critical attitude towards political issues in the evaluation of date Triangulation

can influence people's responses. "*Humanitarian Logistics is politics. So you always have a political filter on it, so if you're going to approach an organization you know there would always be people who are either* [...] *scared of the political system and they don't dare to say anything or they are not allowed to say anything*" (Expert 5). The intra- and inter-organizational power structure should also not be neglected. An overview of the challenges raised by the political environment are summarized in Table 7.

Politically unstable circumstances influence the implementation of case studies. During the preparation phase, attention should be paid to this, and an awareness of the emotional requirements should be encouraged (Wood 2006). In addition, under certain circumstances, priority should be given to experienced researchers for data collection. An intensive examination of the political context is necessary here. Respect for sensitive data should always be ensured. This refers to the phase of case study planning as well as the phases of data collection and reporting (Experts 1 and 2). The political dimension in the context of interviews should not be neglected. An intensive examination in advance helps to ask differentiated questions ("*You have to be very specific in your questions*", Expert 5). A critical attitude should be adopted during the interview and also during the evaluation (Expert 6). In particular, a strong triangulation of the data should be emphasized as a useful tool. The comparison of different statements that were made at various times as well as a comparison with other sources and data helps to create facts. To ensure the reliability of the data, different organizational entities (management level, field workers) can be consulted on the same issue. The evaluation of the collected data material thus can identify contradictions or confirmations. Expert 6 summarizes it as follows: "*You have to be very critical and very often ask the same questions several times in a different way or at a different time or ask different people and triangulate*". An overview can be found in Table 7.

4.4 It's International

Major humanitarian relief missions take place in a highly international context. Up to several hundred organizations from dozens of different countries (Chia 2007) can come together in such operations. Moreover, aid workers can come from different countries and operate in a cultural environment that might be unfamiliar to them and with people whose language they may not speak. This international context also presents challenges for case study research. First and foremost, it makes physical access to data more difficult. Researchers cannot travel to any place in the world to collect data as often as they like. "*The problem is time at least for us [researchers]. For me, it is time that I do not have time to go and have access just [to] find [people in the field]*" (Expert 1). Different languages pose a challenge, especially in interviews. Cultural differences should not be ignored either and may well influence answers and should, therefore, also be taken into account when evaluating data. Expert 6 mentions jargon as another specificity of humanitarian logistics: "*[...] the humanitarian world uses an awful lot of jargon [...] and understanding each other is very difficult*". Expert 6 goes on to explain that this jargon can not only vary between different organizations, but also within an organization where "*they use different words and different concepts for the same thing*". In contrast to the commercial sector, there is rarely a common understanding of concepts or terms. Table 8 summarizes the challenge caused by international aspects.

According to the interviewed experts, certain challenges regarding the strong internationality of humanitarian logistics can be met within the framework of case study research. Regarding the difficult physical access to data, Expert 6 suggests co-operating with local researchers or partners. At the same time, this improves the handling of other challenges. The language barrier in the interview survey may already be solved, and the difficulty of putting the content into the right local and cultural context is thus reduced, because "*they know the language, they know the local conditions*" (Expert 6). The influences of such cooperation extend across all phases of the implementation of case studies. Thus, attention should be paid to this already in the planning phase; positive effects can be achieved in the phases of both data collection and evaluation. To meet the challenge of the different terms and concepts, Expert 5 recommends that intensive preparation takes place: "*I think you need to be very well prepared. You need to understand all of this very well*". Table 8 provides an overview of the recommendations for action.

Table 8 Challenges caused by international issues and recommendations for action

	Challenges	Phase	Recommendations for Action
International	Difficulties regarding physical access Language barriers Cultural differences Humanitarian jargon: variety of terms and concepts	(1)–(4)	Working together with local researchers or partners Preparation across all stages of the research process

4.5 It's Complex

The final characteristic of humanitarian logistics which is mentioned in both literature and the expert interviews is its complexity. The interaction of a dynamic environment with many different stakeholders, the strong influence of political, social, and cultural aspects, and the internationality in which humanitarian aid takes place create an extremely high degree of complexity. On the one hand, this degree of complexity is a strong argument for the application of qualitative methods such as case study research. On the other hand, the complexity of the humanitarian context also creates various challenges that need to be addressed. The various aspects of complexity cannot be ignored in the context of a case study. The completeness of the researcher's perspective is particularly important and is one of the strengths of a case study. "*First of all, you need to know the context*" (Expert 5). Expert 6 states that when "*you make a model* [...] *that ignores safety in an area where you have a lot of fighting going on* [...] *your recommendations will be wrong because you ignored one of the most important things, which is security*". The challenges already arise in the preparation of a case study, especially the case selection and the corresponding definition of case boundaries, which is highlighted by the experts. The structuring of large and complex data sets is a challenge in the analysis phase ("*The amount of data that you get are also very complex to sort out*", Expert 2). In reporting, care should be taken to ensure that the complexity is sufficiently represented. A low degree of generalization and the difficulty of making comparisons can be the consequence here. "*In a normal environment [you already] have to be careful about [...] comparing results and generalizing from them, which I think that is also true for your regular case study research*". In the humanitarian context, "*it might just be an order of magnitude more complicated*" (Expert 6). In Table 9 an overview of the context-related challenges can be found.

Adequately addressing the numerous aspects of complexity is probably the most significant challenge in case study research on humanitarian logistics ("the humanitarian context got to be very complex and dynamic and it has a lot of stakeholders and different objectives and so on", Expert 6). The experts were also able to give individual recommendations for action here, covering the various phases of the case

Table 9 Challenges caused by complexity and recommendations for action

	Challenges	Phase	Recommendations for Action
Complex	The context of humanitarian logistics in general is very complex Various aspects of complexity You need to understand the context very precisely There might be a large amount of data Low degree of generalization and comparison	(1)	Address aspects of complexity from the beginning Special attention paid to case selection and boundaries Experience in both academia and practice
		(2)	Scientific discourse with colleagues
		(3)	Address aspects of complexity adequately

study research process. When selecting the cases in the planning phase, Expert 1 notes that special attention should be paid to the case boundaries. Also, with regard to the evaluation and the presentation of results, particular emphasis should be placed on the adequate description of the complex interrelationships. Gaining experience both in academia and in practice has also been mentioned as an advantage ("I definitely think working inside an organization would make it easier", Expert 5) and certainly contributes to a better understanding of complex interrelations. Furthermore, Expert 6 recommends entering into scientific discourse with colleagues consciously. An overview of the recommendations for action can be found in Table 9.

5 Conclusions

This thesis aimed to produce an overview of the challenges in case study research in the context of humanitarian logistics. For this reason, it was examined how experts from the scientific community discuss the issue. In particular, the following research questions were addressed: What are the challenges in conducting a case study, especially in the context of humanitarian logistics? How can these challenges be categorized? How can these challenges be addressed when carrying out a case study in the context of humanitarian logistics? To answer these questions, the first step was to lay the theoretical foundations for the topics of humanitarian logistics and case study research. This was followed by a description of the applied research methods of literature review and expert interviews. Subsequently, the challenges and corresponding recommendations for action were outlined in the results section of the paper. This was done based on the characteristics of humanitarian logistics.

In summary, the research results show that the fundamental challenges of case study design relate to the lack of methodological knowledge. The lack of understanding exists at a macro (in the general scientific community) and micro (in relation to the individual scientist) level. In addition, the requirements of scientific journals and the bad reputation that the case study method is wrongly attributed make case study development difficult. Four characteristics of humanitarian logistics can describe the specific challenges posed by the particularity of the humanitarian context: first, the dynamic and volatile environment contributes to difficulties. Next, the political, as well as the international context, have to be mentioned. Finally, the high degree of complexity contributes to the fact that unique challenges in case study research in humanitarian logistics cause difficulties. These challenges relate to the different phases of the implementation of a case study. The most important recommendations for action include adherence to methodological rigor throughout the research process, early preparation with regard to financial and time planning, building a network, cooperation with local actors and researchers, promoting confidence-building, formulating questions for specific target groups, critically examining the data material, investigating possible misjudgments, applying the triangulation approach, establishing a methodological distance, understanding methods for structuring the complex data set, obtaining advice from research colleagues, and using mixed methods.

Further research is necessary regarding the specific design of the developed recommendations for action and their systematization within the framework of a detailed procedural model. In this context, quality criteria for case studies in humanitarian logistics could be derived. These would serve both as an assistance for researchers during the implementation and for the subsequent assessment of the quality. A comparison with other research areas is also possible. However, the main purpose of this work is to encourage and support researchers in conducting case studies in the context of humanitarian logistics and to improve their quality.

Limitations regarding the validity of the results arise from the research methodology chosen. Despite careful consideration of the method, a bias in the results due to the subjective interpretation of the data cannot be excluded. The completeness of the research results cannot be guaranteed due to the high complexity and diversity of humanitarian logistics. Furthermore, the challenges identified here and the corresponding recommendations for action only refer to the specifics in the humanitarian context. For a detailed description of the methodological approach and general advice on case study research, explicit reference is made to the corresponding literature. Finally, it should be noted that research on challenges in case study research in the humanitarian logistics context still requires further research. It is certainly possible to supplement the identified recommendations for action. Nevertheless, the strategies identified in this paper can be used to support researchers advancing humanitarian logistics research.

References

Apte A (2009) Humanitarian logistics: A new field of research and action. Found Trends Technol Inf Oper Manag 3(1):1–100

Balcik B, Beamon BM, Krejci CC, Muramatsu KM, Ramirez M (2010) Coordination in humanitarian relief chains: practices, challenges and opportunities. Int J Prod Econ 126(1):22–34

Beamon BM, Balcik B (2008) Performance measurement in humanitarian relief chains. Int J Public Sector Manag 21(1):4–25

Bell J (2014) Doing your research project: a guide for first-time researchers, 4th edn. Open University Press, Berkshire

Blecken A (2009) A reference task model for supply chain processes of humanitarian organisations. PhD thesis. Universität Paderborn. https://d-nb.info/99978434X/34. Accessed 23 Mar 2018

Chia E (2007) Engineering disaster relief. IEEE Technol Soc Mag 26(3):24–29

Chiappetta Jabbour J, Sobreiro A, de Sousa Lopes, Jabbour B, de Souza Campos M, Mariano B, Renwick S (2019) An analysis of the literature on humanitarian logistics and supply chain management: paving the way for future studies. Ann Oper Res 283:289–307

Corbetta P (2003) Social research theory, methods and techniques. Sage, London

Corbin J, Strauss A (1990) Grounded theory research: Procedures, canons and evaluative criteria. Qual Sociol 13(1):3–21

CRED (2019) Natural disasters 2018. https://emdat.be/publications. Accessed 5 Mar 2020

Dubois A, Araujo L (2007) Case research in purchasing and supply management: opportunities and challenges. J Purch Supply Manag 13(1):170–181
Eisenhardt K, Graebner M (2007) Theory building in purchasing and supply management: opportunities and challenges. Acad Manag J 50(1):25–32
Fink A (2014) Conducting research literature reviews: from the internet to paper, 4th edn. Sage, Los Angeles
Gray E (2004) Doing research in the real world. Sage, London
Hein C (2019) Systematization of humanitarian NGOs from a logistical viewpoint: an exploratory study in Germany. In: Logistics management, proceedings of the German Academic Association for Business Research, Halle, 2019, pp 221–237
Hein C, Behrens F, Lasch R (2020) Insights on the costs of humanitarian logistics: a case study analysis. Logist Res 13(3):1–17
Holguín-Veras J, Jaller M, van Wassenhove LN, Pérez N, Wachtendorf T (2012) On the unique features of post-disaster humanitarian logistics. J Oper Manag 30(7–8):494–506
Ishak N, Bakar A (2014) Developing sampling frame for case study. Challenges and conditions. World J Educ 4(3):29–35
Jahre M, Jensen L-M (2010) Coordination in humanitarian logistics through clusters. Int J Phys Distrib Logist Manag 40(8/9):657–674
Karmasin M, Ribing R (2012) Die Gestaltung wissenschaftlicher Arbeiten: Ein Leitfaden für Seminararbeiten, Bachelor-, Master- und Magisterarbeiten sowie Dissertationen, 7th edn. Facultas, Wien
Kovács G, Spens KM (2007) Humanitarian logistics in disaster relief operations. Int J Phys Distrib Logist Manag 37(2):99–114
Kovács G, Spens KM (2009) Identifying challenges in humanitarian logistics. Int J Phys Distrib Logist Manag 39(6):506–528
Kunz N, Reiner G (2012) A meta-analysis of humanitarian logistics research. J Humanit Logist Supply Chain Manag 2(2):116–147
Kvale S (1996) Interviews: an introduction to qualitative research interviewing. Sage, London
Lloyd-Jones G (2003) Design and control issues in qualitative case study research. Int J Qual Methods 2(2):33–42
Malterud K (2001) Qualitative research: standards, challenges, and guidelines. Qual Res Ser 358(1):483–488
Mayring P (2002) Einführung in die qualitative Sozialforschung: Eine Anleitung zu qualitativem Denken, 5th edn. Beltz, Weinheim Basel
McCharthy J, Holland J, Gillies V (2003) Multiple perspectives on the family lives of young people: methodology and theoretical issues in case study research. Int J Soc Res Methodol 6(1):1–23
Merriam S (1998) Qualitative research and case study applications in education. Jossey-Bass, San Francisco
Oloruntoba R, Gray R (2006) Humanitarian aid: an agile supply chain? Supply Chain Manag Int J 11(2):115–120
Ragin C (1992) Cases of what is a case? In: What is a case? Exploring the foundations of social inquiry. University Press, Cambridge
Recker J (2013) Scientific research in information systems: a beginner's guide. Springer, Berlin
Sandwell C (2011) A qualitative study exploring the challenges of humanitarian organisations. J Humanit Logist Supply Chain Manag 1(2):132–150
Schulz SF, Blecken A (2010) Horizontal cooperation in disaster relief logistics: benefits and impediments. Int J Phys Distrib Logist Manag 40(8/9):636–656
Stake E (1995) The art of case study research. Sage, Thousand Oaks
Thomas A, Kopczak LR (2005) From logistics to supply chain management: the path in humanitarian sector. http://www.fritzinstitute.org/pdfs/whitepaper/fromlogisticsto.pdf. Accessed 5 Mar 2020
UNHCR (2020a) Population statistics database. http://popstats.unhcr.org/en/overview. Accessed 5 Mar 2020

UNHCR (2020b) Syria emergency. https://www.unhcr.org/syria-emergency.html. Accessed 5 Mar 2020
van Wassenhove LN (2006) Humanitarian aid logistics: supply chain management in high gear. J Oper Res Soc 57(5):475–489
vom Brocke J, Simons A, Niehaves B, Riemer K, Plattfaut R, Cleven A (2009) Reconstructing the giant: on the importance of rigour in documenting the literature search process. In: ECIS 2009 Proceedings, 161. http://aisel.aisnet.org/ecis2009/161. Accessed 6 Mar 2020
Vega D (2018) Case studies in humanitarian logistics research. J Humanit Logist Supply Chain Manag 8(2):134–152
Webster J, Watson T (2002) Analyzing the past to prepare for the future: writing a literature review. MIS Q 26(2):13–24
Wood E (2006) The ethical challenges of field research in conflict zones. Qual Sociol 9(1):373–386
Yazan B (2015) Three approaches to case study methods in education: Yin, Merriam, and Stake. Qual Rep 20(2):134–152
Yin R (2003) Case study research: design and methods, 3rd edn. Sage, Los Angeles
Yin R (2009) Case study research: design and methods, 4th edn. Sage, Los Angeles

P-median and Maximum Coverage Models for Optimization of Distribution Plans: A Case of United Nations Humanitarian Response Depots

Eligüzel İbrahim Miraç and Özceylan Eren

Abstract The United Nations Humanitarian Response Depot (UNHRD) is a global network of depots located in Ghana, Italy, UAE, Malaysia, Spain, and Panama that procures, manages and transports emergency supplies for the humanitarian community. Managed by the World Food Programme (WFP), UNHRD enables humanitarian actors to pre-position and stockpile relief items and support equipment for swift delivery in emergency situations. According to WFP, 515 different distributions (2,420,258 km in total) are actualized from six UNHRD depots to 88 countries in 2018 to provide 27343 m^3 volume products. Due to working for humanitarian events, maximization of users covered and also minimization of traveled distance have to be taken into consideration for UNHRD network. To do so, two different mathematical models namely maximum coverage and P-median are applied to UNHRD network using the distribution data in 2018. Total traveled distance of 515 shipments is minimized using P-median model and number of covered countries is determined under different distance limits using maximum coverage model. As a result of P-median, the total distance has decreased to 1,003,473 km by 59%. In addition, different scenarios are applied in order to maximize total demand for several coverage areas.

Keywords Humanitarian response depot · Location-allocation · Maximum coverage · P–median · UNHRD

1 Introduction

Improving network design to reduce response time is critical for humanitarian logistics. Humanitarian logistics is about mitigation, preparedness (preparing for disaster including developing competence, prepositioning of stocks etc.), response (during

E. İbrahim Miraç (✉) · Ö. Eren
Department of Industrial Engineering, Gaziantep University, Gaziantep, Turkey
e-mail: miraceliguzel@gmail.com

Ö. Eren
e-mail: erenozceylan@gmail.com

© The Author(s), under exclusive license to Springer Nature Switzerland AG 2020
P. Golinska-Dawson et al. (eds.), *Smart and Sustainable Supply Chain and Logistics – Trends, Challenges, Methods and Best Practices*, EcoProduction,
https://doi.org/10.1007/978-3-030-61947-3_15

a disaster), and recovery (getting back to normal state) (Duran et al. 2013). Typical assistance required in the aftermath of a disaster to stabilize a community after search and rescue is temporary shelter, health support, water and sanitation, food and cooking equipment. The initial goal of humanitarian logistic is to provide search and rescue, shelter, health support and fundamental supplies to live. Most of the humanitarian offices (HOs) tries to achieve both long and short-term operations, in which it is required to manage different supply chains with different warehouses and it ideally should be needed integration of more than one to incline response time and total operating cost (Jahre et al. 2016). Considering the mentioned expects, several location-allocation problems can be applied such as P-median and maximum coverage problems. The location problem which seeks to find *p* facility locations among other nodes to minimize the total distance (or cost) between all demands and their nearest warehouse is expressed as the P-median problem (Serra and Marianov 1998). Maximum coverage problem can be considered as finding the minimum number of warehouse to cover all demand point or to maximize the number of points covered by a given number of warehouses (Daskin 1982, 1983). In 2018, it is estimated that 206.4 million people will be in need due to conflicts in Syria, Yemen and the Democratic Republic of the Congo lead to suffering and promote to the record numbers of migration (Global Humanitarian Assistance 2019). To address the aforementioned objectives, it is examined the one of the HOs called as UNHRD shipments. UNHRD is a global network of hubs which includes procures storage and transporting emergency supplies continuously to the humanitarian community (The United Nations Humanitarian Response Depot (UNHRD) 2018). In addition, UNHRD has six hubs which are strategically located globally, in Italy (Brindisi), Ghana (Accra), Malaysia (Kuala Lumpur), Panama (Panama City), Spain (Las Palmas) and the United Arab Emirates (Dubai). Given six locations for hubs are decided with respect to diversity of transportation and its closeness to areas in need. After the emergency situation, it is required to correspond quickly and supplying a demand consistently. In order to achieve the quick response and consistent supply, uncertain travel times is proposed by deciding the interval of values which is solved by application of robust optimization methods due to guarantee whole determined routes are feasible under the uncertain parameters with predetermined uncertainty set (Balcik and Yanıkoğlu 2020). Another significant issue is the decision making because of facility allocation requirement under different scenarios right after disasters. In order to manage facility allocation after disasters, decision making tool through the spreadsheet developed, which consists of three main parts as; database, decision engine and user interface (Cavdur et al. 2020). In addition, it can be tried to add new facilities in order to achieve cost benefit. One of the examples is adding a new to UNHRD in Kampala and Uganda in East Africa by implementing simulation, network optimization and statistical analyses (Dufour et al. 2018). Another approach to facility allocation proposed by implementing the two-stage stochastic model in order to save from cost and expected total transportation cost especially for temporary medical centers (Oksuz et al. 2020). In addition, there are some heuristic applications by utilizing maximum covering model and one of them is modified genetic algorithm which is applied on locating Malaysia health care facilities in one district (Shariff et al. 2012).

After disaster, victims require the supplies to pursue vital activities and some of them need an immediate assistance such as delivery of medical supply. Therefore, relief network design is presented for pharmaceutical items by applying bi-objective min-max robust model by considering the perishability of item, mobility of facilities and benefits from the cooperative coverage mechanism (Akbarpour et al. 2020). There are several mathematical models that applied on humanitarian logistics. In order to understand mathematical models for humanitarian logistics, the study is conducted which includes details on goals, constraints, and structures of available mathematical models with solution approaches (Özdamar and Ertem 2015). In this chapter, the main objective is to provide shortest time response to emergency calls and the applying aforementioned models (P-median and Maximum coverage) to solve HO-related location problems in order to minimize the time required for meet with demand or minimizing the distance between a demand point and its nearest warehouse. To the best knowledge of authors this is the first study that implements P-median and maximum coverage problems to UNHRD. Structure of this chapter is as follows. Firstly, models and application part is presented, which includes the general information about the case and formulation of models. Secondly, Applications of models are given and results demonstrated. Finally, conclusion is given.

2 Case and Mathematical Models

UNHRD are logistics hubs strategically placed close to disaster-prone areas across the world which is given in Fig. 1 (Ireland & The UN Humanitarian Response

Fig. 1 A map of six hubs

Depot 2020) in order to provide quick response and readiness in the case of emergency. These hubs and information about the hubs are given as (United Nations Humanitarian Response Depot 2020):

1-Ghana (Accra)

The hub in Accra is placed in the Kotoka International Airport and located to support emergency situation specifically for the West and Central African regions.

2-Italy (Brindisi)

UNHRD in Brindisi has significant place in emergency support and response worldwide, which founded in 2000

3-Spain (Las Palmas)

In 2014, UNHRD in Las Palmas is the new addition to the UNHRD Network in order to support the World Food Programme's Ebola response operations, which utilized as a transshipment and staging.

4-Malaysia (Kuala Lumpur)

The facilities provides open and closed storage space. Also, it facilitates cold storage and guest houses.

5-Panama (Panama City)

The facilities provides open and closed storage space. Also, it facilitates cold storage and guest houses.

6-United Arab Emirates (Dubai)

The facilities provides open and closed storage space. Also, it facilitates cold storage and guest houses.

The aim of this study examines in which hubs and in which ways the demands of the countries in need of assistance in these events took place in 2018 were delivered by UNHRD and looking for an alternative route to the deliveries made. It aims to provide services to people in need as soon as possible by planning to minimize the total transport distance. In order to achieve given objectives six hubs are placed which are demonstrated in Fig. 1. First of all, UNHRD 2018 activity reports have been used to determine which countries, with which transport type, how much volume of product and by which hub is sent. Secondly, the distances between the delivering hubs and the delivering countries were calculated according to the type of transport. Air and land routes are calculated as kilometers on Google Maps. The sea route was calculated as nautical mile in seadistance.org and converted to kilometers. Finally, P-median and maximal coverage models were applied and the results were shown.

The shipment data from 2018 provide by the UNHRD totaled 578 records which are retrieved from the official web-site of UNHRD (The United Nations Humanitarian Response Depot (UNHRD) 2018). These data included the shipment volume, country, transportation type, partner, date, hubs, and other information. These data include 6 hubs, 91 countries and four different types of transport: air, sea, land and multimodal. Demand points of the countries, locations of hubs, distance between each country and each hub were calculated according to transportation type. Air and land transports were calculated in kilometers by using Google Maps. Sea distance was calculated as nautical mile on seadistance.org and converted to kilometers. Since 63 of 578 deliveries were delivered in a multimodal form, it is not included in proposed study due to lack of retrieving information. Therefore, it is continued with 27343 m^3 delivered to 88 countries by considering remaining 515 deliveries.

Remaining data indicates 515 delivery volumes with 27343 cubic meters at total. With respect to Figs. 2 and 3 generated by retrieving data from (The United Nations Humanitarian Response Depot (UNHRD) 2018), 61% of the total volume is supplied from Dubai with 279 shipments, 20% from Accra with 91 deliveries, 18% from

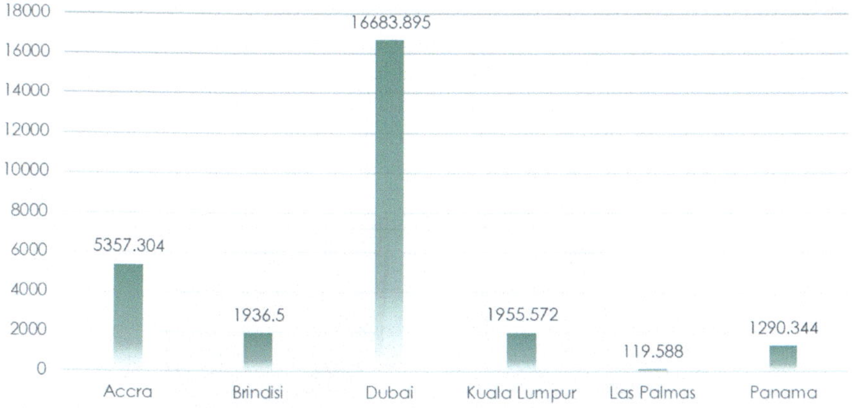

Fig. 2 Volume of each hub

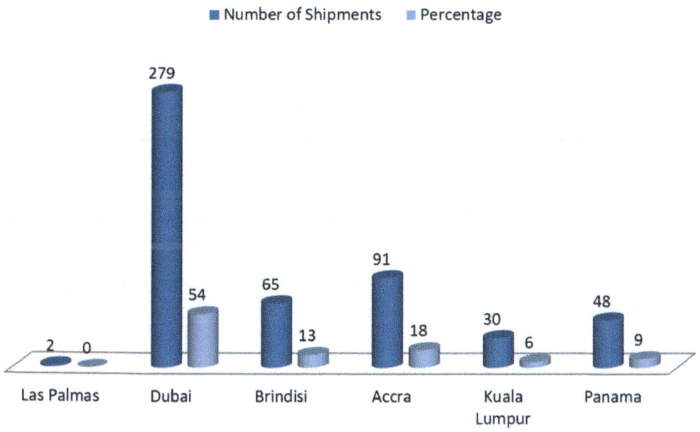

Fig. 3 Number of hubs shipment

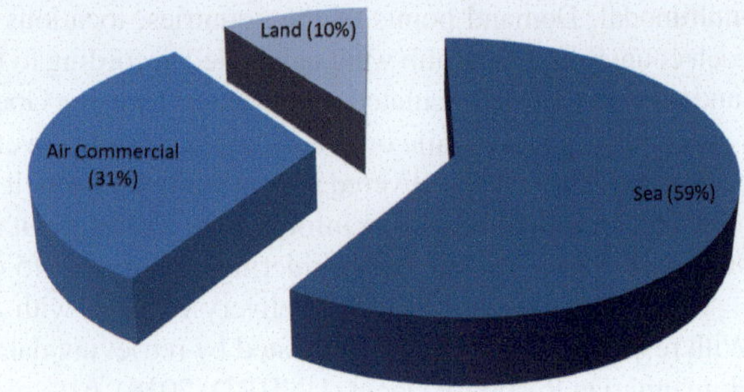

Fig. 4 Distribution of volumes by type of transportation

Panama with 48 shipments (9%), Brindisi 65 shipment (7%), Kuala Lumpur 30 shipment (4.57%) and Las Palmas 2 shipment (0.43%).

UNHRD delivered shipments with four different transportation types. Data of related figures (Figs. 4, 5) and tables (Table 1, 2, 3) gathered from UNHRD 2018 report (The United Nations Humanitarian Response Depot (UNHRD) 2018). However, multimodal form is not utilized in this study. Therefore, it is conducted

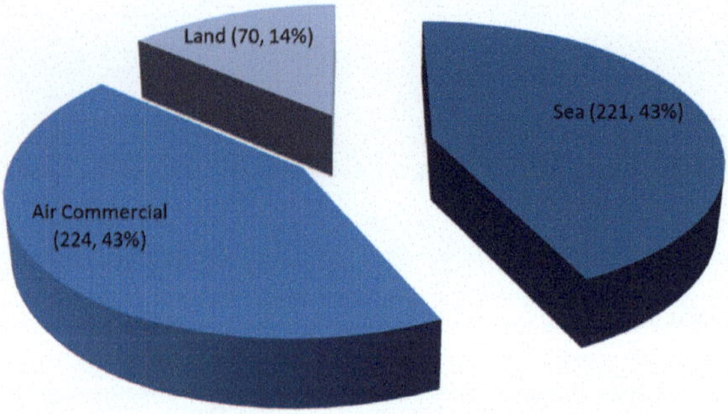

Fig. 5 Distribution of shipments by type of transportation

Table 1 Number of shipment

	Countries reached	Number of shipments	Volume (m^3)
With multimodal	91	578	32463
Without multimodal	88	515	27343

Table 2 Volume of per shipment

	Air	Sea	Land	Total
Number of shipment	224	221	70	515
Volume (m^3)	8535.4	16112.2	2695.4	27343
Volume/Shipment	38.10	72.91	38.51	53.09

Table 3 Volume per hubs shipment

	Dubai	Brindisi	Accra	Kuala Lumpur	Panama	Las Palmas	Total
Number of shipments	279	65	91	30	48	2	515
Volume (m^3)	16683.90	1936.50	5357.70	1955.57	1290.34	119.59	27343.00
Volume/Shipment	59.80	29.79	58.88	65.19	26.88	59.80	53.09

3 different delivery methods as air, land and sea. When total volume of deliveries are taken into consideration, it can be expressed from the Fig. 4 that 59% delivery managed by sea, 31% delivery managed by air and 10% delivery managed by land.

The Table 3 shows the number of deliveries delivered by each hub and the total volume carried by each hub. Volume per delivery is calculated. Although 59% of the total volume was delivered by sea, 221 deliveries made by sea are accounted for 43% of 515 deliveries as it is shown in Fig. 5. From the Fig. 5 it can be expressed that contrary situation occurs in air delivery with 224 deliveries that is 43% of total number of delivery, but 31% of the volume was delivered by air. The remaining 71 deliveries were made by land (%13.8) and accounted for 10% of the volume.

The Fig. 6 shows the distances of each hub's deliveries. The distance of each delivery is calculated according to the mode of transportation. The distances of air and land deliveries are measured with Google Maps. Measurements of sea deliveries are made with sea-distances.org. After all measurements have been completed, the total distance is found as 2420258.414 km.

It have been seen that 70 deliveries is managed by land transport type in Fig. 5. These 70 deliveries take place in 18 different countries. The map in Fig. 7 shows deliveries is done by which hubs to which one of the 18 countries.

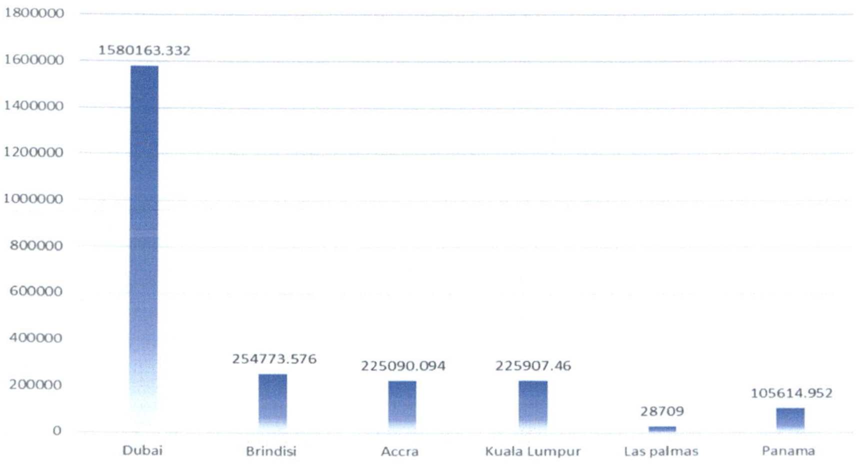

Fig. 6 Total distance of hubs (km)

Fig. 7 A map of land shipment

224 deliveries managed through Air transportation as it is demonstrated in Fig. 5. These 224 deliveries took place in 65 countries. The map in Fig. 8 shows which hubs meet the deliveries to presented countries by air shipments.

It can be seen that there are 221 deliveries sent by Sea transport type in Fig. 5. These 221 deliveries took place in 71 different countries. The map in Fig. 9 shows which hub makes deliveries to which country.

2.1 Location-Allocation Models

In this study, P-median and Maximal Coverage models are utilized. These models include delivery categories that consider hub locations, country delivery points, demand quantities, and objective function. The objective of facility location problems is to find a place to locate a facility in order to minimize the total setup cost and the total cost of transportation between clients and facilities. The objective of P-median problem is to determine the locations of p facilities and their assigned clients in order to minimize the total cost of transportation between clients and facilities (Dantrakul et al. 2014). Covering problems are to find the minimum number of facilities to cover all clients or to maximize the number of clients covered by a given number of opened facilities.

Fig. 8 A map of air shipments

Fig. 9 A map of sea shipments

2.1.1 P-median Model

The p-median problem was introduced by Hakimi (1964). With the given number of p, the objective is to minimization of total transportation cost between clients and the p facilities. The uncapacitated P-median problem can be solved in polynomial time for fixed values of p but this problem is NP-hard for variable values of p (Garey et al.

1990). Many heuristics and exact methods have been proposed to solve the problem during 1979–2005, as reviewed by Mladenović et al. (2007).

Methods for solving P-median problems are similar to methods for solving facility location problems such as variable neighborhood search (Hansen et al. 2009), simulated annealing (Brimberg and Drezner 2013), greedy heuristic (Almeida and Araujo 2010) and etc.

Sets and indices:

i ∈ I set of demand note

j ∈ J set of candidate facility

k ∈ K set of transportation type

Parameters:

P = available number of facilities

d_{ijk} = distance between locations i, j and k

w_i = weight of demand i

Decisions Variables:

$y_j = 1, if\ potential\ hub\ j\ is\ opened\ (\forall j \in J); 0\ otherwise$

$x_{ijk} = 1,\ if\ country\ i\ assigned\ to\ hub\ j\ and\ transportation\ k\ (\forall i \in I, \forall j \in J, \forall k \in K); 0 otherwise$

Objective Function:

$$minZ = \sum\nolimits_{j \in j} \sum\nolimits_{i \in I} \sum\nolimits_{k \in K} d_{ijk} W_i x_{jik} \tag{1}$$

Constraints:

$$\sum\nolimits_{j \in J} \sum\nolimits_{k \in K} x_{jik} = 1 \quad \forall i \in I \tag{2}$$

$$x_{jik} \le y_j \quad \forall i \in I, \quad \forall k \in K \tag{3}$$

$$\sum\nolimits_{j \varepsilon J} y_j = P \tag{4}$$

$$y_j, x_{jik} \in \{0.1\} \quad \forall i \in I, \quad \forall j \in J, \quad \forall k \in K \tag{5}$$

The objective function (1) minimizes the total demand-weighted distance between each demand node. Constraint (2) requires each demand node i to be assigned to exactly one facility j and one transportation type k. Constraints (3) link the location variables (x_j) and the allocation variables (y_{ij}). They state that demands at node i can only be assigned to a facility at location j ($y_{ij} = 1$) if a facility is located at node j ($x_j = 1$). Constraint (4) states that exactly p facilities are to be located. Constraints (5) are the standard integrality conditions.

2.1.2 Maximal Coverage

The maximal covering location problem (MCLP) considers the objective of locating a given number of facilities to maximize the covered number of demand nodes and demand nodes are expected be covered entirely if nodes are in the range of critical distance of the facility, otherwise it is assumed to be not covered (Karasakal and Karasakal 2004). The optimal result of a MCLP is depends on pre-decided critical distance, decision on a critical distance value without altering coverage may lead "fully covered" to "not covered" and erroneous results. Therefore, it is allowed the coverage alter "covered" to "not-covered" in given range that is called as level partial coverage (Karasakal and Karasakal 2004). In addition, mixed integer linear programming that aims to cover maximum demand and objective function which aims to cover maximum weighted demand by providing minimum average distance to uncovered demand is presented as the solution approaches (Haghani 1996). Maximum coverage model is also utilized in daily life areas such as deciding on location of stations. One of the instances is bicycle sharing stations which utilizes the optimization that comprises the aim of maximizing covered demand and available budget as a constraint (Frade and Ribeiro 2015). Another instance is the placing the stations for university campus and it utilizes the model that has an objective of deciding coverage capability of pre-determined number of stations for demand points (Mete et al. 2018). Moreover, Maximum coverage model is utilized for humanitarian logistic operation optimization. For instance, optimization of number of relief centers and their capacities is conducted for flood victims in Malaysia (Hashim et al. 2017). Maximum Coverage model for this study is given as follows (Church and ReVELLE 1974):

Sets and indices:

I = the set of demand nodes

J = the set of facility sites

S = the distance beyond which a demand point is considered "uncovered" $N_i\{j \in J/d_{ij} \leq S\}$ the nodes j that are within a distance of S to node i

Parameters:

P = number of facilities

d_{ij} = shortest distance between locations i, j

a_i = population to be served at demand node i

Decisions Variables:

a_i= *demand at node i*

$x_j = 1\, if\ a\ facility\ sited\ at\ the\ jth\ node\ (\forall \in J);\ 0\ otherwise$

$y_i = 1\, if\ node\ i\ is\ covered\ by\ one\ or\ more\ facilitiesstationed\ within\ S;\ 0\ otherwise$

Objective Function:

$$max\,Z = \sum\nolimits_{i \in I} a_i y_i \tag{6}$$

Constraints:

$$yi \leq \sum\nolimits_{j \in N_i} x_j \tag{7}$$

$$\sum\nolimits_{j \in J} x_j = p \tag{8}$$

$$y_i,\ x_j \in \{0,\ 1\} \quad \forall i \in I \quad \forall j \in J \tag{9}$$

The objective function (6) maximizing the sum of covered elements in the sets in which they are covered. Constraint (7) if $y_i > 0$ then at least one set $j \in N_i$ is selected. Constraint (8) states that exactly p facilities are to be located. The sign constraints are shown in Constraint (9).

3 Results

In this section, distance between countries and each hub are examined and the result is gathered by using mathematical models (P-median and Maximal Coverage) described in the previous section. Finally, LINGO 14.0 software results and proposed mathematical model's results are compared. In this section, two different location-allocation models which are P-median and Maximal Coverage, applied to the UNHRD Logistics Service. The cases from Ghana (Accra), Italy (Brindisi), Spain (Las Palmas), Malaysia (Kuala Lumpur), Panama (Panama City), United Arab Emirates (Dubai) to 88 countries are considered. Distances from each of these six points to 88 countries were measured according to transport types. We took the run on a server with 3.00 GHz Intel Core processor and 2 GB of RAM. The computation time required to solve the models are less than 10 s.

3.1 P-median Solution

After obtaining distances from each hub to countries, P-median is implemented to assign UNHRD depots and demand points of countries by aiming the total transportation distance minimization. The obtained results of PMP by LINGO 14.0 are given in Tables 4, 5, 6 and illustrated in Figs. 10, 11, 12.

Table 2 demonstrates the situation before the application of P-median and Maximal Coverage which is 224 transports took place by air (a total volume of 8535.4 m^3 and 38.1 m^3 per shipment), 221 transports managed by sea (a total volume of 16112.2 m^3, and 72.91 m^3 per shipment) and the remaining 70 transports are carried out by land (a total volume of 2695.4 m^3 and 38.51 m^3 per shipment). When the results from application of p-median are analyzed, which is demonstrated in Table 4, it is seen that the number of deliveries for air transportation is increased by 32% and the number of deliveries for sea and land transportation are decreased by 7.23 and 80% respectively.

Figures 7, 8, 9 illustrates the situation before the application of focused models. Initially, the longest distance in the distribution network is measured as 25789.1 km. After applying the p-median model, this distance fell to 7719.4 km. The Fig. 10 shows 515 deliveries to 88 countries from each hubs after the implementing the p-median model. Figure 10 also shows how many deliveries are made from each hub to each country.

Initial total distance of the deliveries is measured as 2420258.414 km. As a result of the p-median model application, total distance of optimized deliveries is updated to 1003473.646 km by 58% decrease. Before the improvement, as it is shown in Fig. 3, the demands (279 deliveries) are mostly supplied by Dubai. However, most of the demands (206 deliveries) are supplied by Accra after improvement.

In a nutshell, there is a significant change in activity of hubs. The considerable change is seen in Dubai hub activity. As a result of the model, a significant reduction in total distance has been achieved and deliveries are made from the hub to the closest requesting country. The number of deliveries carried by Accra has increased.

As it is shown in Fig. 12, the volume transported from Dubai decreased almost by half, while the volume transported from Accra doubled. Other hubs showed small changes.

According to the results of our model presented in Table 6, the hub point that meets the demand of each country, the transportation point of the deliveries, the hub point that meets the demand, the distance between the country and finally the number of deliveries made to that country are indicated. As it is expressed previously, there

Table 4 P-median number of transportation and volume solution

Transportation	Number of shipment	Volume (m^3)	Average volume
Air	296	13409.8	45.3033
Sea	205	12927.7	63.0619
Land	14	1005.71	71.8364

Table 5 Average volumes of each shipment of hubs and distances

Transportation	Accra	Dubai	Brindisi	Las Palmas	Kuala Lumpur	Panama	Total
Distance	392423.20	310868.66	46146.38	14530.80	125738.85	113765.74	1003473.64
Volume (m^3)	11823.57	8701.49	1732.42	444.46	2804.92	1836.32	27343.21
Number of shipment	206	127	57	8	51	66	515
Average volume	57.396	68.515	30.39	55.55	54.99	27.82	

Table 6 P-median assignments

Country	Hub	Mode	Demand (m^3)	Distance (km)	# of Shipments	Country	Hub	Mode	Demand (m^3)	Distance (km)	# of Shipments
Afghanistan	Dubai	Sea	65.67	1281.58	5	Libyan	Brindisi	Air	271.42	977.83	9
Angola	Accra	Air	11.83	2190.22	1	Liechtenstein	Brindisi	Sea	0.37	1	1
Antigua and Barbuda	Panama	Air	23.75	2105.80	2	Madagascar	Dubai	Air	67.41	4928.80	3
Armenia	Brindisi	Sea	41.80	2572.42	1	Malawi	Accra	Air	23.57	4535.25	2
Bangladesh	Kuala Lumpur	Sea	1752.17	2427.97	22	Malaysia	Kuala Lumpur	Air	119.82	1	6
Barbados	Panama	Air	65.97	2216.82	3	Mali	Las Palmas	Sea	132.98	1553.82	2
Benin	Accra	Sea	96.11	285.20	3	Marshall Island	Kuala Lumpur	Air	0.18	7719.40	1
Bhutan	Kuala Lumpur	Sea	54.41	2674.28	3	Mauritania	Accra	Air	22.52	1095.80	3
Brazil	Panama	Air	23.23	4435.20	2	Mozambique	Accra	Air	266.30	4975.77	4
Burkina Faso	Accra	Sea	253.25	479.66	6	Myanmar	Kuala Lumpur	Sea	9.83	2146.46	1
Burundi	Accra	Air	234.63	3423.49	6	Namibia	Accra	Sea	20	3539.17	1
C. African Rep.	Accra	Sea	127.70	1111.20	4	Niger	Accra	Sea	1411.12	285.20	17
Cameron	Accra	Sea	958.07	1111.20	8	Nigeria	Accra	Sea	1787.67	394.47	23
Chad	Accra	Sea	137.30	1111.20	8	Pakistan	Dubai	Air	4.75	1887.70	2
Comoros	Dubai	Sea	466.33	4942.98	4	Palestinian	Brindisi	Air	0.01	1785	1

(continued)

Table 6 (continued)

Country	Hub	Mode	Demand (m^3)	Distance (km)	# of Shipments	Country	Hub	Mode	Demand (m^3)	Distance (km)	# of Shipments
Congo	Accra	Air	103.30	2033.48	5	Panama	Panama	Sea	30.50	1	9
Congo Demo. Rep.	Accra	Sea	3463.13	1902.00	42	Papua New Guinea	Kuala Lumpur	Sea	27.18	5983.81	2
Côte d'Ivoire	Accra	Land	5.96	536	1	Peru	Panama	Air	127.67	2557.7	5
Cuba	Panama	Air	293.92	1592.90	10	Philippines	Kuala Lumpur	Air	44.03	2741	3
Djibouti	Dubai	Air	256.32	1938.70	9	Rwanda	Accra	Air	172.34	3469.44	12
Dominica	Panama	Air	26.67	2118.66	4	Senegal	Las Palmas	Sea	191.90	1553.82	4
Ecuador	Panama	Air	13.71	1015	2	Sierre Leone	Accra	Air	45.00	1475.91	2

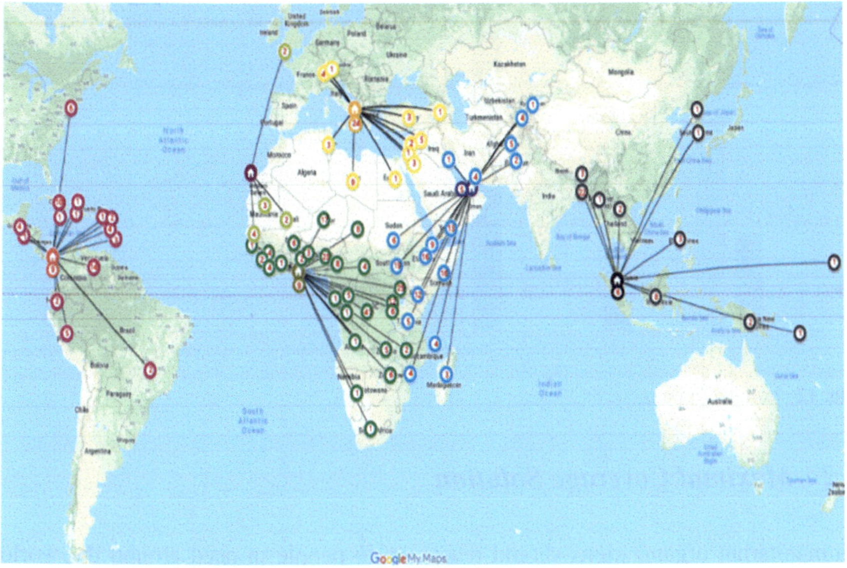

Fig. 10 P-median solution map

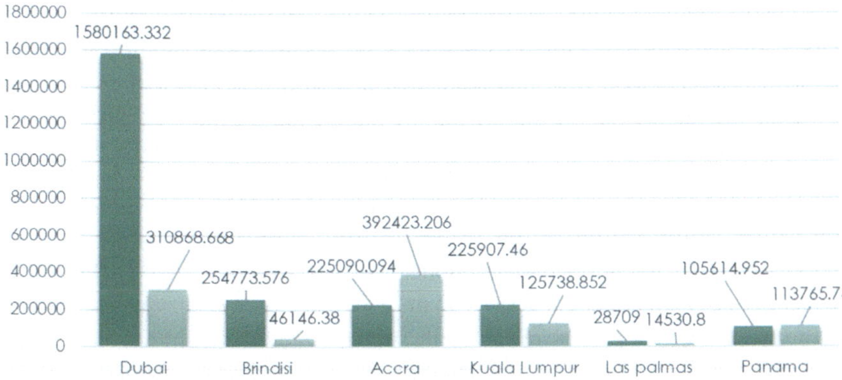

Fig. 11 Distance of each hub

are 515 deliveries which are not shown all in Table 6. From the results, it can be said that air transportation is the mostly used type, and the Accra is the hub where demand is mostly met.

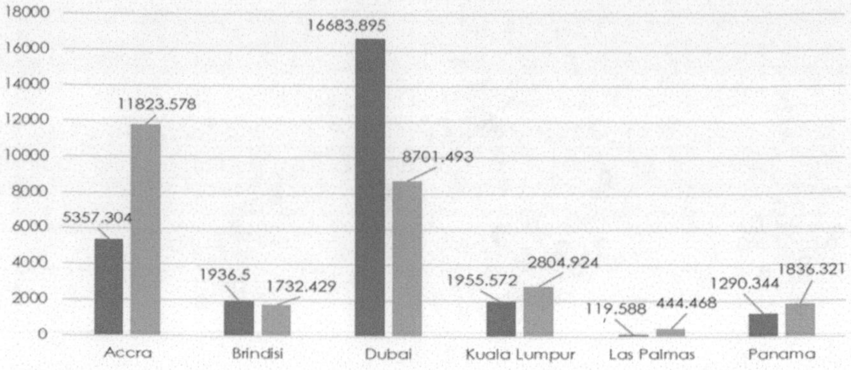

Fig. 12 Volume of hubs

3.2 *Maximal Coverage Solution*

Humanitarian organizations should reach out to people in need around the world as much as possible. UNHRD has embraced it and tries to reach every part of the world in minimum time. Minimization is done on the total distance of UNHRD 2018 deliveries by applying P-median. In this section, we applied 3 different distance constraints such as 3000, 4000 and 5000 km using UNHRD (2018) data and measured the total covered demand (Table 7). Looking at the results prior to optimization, the total distance of 515 deliveries was calculated as 2420258.414 km and the average distance for each delivery was 4699.7 km. After the application of P-median, the total distance was calculated as 1003473.646 km and the average distance decreased to 1948.5 km.

First of all, it is arranged retrieved data separately for maximal coverage model according to decided distance constraints. Firstly, distance constraint is set to 5000 km. By LINGO 14.0 software application, it is observed that 27211.06 m^3 of demand is met. The remaining 131.941 m^3 and 6 countries are uncovered and the rate of uncovered countries is 7%. The uncovered countries are Solomon Islands, Papua New Guinea, South Africa, Marshall Islands, North Korea, and South Korea. Secondly, it is edited data by setting a 4000 km coverage area for the maximal coverage model. In this scenario, it is observed that 26219.445 m^3 of our demand is met. The remaining 1123.555 m^3 demand and 12 countries are excluded. These

Table 7 Maximal coverage solutions

Coverage area (km)	Total demand	Meet demand (m^3)	Missing demand (m^3)	Number of covered country	Number of uncovered country	Uncovered rate (%)
3000	27343	23474.33	3868.66	69	19	22
4000	27343	26219.44	1123.55	76	12	14
5000	27343	27211.05	131.94	82	6	7

Fig. 13 Maximal coverage assignments for 5000 km

countries are shown in Fig. 14. Finally, maximal coverage model is implemented with 3000 km coverage area. In final scenario, it is observed that 23474.336 m^3 of demand is met. The remaining 3868.664 m^3 demand and 19 countries are excluded. Unfortunately, rate of uncovered countries are consists of 22%. The uncovered countries are Solomon Islands, Papua New Guinea, South Africa, Marshall Islands, North Korea, South Korea, Brazil, Comoros, Madagascar, Malawi, Zimbabwe, Mozambique, Burundi, Namibia, Rwanda, Tanzania, Uganda, United States, and Zambia. As coverage distance declined, the country that is not covered increased dramatically. New solutions should be tried for this situation. These countries are shown in Fig. 15.

Figure 13 for 5000 km coverage area, purple colors indicate the countries covered by Panama. Also, green ones show Ghana, yellow ones indicate Italy, blue ones illustrate UAE, olive ones show Las Palmas de G.C and Gray ones Malaysia. Black dots indicate countries that cannot be assigned. As it can be seen from Fig. 13, 82 countries are covered and 6 are not covered.

Figure 14 indicates the countries covered for 4000 km coverage, as in Fig. 13 the same colors represent the same hubs. From Fig. 14, it can be deducted that 76 countries are covered and 12 are not covered.

Figure 15 indicates the countries covered and not covered for 3000 km coverage. In Fig. 15 69 countries are covered and 19 countries are not.

4 Conclusion

The aim of this study is to minimize the total distances of 515 deliveries to 88 countries that request from 6 warehouses owned by UNHRD. These deliveries took place in 3 different ways: air, land and sea. Data on deliveries are taken from the 2018 activity

Fig. 14 Maximal coverage assignments for 4000 km

Fig. 15 Maximal coverage assignments for 3000 km

reports of UNHRD and the total distance of 515 deliveries is 2420258.414 km and the total volume is 27343 m^3. 224 of these deliveries were made by air and 8535.4 m^3 of the volume was met in this way. 221 of the remaining deliveries were made by sea and 16112.2 $m^3$of the remaining volume was met. Finally, 70 transports were realized by land and 2695.4 m^3 of the total volume was met in this way. Firstly, distances from 6 hubs to 88 countries were measured on Google Maps and seadistance.org according to the delivery type. Secondly, P-median mathematical model is applied to the total distance and the demand's longest by aiming distance minimization. As a result of P-median, our total distance from 2420258.414 km has decreased to 1003473.646 km.

In this case, 296 of 515 deliveries were made by air and 13409.8 m^3 of the total volume was met. 205 deliveries were made by sea and 12927.7 m^3 of the total volume was met. The remaining 14 deliveries and 1005.71 m^3 volume were provided by road. Finally, the Maximal Coverage model was applied in 3 different distances to increase the total covered demand. A total of 69 countries were covered when 3000 km was applied and 23474.336 m^3 of the total demand was met. When 4000 km is applied, 76 countries are covered and 26219.445 m^3 of demand is met. When we reached 5000 km, 82 countries were covered and 27211.059 m^3 of the total demand was met. In order to cover all countries and meet all the demand, 7719.4 km is required. For the future studies, set covering approach should be implemented in order to see performance of other methods and location analysis can be conducted for new establishments for the hubs. Secondly, new locations for potential new hubs should be investigated using geographic information systems. Lastly, a web-based decision support system should be developed to provide dynamic solutions for humanitarian logistic operations.

References

Akbarpour M, Ali Torabi S, Ghavamifar A (2020) Designing an integrated pharmaceutical relief chain network under demand uncertainty. Transp Res Part E Logist Transp Rev 136(January):101867. https://doi.org/10.1016/j.tre.2020.101867

Almeida WG, Araujo E (2010) Metaheuristic search procedure based on fuzzy graph for p-facility problems. In: 2010 IEEE World congress on computational intelligence WCCI 2010, pp 1–5

Balcik B, Yanıkoğlu İ (2020) A robust optimization approach for humanitarian needs assessment planning under travel time uncertainty. Eur J Oper Res 282(1):40–57

Brimberg J, Drezner Z (2013) A new heuristic for solving the p-median problem in the plane. Comput Oper Res 40(1):427–437

Cavdur F, Sebatli-Saglam A, Kose-Kucuk M (2020) A spreadsheet-based decision support tool for temporary-disaster-response facilities allocation. Saf Sci 124(January):104581. https://doi.org/10.1016/j.ssci.2019.104581

Church RL, ReVELLE C (1974) The maximal covering location problem. Pap Reg Sci Assoc 32(1):101–118

Dantrakul S, Likasiri C, Pongvuthithum R (2014) Applied p-median and p-center algorithms for facility location problems. Expert Syst Appl 41:3596–3604

Daskin M (1982) Application of an expected covering model to emergency medical service system design. Decis Sci 13(3):416–439

Daskin M (1983) A maximum expected covering location model: formulation, properties, and heuristic solution. Transp Sci 17:48–70

Dufour É, Laporte G, Paquette J, Rancourt M–È (2018) Logistics service network design for humanitarian response in East Africa. Omega 74:1–14

Duran S, Keskinocak P, Ergun Ö, Swann JL (2013) Humanitarian logistics: advanced purchasing and pre-positioning of relief items. Handbook of global logistics. Springer, New York, pp 447–462

Frade I, Ribeiro A (2015) Bike-sharing stations: a maximal covering location approach. Transp Res Part A Policy Pract 82:216–227

Garey MR, Johnson DS (1990) A guide to the theory of NP-completeness. Comput Intractability 37–79

Global Humanitarian Assistance (2019) GHA Annual Report Global Humanitarian Assistance Report. http://www.globalhumanitarianassistance.org/report/gha-report-2015

Haghani A (1996) Capacitated maximum covering location models: formulations and solution procedures. J Adv Transp 30(3):101–136

Hakimi S (1964) Optimum location of switching centers and the absolute centers and medians of a graph. Oper Res 12:450–459

Hansen P, Brimberg J, Urošević D, Mladenović N (2009) Solving large p-median clustering problems by primal-dual variable neighborhood search. Data Min Knowl Discov 19(3):351–375

Hashim NM, Shariff SSR, Deni SM (2017) Capacitated maximal covering location allocation problem during flood disaster. Adv Sci Lett 23(11):11545–11548

Ireland & The UN Humanitarian Response Depot (2020). https://www.irishaid.ie/what-we-do/responding-to-emergencies/rapid-response-initiative/ireland-and-the-un-humanitarian-response-depot/

Jahre M et al (2016) Integrating supply chains for emergencies and ongoing operations in UNHCR. J Oper Manag 45:57–72. https://doi.org/10.1016/j.jom.2016.05.009

Karasakal O, Karasakal E (2004) A maximal covering location model in the presence of partial coverage. Comput Oper Res 31:1515–1526

Mete S, Cil ZA, Özceylan E (2018) Location and coverage analysis of bike-sharing stations in university campus. Bus Syst Res 9(2):80–95

Mladenović N, Brimberg J, Hansen P, Moreno-Pérez JA (2007) The p-median problem: a survey of metaheuristic approaches. Eur J Oper Res 179(3):927–939

Oksuz MK, Satoglu SI (2020) A two-stage stochastic model for location planning of temporary medical centers for disaster response. Int J Disaster Risk Reduct 44:101426. (August 2019). https://doi.org/10.1016/j.ijdrr.2019.101426

Özdamar L, Ertem MA (2015) Models, solutions and enabling technologies in humanitarian logistics. Eur J Oper Res 244(1):55–65

Serra D, Marianov V (1998) The p-median problem in a changing network: the case of Barcelona. Locat Sci 6(1–4):383–394. May–December 1998

Shariff SSR, Moin NH, Omar M (2012) Location allocation modeling for healthcare facility planning in Malaysia. Comput Ind Eng 62(4):1000–1010

The United Nations Humanitarian Response Depot (UNHRD) (2018). https://unhrd.org/hub

United Nations Humanitarian Response Depot (2020) World food programme. https://www.wfp.org/unhrd

International Logistics Criteria of Medicinal Products for Human Use: How Do They Relate to Cargo Logistics Providers?

Shiang-Min Meng

Abstract This study explores the logistics criteria of medicinal products with logistics outsourcing. It examines the subjective decisions of the managers at medicinal product firms with regard to the outsourcing of logistics work to cargo logistics providers. The research targets outsourcing logistics decision makers from firms that manufacture medicinal products for human use in Taiwan. The empirical evidence supports five key logistics criteria factors: documentation, equipment, technology, certification, and training. The four key logistics outsourcing decision factors examined are quality, freight, service, and delivery. A significant relationship between the logistics criteria of the cargo logistics providers and the outsourcing logistics decisions of the medicinal products firms is also identified. The study provides an empirical validation of the two-factor construct to develop survey scales for international logistics criteria and client decisions with the establishment of a standard questionnaire, which uses 47 successful questionnaires and follows the Delphi method. Factor analysis was used on 40 criteria between the medicinal products firms and cargo logistics providers. The Pearson's coefficients on the correlations applied the two factors of logistics criteria and outsourcing decisions.

Keywords International logistics criteria · Client outsourcing decision · Medicinal products for human use · Cargo logistics providers

1 Introduction

The global COVID-19 coronavirus outbreak has seriously hampered the delivery of medicinal products by cargo logistics enterprises. Global logistics companies are essential service providers as they enable medicinal product enterprises to outsource the on-time delivery of their products. In the past decade, Taiwan's medicinal products industry has increased its competitive advantages in terms of production through

S.-M. Meng (✉)
Department of International Logistics and Transportation Management, Kainan University, Taoyuan City, Taiwan
e-mail: eddymeng@mail.knu.edu.tw

© The Author(s), under exclusive license to Springer Nature Switzerland AG 2020
P. Golinska-Dawson et al. (eds.), *Smart and Sustainable Supply Chain and Logistics – Trends, Challenges, Methods and Best Practices*, EcoProduction,
https://doi.org/10.1007/978-3-030-61947-3_16

outsourcing its deliveries to external cargo logistics providers. The outsourcing of a company's logistics functions represents an effective allocation of their resources and reconfigured its supplier network to build long-term partnerships. Organizations have traditionally focused their attention on controlling their own costs to increase profitability by concentrating on improving the business processes that the organization controls independently (Ergun et al. 2007). The medicinal products transportation process in different regions has required giving special consideration to the aspects of long-range transportation by trucks, ships, and air cargo. There has been a need to control the temperature, humidity, and other factors during the transportation process. The outsourcing of key logistics activities has been one of the most important trends in medicinal companies, and the logistics service criteria affect the outsourcing firm's logistics perception, consequent behavior, and partnership selection (Meng 2014). The logistics criteria and outsourcing decision factors have led to a growing body of research to define their essential characteristics. Du et al. (2005) pointed out that product customization has been recognized as an effective means to meet individual client's needs, and customers can be actively involved in the product customization. Logistics outsourcing controls a firm's differentiation choices on logistics customization and enables executives to pursue improved efficiency. There are many implications for service value of logistics outsourcing in terms of the customization of services (Remko 2000). Fugate et al. (2006) stated that an effective supply chain flow requires the creation of synergistic relationships between the supply and distribution partners in order to maximize service and provide a profit. These decisions are largely determined by the political and organizational factors associated with the regional autonomy, geographical factors, climate, travel distances, infrastructure, roads, and electricity supply in which the population is distributed. Logistical outsourcing issues are included in the strategic agenda because they can have a positive influence on company performance (Tracey 2006). Although logistics studies have become a key component of research work, they have invariably lacked a focus on organizations' logistics decisions in respect of medicinal products for human use.

2 Logistics on Medicinal Products for Human Use

The Pharmaceutical Inspection Co-operation Scheme (PIC/S) sets out medicinal products specifications for good distribution practices (GDP), which continue the thrust of the good manufacturing practices (GMP) established in the scheme. A manufacturer of medicinal products for human use is required to comply with the local legislation. This legislation specifies the application of GMP for medicinal products exclusively intended for export. A manufacturing license, the GMP certificate, is issued, which signifies that laboratory control of the pharmaceutical products has been conducted (Tomić et al. 2010). The quality of international pharmaceutical products continues to improve, and the quality control regime spans the entire process from production to sales; that is, the drug quality management covers the entire life cycle of the drugs and includes stringent requirements for the procedures. The quality control of international pharmaceutical companies has continued to improve. The drug quality management process covers the entire life cycle of drugs, and since the GDP came into being, it has extended to drug distribution. Over 130 countries have adopted the GDP specifications. These specifications cover the management required by the wholesale distributors in respect of risk management, inspection, storage, and transport. The distribution processes under GDP are consistent with quality assurance principles (Meng et al. 2019). Thus, logistics providers are to be able to meet the needs of their medicinal clients by acting as their geographical partner to propagate the spread and range of their clients' service criteria. Through factor and correlation analysis, this study designs valuable criteria to align with the needs of their outsourcing medicinal companies to enable them to perform some or all of the logistics activities. These will evolve into tailor-made services to meet the diverse needs of their clients. Chikumba (2009) indicated that medicinal products logistics operators need to get additional spatial information, understand the needs of the health facilities, and take into account environmental factors in order to monitor the flow of medicinal products at each health facility. Central logistics models are primarily coordinated by arranging for the products to be received first at their central medical stores and then distributed to intermediate stores and ultimately to the final service points. In a decentralized system the intermediate stores are responsible for receiving, storing, and distributing the end products, and they are also responsible for procurement of the production inputs. The active pharmaceutical ingredient (API) specifications need to comply with the GDP regulations, which comprise procurement, importation, storage, supply, distribution, export activity, processing, packing, reverse logistics, and warehousing, as well as the rules for contract agents, brokers, traders, and others involved in the supply chain that are not the original manufacturers' suppliers (APIC 2017). APIs are the parts of any drug that are used in the formulation of the finished product to produce the intended effects of the drug. Based on the drug GMP, the definition of API refers to the components of a drug by reference to its physical and chemical properties, its biological treatment, the technology used in manufacturing to produce the necessary pharmacologically effective active chemical ingredient. The active chemical ingredients commonly used are drugs manufactured in the production of biotechnology products.

With medicinal products for human use, international logistics providers play an important part in aligning GDP and GMP to ensure the stability of the quality during the transport or storage of the products. They need to monitor the temperature and humidity to ensure they meet the requirements of the national health authority, and to do so, they need to develop customized criteria for pharmaceutical logistics (Meng et al. 2019). For cargo logistics providers involved with medicinal products, the main functions concern the degree of centralization and the number of storage points through which the product will pass before being delivered to the recipient. Roy et al. (2009) argued that the drug supply available must not only meet with the requirements of the population and the estimated demand for the pharmaceutical items, but it also scientifically meet the quantity and quality of the medicinal products demanded. The overall API logistics links are not allowed to have errors that could compromise the quality of the drugs during transportation. Accordingly, a strict set of comprehensive service standards for drug delivery have been set up to reduce the incidence of risks and accidents in order to maintain a consistent quality of the drugs and to ensure there are no adverse effects on any of the key factors in the delivery process.

3 Delphi Method and Questionnaire Design

The data for this study were collected from a survey. To determine the questionnaire items, it was crucial to ensure their content validity and survey instrumentation accuracy. A standard questionnaire using the Delphi method was established. The questions were based on expert opinions and suggestions from pharmaceutical industry representatives, academics, and those with advanced knowledge in related fields. The first Delphi step involved experts and professors determining the number of experts required for each group. They decided on one from international ocean freight transportation, one from international air freight transportation, one professor from a university, and one product executive from a pharmaceutical company to provide guidance on the principals of good distribution practices in respect of active substances for medicinal products for human use. The Meng et al. (2019) studies on medicinal criteria were divided into a total of 40 measures of logistics criteria and client decision-making, as shown in Table 1.

The research target for this study is the actual outsourcing logistics decision makers in respect of medicinal products for human use in Taiwan. The results of the pretest of all 40 items were very good, so the survey was distributed to 60 logistics outsourcing decision makers and was carried out during June to September 2019. A total of 52 successful questionnaires were returned, but five of these were discarded because of incomplete information. Accordingly, there were 47 usable responses for an overall response rate of 78.3%. Statistical analyses and factor analysis were conducted using SPSS. First, factor analysis was used to investigate any separate underlying factors and to reduce redundancy. Next, Pearson correlations were used to assess the bivariate relationships. The Cronbach's α values were statistically determined to provide a summary measure of the inter-correlations that existed. In this study, the Cronbach's α values, which are depicted in Tables 2 and 3, show that the factor dimension attributes with respect to the logistics criteria and outsourcing decisions are all greater than or equal to 0.6. This shows that each dimension achieved reliability.

4 Empirical Results

To simplify the analysis structure, this study conducted two factor analyses of the forty items in the logistics criteria and outsourcing decisions lists. This study adopted the principal components to extract the factors whose Eigenvalue was greater than 1

Table 1 Measurements of international logistics criteria and client decisions

Measurements of logistics criteria	Measurements of outsourcing decisions
1. Self-check	1. Medicinal products for human use distribution regulations
2. Personnel service	2. Drug delivery processes files need to be saved properly and could date back
3. Premises and equipment	3. GDP/GMP specifications in respect of compliance document management
4. Complaints	4. Delivery documents are immediately returned
5. Containers and labels	5. Sufficient manpower
6. Organization and management	6. Temperatures in warehouses are periodically calibrated
7. Vehicles and equipment	7. Workplace environment is clean
8. Sending and receiving	8. Drug distribution has excellent specifications (GDP) (blue-chip vendors)
9. Repackaging and relabeling	9. Provide written procedures for processing jobs under abnormal conditions
10. Product returns	10. Has dangerous goods expertise
11. Contract activities	11. Has expertise in highly potent and toxic drugs
12. Transport	12. Emerging market delivery route planning
13. Recalls	13. Capacity for risk assessment for transport
14. Self-audit	14. Drug storage conditions reflect GMP/GDP knowledge
15. Self-inspection	15. Reverse logistics capacity; able to handle rejection and recalls of pharmaceutical products
16. Information transfer	16. Workers receive the necessary GMP/GDP training regularly
17. Shipping	17. Acquired Partners in Protection (PIP) certification
18. File management	18. Obtained United States Business Coalition against Terrorism (C-TPAT) certification
19. Quality systems	19. Obtained Authorized Economic Operator (AEO) certification
20. Final provisions	20. Transportation vehicles cleaning procedures

and then used the varimax of orthogonal rotation to obtain the rotated coefficients. Finally, factors four and five were extracted. The accumulated variance explained was 72.13% in respect of the logistics criteria and 68.42% for the outsourcing decision; the analysis results of the individual factors of the logistics criteria and outsourcing decisions are illustrated in Tables 2 and 3, respectively.

Factor analysis was conducted to classify the identified client's outsourcing decisions into several critical dimensions. The results identify equipment as the most important dimension of the logistics criteria, whereas quality is the most important dimension of the outsourcing decisions.

Table 2 Factor analysis of logistics criteria

Factor dimensions of logistics criteria	Factor loading	Cronbach's α	Eigenvalue	Explanation rate (accumulated)
DOCUMENTS - File management - Contract activities - Self-inspection - Self-audit	**0.59** **0.74** **0.58** **0.66**	**0.72**	**3.68**	**23.82% (23.82%)**
EQUIPMENT - Premises and equipment - Containers and labels - Vehicles and equipment - Repackaging and relabeling - Shipping - Transport	**0.66** **0.54** **0.69** **0.71** **0.48** **0.51**	**0.77** **0.71**	**2.91** **2.48**	**16.2% (41.18%)** **17.44% (57.46%)**
TECHNOLOGY - Sending and receiving - Information transfer - Self-check - Final provisions	**0.58** **0.76** **0.59** **0.53**	**0.61**	**2.98**	**11.63% (74.41%)**
CERTIFICATION - Recalls - Product returns	**0.78** **0.71**	**0.67**	**2.43**	**9.19% (71.21%)**
TRAINING - Personnel service - Organization and management - Quality systems - Complaints	**0.69** **0.46** **0.44** **0.68** **0.54**			

In order to understand the relationships between the factors of the logistics criteria and outsourcing decision indicators, the study used Pearson's correlation analysis. The results are shown in Table 4.

The empirical results show that there are significant relationships between the factors of the logistics criteria and outsourcing decisions. The detailed analyses and corresponding relationships are stated below.

1. A significant relationship exists between "Documents" and "Quality" as well as "Documents" and "Delivery."
2. A significant relationship exists among "Equipment," "Technology," and "Certification" and among "Quality," "Freight," "Service," and "Delivery."
3. There are significant relationships between "Training" and "Quality" and "Service" and "Delivery."

Table 3 Factor analysis of outsourcing decisions

Factor dimensions of outsourcing decisions	Factor loading	Cronbach's α	Eigenvalue	Explanation rate (accumulated)
QUALITY - Temperatures in warehouses are periodically calibrated - Drug delivery processes files need to be saved properly and could date back - Medicinal products for human use distribution regulations - Sufficient manpower - Workplace environment is clean	**0.69** **0.77** **0.74** **0.64** **0.58** **0.71**	**0.71**	**7.94**	**37.54%** **(37.54%)**
FREIGHT - Transportation vehicles cleaning procedures - Provide written procedures for processing jobs under abnormal conditions - Workers receive the necessary GMP/GDP training regularly - Low cost provider - Acquired Partners in Protection (PIP) Certification	**0.59** **0.57** **0.73** **0.58** **0.54**	**0.73** **0.87**	**2.41** **1.68**	**16.77%** **(54.45%)**
SERVICE - GDP/GMP specification of compliance document management - Drug storage conditions reflect GMP/GDP knowledge - Expertise in highly potent and toxic drugs - Obtained Authorized Economic Operator (AEO) certification - Has dangerous goods expertise - Obtained United States Business Coalition against Terrorism (C-TPAT) certification	**0.69** **0.58** **0.48** **0.57** **0.81** **0.78** **0.58**	**0.61**	**1.62**	**13.41%** **(75.86%)**

(continued)

Table 3 (continued)

Factor dimensions of outsourcing decisions	Factor loading	Cronbach's α	Eigenvalue	Explanation rate (accumulated)
DELIVERY - Delivery documents are immediately returned - Drug distribution has excellent specifications (GDP) (blue-chip vendors) - Reverse logistics capacity; able to handle rejection and recalls of pharmaceutical products - Emerging market delivery route planning	**0.54** **0.69** **0.43** **0.58**			**8.36% (72.18%)**

Table 4 Pearson's coefficients on logistics criteria and outsourcing decisions

		Logistics criteria				
Outsourcing decision	Factors	Documents	Equipment	Technology	Certification	Training
	Quality	0.24**	0.18**	0.59**	0.35**	0.25*
	Freight	0.18	0.27**	0.64**	0.31**	0.06
	Service	0.04	0.36**	0.57**	0.48**	0.31*
	Delivery	0.53**	0.56**	0.48**	0.28**	0.27**

** Significance level $p < 0.01$; Significance level $*p < 0.05$

4. "Equipment,"Technology,"and "Certification" in operations are potential logistics factors to connect with the outsourcing decisions. The results showed that the logistics criteria have a significant relationship with the outsourcing decisions.

5 Conclusions

This study conceptually defined the domain factors and, using the Delphi method, designed 40 survey items to empirically validate and explore the relationships between the service criteria and outsourcing decisions on the medicinal products examined. The results align with those of Gouveia et al. (2015), who surveyed European participants with respect to GMP guidelines in which quality, security, and effectiveness at exceptional levels are applicable to the manufacturing of health products. By providing positive logistics criteria to the client, the potential negative service failure associated with the logistics decision is reduced. This suggests that cargo logistics providers should place their emphasis on the documents, equipment, technology, certification, and training factors. The logistics criteria are the matters that drive the clients' outsourcing intentions. Understanding the behavioral intention on outsourcing is the main factor to influence clients' decisions (Meng et al. 2010). This study demonstrated that equipment is the most important criterion and that firms should take it into account when developing their strategy. The study also presented the strategic management that is necessary regarding the decision factors of quality, freight, service, and delivery. In summary, this study identified the essential factors that need to be comprehensively evaluated to provide an understanding of the logistics criteria and outsourcing decisions. We expect that it can be used as a reference for strategic planning carried out by cargo logistics providers. The results revealed that the proposed two factors have psychometric properties that are desirable to clients when they make decisions, and that these factors can improve efficiency and core competency.

References

APIC (2017) Active pharmaceutical ingredients committee: how to do document: GDP for API. European Chemical Industry Council

Chikumba PA (2009) Application of geographic information system (GIS) in drug logistics management information system (LMIS) at district level in Malawi: opportunities and challenges. In: International conference on e-infrastructure and e-services for developing countries, e-infrastructures and e-services on developing countries, pp 105–115

Du X, Jiao J, Tseng MM (2005) Understanding customer satisfaction in product customization. Int J Adv Manuf Technol 31:396–406

Ergun O, Kuyzu G, Savelsbergh M (2007) Shipper collaboration. Comput Oper Res 34:1551–1560

Fugate B, Sahin F, Mentzer JT (2006) Supply chain management coordination mechanisms. J Bus Logist 27:129–162

Gouveia BG, Rijo P, Gonçalo TS, Reis CP (2015) Good manufacturing practices for medicinal products for human use. J Pharm Bioallied Sci 7:87–96

Meng SM (2014) Logistics image of outsourcing clients in the wireless telecommunications industry: how is it related to the service value of air cargo logistics provider? Int J Asian Soc Sci 4:940–955

Meng SM, Liang GS, Lin K, Chen SY (2010) Criteria for services of air cargo logistics providers: How do they relate to client satisfaction? J Air Transp Manag 16:284–286

Meng SM, Yang HY, Dai J (2019) International logistics criteria on medicinal products for human use. Int J Asian Soc Sci 9:148–168

Remko IVH (2000) The role of third-party logistics providers in mass customization. Int J Logist Manag 11:37–46

Roy C, Das JK, Jha HK, Bhattacharya V, Shivdasani JP, Nandan D (2009) Logistics and supply management system of drugs at different levels in Darbhanga District of Bihar. Indian J Public Health 53:147–150

Tomić S, Filipović SA, Ilić MA (2010) Good manufacturing practice: the role of local manufacturers and competent authorities. J Artic Arch Ind Hyg Toxicol 61:425–436

Tracey M (2006) The role of logistics in strategic management. Int J Integr Supply Manag 2:356–382

Human Factor in Industry 4.0—Perception of Competences of Graduates and Employees

Anna Stasiuk-Piekarska

Abstract The concept of Industry 4.0 is focused on the production of personalized products in time and cost which are approximate to mass production. The generated technical and organizational solutions force enterprises to change the role of their employees. He stops to be a contractor and becomes a supervisor of the self-initiating production process. The chapter presents the research about the perception of competences of employees and graduates by the enterprises in the context of the discussed strategy of Industry 4.0. This is a part of a wider research carried out using the CATI method on a group of 108 large industrial enterprises located in Poland. They allowed to find out not only how the idea of Industry 4.0 is understood, but also what is necessary to implement it. It gives a chance to formulate guidelines to which areas the attention should be paid to build the competences of future and current staff of industrial enterprises—including logistics—not only in Poland but also in the world.

Keywords Industry 4.0 · Employee's competences · Industrial enterprise

1 Introduction

Enterprises that make a decision to compete on the global market look for ways to reduce costs and acquire commissions to survive on the market. The resulting, inevitable changes are often referred to as industrial revolutions. Since 2013, we have been watching how the world is preparing and attempting to introduce the Fourth Industrial Revolution, called the Industry 4.0 (Dr. Wieselhuber & Partner GmbH, Fraunhofer-Institut für Produktionstechnik und Automatisierung IPA 2015). The concept was proposed by the German government as a direction in development and support for the competitive efforts of German enterprises. The new method of production forces the existing and potential employees, meaning staff educated by

A. Stasiuk-Piekarska (✉)
Faculty of Engineering Management, Poznan University of Technology, 60965 Poznan, Poland
e-mail: anna.stasiuk-piekarska@put.poznan.pl

© The Author(s), under exclusive license to Springer Nature Switzerland AG 2020

P. Golinska-Dawson et al. (eds.), *Smart and Sustainable Supply Chain and Logistics – Trends, Challenges, Methods and Best Practices*, EcoProduction,
https://doi.org/10.1007/978-3-030-61947-3_17

universities, to acquire and use new skills. The author elected to combine considerations on the competences of the existing and future employee, because the need to expand skills and undertake training on communication and the so-called soft competences exists for the both groups of employees. This creates a special education need, especially among the engineering staff, who must not only prove their technical skills and expertise, but also be willing to introduce changes and drive self-development. For higher education institutions, on the other hand, this represents an area for scientific research and where educational activities can be expanded and updated.

The Industry 4.0 concept aims to accelerate the emergence of smart factories, which should be based on network cooperation with the use of informational and communication technologies for the purpose of combining machinery, processes, systems, products, clients and suppliers (Bembenek 2017). An important role here is played by employees, who receive information at any time and from any place (Wyrwicka and Mrugalska 2017), allowing the production of personalized products, the so-called mass customization. This idea defines a broad spectrum of innovations in IT, production technology and material sciences that impact the implementation of the concept of Industry 4.0. Special attention was paid to challenges resulting from Spath et al. (2013):

- handling large amounts of data,
- the businesses' ability to introduce innovations,
- flexibility of production processes.

These issues may be further expanded with the complexity of enterprise management and ensuring data safety (Spath et al. 2013). As a result, new tasks must be carried out by employees, which, in turn, forces them to acquire different skills and competences.

The use of new technologies gives access to all working parameters in real time and allows analysing massive data bases. This makes it possible to use new steering and control methods with the application of mobile devices and technologies. Consequently, the merging of the real and virtual worlds is observed. An expansion of the traditional concept of the man-machine array (Bendkowski 2017).

2 Employees in Industry 4.0

When arranging the operations of an organisation, one should remember about the individuals tasked with handling it, which most often means the employees themselves. According to Bendkowski, two models of enterprises can be distinguished when analysing work organisation in a smart factory. The first model is based on structures with changing personnel composition that supports the process of learning on a job position and give the units operational flexibility and a high degree of autonomy. Their structure will be based on self-organising activity networks whose purpose is the completion of a specific task. Another organisational model will consist in an enterprise based on polarisation of qualifications, one that supports a far-reaching work division that is a combination of decentralisation and expansion of the scope of duties, while maintaining a strict division of work and standardisation. The creation of new structures and a different organisation of work may lead to the creation of problems at the intersection of humans, technologies and organisation and their relations to socio-technical systems. This constitutes a determinant for commencing analyses within the area of the impact of technology on shaping social reality. The result is the necessity to mutually adjust the technology and the requirements of organisational structures and work quality criteria (intellectual development, growth of the employees' competences and knowledge, among else) (Bendkowski 2017).

The ASTOR Reports (2017) on Industry 4.0 Engineers contains research that points to the perception of indispensable competences by employees. The necessity of training is noted, on the so-called "hard" skills related to technical competences (new automation systems, programming skills, SAP training), financial and organisational competences, cost accounts, work and production organisation, project management and self-management in time. Among the so-called "soft" competences, the respondents would most often point to human resources management, skills useful for contacting clients and negotiations, acquiring new clients, sharing and transferring competences, competences related to communication, creativity and leadership. At the same time, the report underscores that competences on which cyber security depends will be the basic competences for organisations operating in cybernetic systems (ASTOR 2017). The employees will cease to be executors, but will become work supervisors, so it seems important to combine the activities on the verge of "cybernetics" with, for example, working with robots or using IT and programming tools (2016). Many authors have noted that employees in companies based on the Industry 4.0 concept must not only have expert competences, but they should also be open to varying tasks and low levels of stability and task repetitiveness. They must be ready to process and analyse data from multiple sources, with the use of Big Data tools. This creates the need to know how to choose what is important to analyse a given problem, to evaluate the reliability of data and to draw conclusions (Piątek 2018) It is also noted that the competences of the future will have the client's needs in their centre. Among the analysed features the following are underscored (Szczerblewska 2017):

- client orientation,
- flexibility,
- critical thinking skills,
- openness to digitisation,
- readiness and openness to change,
- easy adaptation to new conditions and environment,
- team work skills,
- comprehensiveness and integrity,
- ability to work in varying cultural conditions.

Hecklau et al. also mention the ability to transfer knowledge, media skills, understanding IT security or tolerance of ambiguity (Hecklau et al. 2016).

The omnipresent computerisation will also lead to increased complexity of production processes. It is also assumed that the basic competences will include the ability to solve complex problems and to learn at work. Smart factories will also require employees who can collaborate with others in the pursuit of a common goal within the value creation process. A view present in literature posits that production workers will have to improve their skills in the area of flexible manufacturing (among else to acquire and improve digital competences in manufacturing, assembly and auxiliary areas related to preparing and planning production, quality management and logistics). They are also expected to have competences allowing to fulfil the process owner role—sponsoring and designing it, managing changes, introducing improvements. Employees should also strive to take the role of process managers, whose duties included planning and coordinating activities related to carrying out the process, monitoring it and reporting it. The need to have skills in counteracting any disruptions that may stem from the growing complexity of manufacturing processes is also seen (Bendkowski 2017).

Analyzing the above, it is worth noting that Polish universities educate more and more graduates—currently 27.2% of Poles have a university degree (the average in the EU is 28.7%). However, Polish graduates still lack the qualifications needed at work. Poland achieved a result of only 3.9 according to the Global Competitiveness Index scale (where 7 is the highest assessment of matching students' competences to their future work). The average for EU countries in this respect is 4.897. This may suggest that it is only at work that young Poles acquire useful competences. What may be noted here is the role of employers in educating staff based on the skills gained from the university. The increasing degree of automation causes an increase in demand for employees with high technological and digital skills. Higher cognitive skills and socio-emotional skills also seem important. Both the education sector and the companies themselves will have to help employees adapt to this demand. In this context, retraining strategies are important, including lifelong learning and formal employee training. In Poland, adults do not often continue their education. (McKinsey and Company 2019).

3 Evaluation of Competences of Graduates and Employees

Research methodology

The survey was performed using CATI method, on a sample of 108 respondents. They were performed as a part of the design works of financing the Youth Staff of the Faculty of Engineering Management of the Poznan University of Technology (No. 11/141/DSMK/0586). The survey was completed at the turn of November and December 2019. Large production enterprises (with over 250 employees) located in Poland were selected for the survey. The questionnaire featured open-ended and close-ended questions. The majority of respondents were employed as specialists,

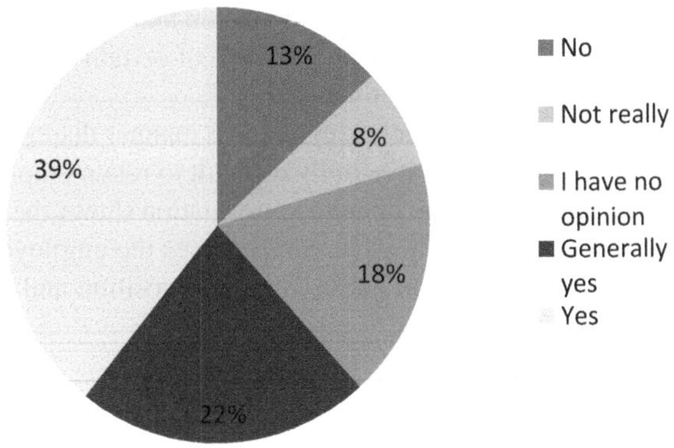

Fig. 1 Summary list of the survey participants' responses to the question: "Do the existing employees' competences allow the implementation of the Industry 4.0 idea?"

head managers or directors (more information: Stasiuk-Piekarska and Mrugalska 2020).

The presented results apply to viewing the competences of employees of the examined organisations and the respondents' evaluation of matching the students' competences for the purpose of implementing the Industry 4.0 strategy and they constitute a part of the conducted examination.

Competences of employees

The first questions aiming to evaluate viewing the possession of competences by individuals from the organisation's environment (employees or graduates in the labour market) was the request to evaluate the statement "Do the existing employees' competences allow the implementation of the Industry 4.0 idea?". To better illustrate the results, the responses were rated according to Likert scale (1–5), where: 1—no, 2—not really, 3—I have no opinion, 4—generally yes, 5—yes. The results are presented in Fig. 1.

The responses to the open-ended question "Do the existing employees' competences allow the implementation of the Industry 4.0 idea?" show that 19 in 108 respondents have no opinion in the examined area.

The responses of 24 survey participants (22% of the surveyed) were considered statements close to "generally yes". The respondents had doubts expressed by such statements as "the managerial and engineering staff yes, but not necessarily the older employee", "depending on the field" or declarations related to training and improving qualifications of their employees. The responses of 42 survey participants were qualified as the confirmation that the employees of the examined organisations have competences allowing the implementation of the Industry 4.0 idea. The respondents in this group used rather decisive language in their evaluation, with expressions such as "yes" and "definitely yes", assessing them as good.

14 respondents negatively evaluated the competences of the employees of their respective enterprises in relation to satisfying the requirements of the Industry 4.0 concept. 8 survey participants evaluated this area as rather unfitting to the requirements of the examined idea. Some of the surveyed noted the lack of financing for

training, however, some of them assured that their organisation would surely support such activities in the future. The lack of certain competences is also identified, but the specific skills are not named.

One of the respondents replied in a manner disregarding the topic, stating that at the time of reply it is generally difficult to retain employees.

The analysis of the obtained information shows the necessity to not only reiterate the request to identify what competences the employees miss, but also the need to analyse the correlation between the job position and the level of assessment of the employed staff.

Shaping the competences of future staff

Another examined issue that was supposed to examine the view of having competences by the individuals in the organisation's environment (employees or graduates in the labour market) was the request to assess the statement "In your opinion, do universities educate staff that supports the development of automation and production customisation?". Similarly to the analysis of responses to the previous question, to better illustrate the results, the responses were rated according to Likert scale (1–5), where: 1—no, 2—not really, 3—I have no opinion, 4—generally yes, 5—yes. The results are presented in Fig. 2.

To the question "In your opinion, do universities educate staff that supports the development of automation and production customisation?" 11 respondents (10% of the surveyed) had no opinion with regard to the examined issue, citing the lack of contact with graduates or the relatively small comparison group ("I don't know, I recently hired 1 person and got good results, but this is not enough").

10 respondents (10% of the surveyed) believes that universities generally prepare the graduates to support the activity of businesses in terms Industry 4.0. The used wording was related to the respondents' own experience ("I have graduates several years ago, but I think that [the situation] is improving every year and that the education gets better; the universities are developing and introducing practical classes") and references were made to collaboration with selected academic units. ("We are now seeing improvements in mutual relations and attempts to share skills, knowledge and growth"). The positive side is that as many as 51 of the survey participants

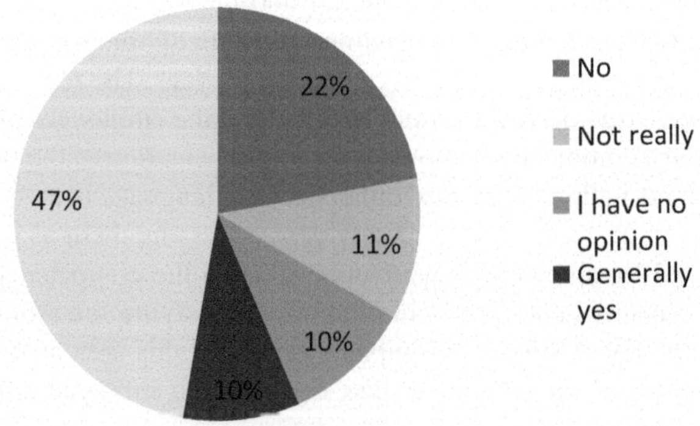

Fig. 2 Summary list of the survey participants' responses to the question: "In your opinion, do universities educate staff that supports the development of automation and production customisation?"

assessed the examined issue whether universities educate staff that supports the development of production automation and customisation—as "yes", "definitely yes" or "absolutely yes". Still, even these opinions came with reservations that there are few such universities or that the graduates lack practical knowledge and on-the-job practice.

24 of the 108 respondents underscored that universities do not or definitely do not educate their graduates adequately for the needs of Industry 4.0. This group also included respondents who used phrases such as "the university is visibly delayed in terms of the latest technologies" or "the knowledge is there, but there is motivation for work" among the graduates. The surveyed would also use phrases like "not enough and the availability of such personnel is inadequate". In relation to the examined issue, 12 respondents believe that there is generally no support for the education supporting Industry from the universities or that there is not enough of it. Moreover, one opinion stated that the graduates have high financial demands.

We may consider it a positive that the respondents make references to working with universities and building mutual win-win relations. On the other hand, we may consider it an alarming trend that graduates are assessed to have low motivation for work, but this may result from the generational gap and the entering of millennials into the labour market. In this regard, universities can support not only their own graduates, but also Gen Xers and Gen Yers, with the purpose to improve human collaboration. We should also not forget about the use of the newest technologies and teaching them to the students.

4 Conclusions

This chapter serves as an initial review of literature with regard to competences seen as necessary for implementing and functioning of companies based on the Industry 4.0 concept. Then, the chapter presents the results of examination of the assessment of the degree to which such competences are possessed by employees of large, industrial enterprises operating in Poland and by graduates of higher education institutions. Significant optimism can be seen among the respondents, as evidenced on the conducted examinations. Nearly 60% of the respondents assessed that the employees of their respective organisations have or generally have competences necessary to implement the concept of Industry 4.0. Nearly 70% of the surveyed evaluated positively or largely positively the competences of graduates in terms of supporting customised and automated production in their enterprises. We should remain mindful that there is no way to develop business without collaboration with and support from the external environment, also in the areas covered by the educational system.

The conducted examination constitutes a foray into the evaluation of the readiness of enterprises to introduce the Industry 4.0 concept. At the same time, we see the need to deepen the research, if only to address the remarks of the respondents on which competences they consider to be key in terms of developing business according to the analysed strategy.

Summing these considerations up, a conclusion can be made that the analysed area may constitute a certain group of determinants for the implementation of modern strategies, such as Industry 4.0. However, in order to be able to obtain highly customised products, it is not only necessary to declare the openness to collaborate with clients and suppliers, but also a more extensive preparedness with regards to production management that coves, among else, digitisation and automation of processes.

Acknowledgements Research financed by the Faculty of Engineering Management—Young Personnel Statutory Activities (11/141/DSMK/0586). Publication financed by the Faculty of Engineering Management—Statutory Activities (11/142//SBAD/1105).

References

ASTOR (2017) Inżynierowie Przemysłu 4.0 (Nie)gotowi do zmian?. https://www.astor.com.pl/images/Industry_4-0_Przemysl_4-0/ASTOR_Inzynierowie_4.0_whitepaper.pdf. Accessed 29 Feb 2020

Bembenek B (2017) Klastry Przemysłu 4.0 w zrównoważonej gospodarce opartej na wiedzy. In Prace naukowe Uniwersytetu Ekonomicznego we Wrocławiu nr 491. https://www.dbc.wroc.pl/Content/39466/PDF/Bembenek_Klastry_Przemyslu_4_0_w_Zrownowazonej_Gospodarce_2017.pdf. Accessed 01 Jan 2020

Bendkowski J (2017) Zmiany w pracy produkcyjnej w perspektywie koncepcji „Przemysł 4.0". In Zeszyty Naukowe Politechniki Śląskiej 2017, seria: Organizacja I Zarządzanie Z. 112 Nr Kol. 1990, pp 21–33. https://yadda.icm.edu.pl/baztech/element/bwmeta1.element.baztech-02e3754a-33a3-4262-8691-7b2d31ae63b3. Dostęp: 03 Mar 2020

Dr. Wieselhuber & Partner GmbH, Fraunhofer-Institut für Produktionstechnik und Automatisierung IPA (2015) Geschäftsmodell-Innovation durch Industrie 4.0 Chancen und Risiken für den Maschinen- und Anlagenbau. https://www.wieselhuber.de/migrate/attachments/Geschaeftsmodell_Industrie40-Studie_Wieselhuber.pdf. Dostęp: 03 Mar 2020

Hecklau F, Galeitzke M, Flachs S, Kohl H (2016) Holistic approach for human resource management in Industry 4.0. Procedia CIRP 54:1–6. 6th CLF-6th CIRP conference on learning factories. https://core.ac.uk/reader/81956090. Accessed 30 Apr 2020

McKinsley and Company, Polska 2030 (2019) Szansa na skok do gospodarczej ekstraklasy. https://www.mckinsey.com/pl/~/media/McKinsey/Locations/Europe%20and%20Middle%20East/Polska/Raporty/Polska%202030/Raport%20Polska%202030%20McKinsey%20Forbes.ashx. Accessed 30 Apr 2020

Piątek Z (2018) Jak rozwijać kompetencje Inżyniera 4.0. https://przemysl-40.pl/index.php/2018/04/11/jak-rozwijac-kompetencje-inzyniera-4-0/. Accessed 01 Mar 2020

Spath D (Hrsg) et al (2013) Fraunhofer-Institut Für Arbeitswirtschaft und Organisation IAO, Produktionsarbeit der Zukunft–Industrie 4.0. https://www.iao.fraunhofer.de/images/iao-news/produktionsarbeit-der-zukunft.pdf. Accessed 05 Mar 2020

Stasiuk-Piekarska AK, Mrugalska B (2020) Możliwości wdrożenia Przemysłu 4.0 w obszarze technologii w polskich przedsiębiorstwach przemysłowych. In: Knosala R (ed) Inżynieria zarządzania 2, Polskie Wydawnictwo Ekonomiczne, w druku

Szczerblewska A (2017) Pracownik 4.0, czyli jakich kompetencji wymagają ciągłe zmiany?. https://przemysl-40.pl/wp-content/uploads/2017/07/konferencja_MR_2017_4.pdf. Accessed 03 Mar 2020

Wyrwicka MK, Mrugalska B (2017) "Industry 4.0"-towards opportunities and challenges of implementation. In: DEStech transactions on engineering and technology research, 24th international conference on production research (ICPR 2017), pp 382–387. https://www.dpi-proceedings.com/index.php/dtetr/article/view/17640/17146. Accessed 15 Dec 2019

Management of the Human Capital of the Persons with Disabilities in Logistics Organizations

Krzysztof Czyrka and Józef Frąś

Abstract For many years the low level of awareness on the subject of the meaning of the social and economic human capital of the disabled persons in many economies has led to annihilation of the actions in the area of its exploration, development and using it with the result which is negative not only for the disabled persons themselves, but also budget of these countries whose discreditable instance is Poland. The first goal of the article is to indicate the essence and importance of managing human capital of disabled people. The second goal of the article is to point out recommendations (based on research) for better management of people with disabilities in logistics companies. The study was applied logit models as a tool reinforcing the decisive processes of management of the human capital of the persons with disabilities. One shall claim that except for the fact of bringing to life the satisfactory array of tools of support of employment of the disabled persons, in the result of the lack of the properly conducted informative policy there takes place a considerable limitation of their role and function.

Keywords Management · Logistics companies · Labor market · Disabled people

1 Introduction

In the light of contemporary challenges of the demographic and economic areas of Europe a significant meaning commences to have the notion of the human potential of the people with disabilities. The use of intellectual capital of the disabled is not only benefits for the disabled themselves but also for the labor market and the state budget. The logistics industry market seems particularly interesting because it is characterized by rapid technological progress and large staffing needs.

K. Czyrka
The Jacob of Paradies University, Gorzow, Poland
e-mail: kczyrka@interia.pl

J. Frąś (✉)
Poznan University of Technology, J. Rychlewskiego 2, 60-965 Poznan, Poland
e-mail: jozef.fras@put.poznan.pl

© The Author(s), under exclusive license to Springer Nature Switzerland AG 2020
P. Golinska-Dawson et al. (eds.), *Smart and Sustainable Supply Chain and Logistics – Trends, Challenges, Methods and Best Practices*, EcoProduction,
https://doi.org/10.1007/978-3-030-61947-3_18

2 Methodology

Within the framework of the conducted research the quality methods were used, in addition to the quantity ones. Within the quality research the desk research was implemented and what is more, individual extended interviews. However, in the framework of the quantity methods, the questionnaire was used.

Stage 1: Desk research

This stage encompassed the theoretical analysis of the scrutinized notions and review of the literature on the topic of the state and situation within the framework of benefiting from the human potential of the disabled persons especially in the context of the logistics companies. The analysis was conducted with the use of the available publications and enclosed the following issues: theoretical studies on the operationalization of substantial terms connected with management of the human capital of the persons with disabilities on the logistics companies, synthetic studies on the implementation of methods and tools of so called 'good practice' in the range of management of the human capital of the persons with disabilities in enterprises, studies on the available statistic data, expertize and reports concerning the state and size of employment of the disabled persons in Poland, studies on the available statistic data, expertize and reports concerning the state and size of employment of the disabled persons in the EU, the research on materials and desk documents directed on the verification of barriers in the range of benefiting from the potential of disability in enterprises of the logistics companies.

Stage 2: Individual in-depth interviews

Due to a wide research area—the territory of the whole Poland the individual in-depth interviews (IDI) were conducted in the chosen enterprises of the lubuskie and zachodniopomorskie voivodships. The goal of the conversation was to receive as much of pivot information from the respondents on the topic of the issue of management of human capital of the disabled persons as plausible. The interviews were conducted among the decision-makers of the researched entities as well as the experts connecting the theoretical knowledge with practice. In this way there were diagnosed core resources of information in the scope of the conducted research.

Stage 3: The questionnaire

Realization of the accepted research method in the third stage required the construction of the research tool—a questionnaire. In order to check the correctness of the provided method of the research and the tools used there was conducted a pilot research which was supposed to verify the correctness of the construction of the research tool—the questionnaire and the way of collecting the data and the implementation of the possible corrections. The pilot research encompassed 16 randomly chosen firms from the list of the companies subjected to the core survey. Having checked the results of the pilot research and taking into account the remarks of the explored enterprises concerning the survey, the author undertook the core research.

The correlated questions in the questionnaire were subjected to the issues stemming from the realization of an average research goal and the verification of the research thesis. The questionnaire concerned the quantity and quality traits, encompassed close-ended questions in which a wide range of statements were cited and the respondent was asked to mark their answer according to a scale consisting of a few points and which was also a bipolar one, ranging from a full approval to a total disapproval.

3 Results/Methods Testing/Model Validation

The effectiveness of benefiting from the human potential of the persons with disabilities is explored and scrutinized most often in reference to the macro economic factors. On the basis of the elaborate study touching upon literature of the subject and synthetic exploration presented in the first and the second chapter of the very dissertation one may state that truly significant macroeconomic factors which constitute barriers and problems in using the human capital of the disabled persons are: unstable and complex legal regulations in the range of employment of the disabled persons, unstable and complex conditions of financial support, low efficiency and purposefulness of expenditures of the supporting funds in the range of promoting the employment of the disabled persons, generally low professional activity of the disabled persons as a group and individuals on the logistics companies. It is confirmed by the results of the research conducted by the author of the following thesis within the framework of the project 'Management of the human potential of the disabled persons in the organizations on the example of the Protected Labor Plant of the Western Poland' in 2011 (https://docplayer.pl/3206536-Co-to-jest-sodir-system-obslugi-dofinansowan-i-refundacji.html 2019). However, apart from the undoubtful core role of the barriers mentioned above and being the obstacles of management of the human potential of the disabled persons, one rarely indicates here internal determinants (lying in conditions of a company) of a low effectiveness of benefiting from the human potential of the people with disabilities. That is why the problem of benefiting from the human capital of the disabled persons undoubtedly does not lay solely in the sphere of the macro economic factors but also in the great extent depends on the readiness of enterprises of the logistic companies to employ the disabled persons and is apprehensive about: openness to employ the persons with disabilities (directing oneself with stereotypes and prejudices), possessing the decent knowledge allowing to benefit from the available tools of the financial support, readiness to conduct technological and organizational changes if the need occurs, possessing the essential knowledge and skills to implement methods and tools of managing the human capital, when taking into account the persons with disabilities.

Bearing in mind the fact that it is well worth either in this perspective to have a look at the received results of the research. The basic factor which conditions the readiness to employ the persons with disabilities is the inclusion in the strategy of gaining and management of the human capital of the firm the policy of employment of the disabled persons. It is exclusively about forecasting the possibilities of employment of the persons with disabilities, but practically taking into account this probability in the processes of management of the human resources, so in:

- procedures of recruitment, selection and nominating,
- policy of professional development and trainings,
- policy of appraisal,
- policy of motivation,
- interpersonal relations of organization.

Openness in the prism of problematics of employment of the persons with disabilities understood as breaking the stereotypes, prejudices and constraints connected with them at the potential employers is the primary and fundamental means, however cannot be treated as a sole category and what is even more, the ultimate one. Moreover, one needs to mark that openness in itself in the sphere of practical management does not imply anything. Without taking into account practical spheres as the results impose, there occur much trouble in the area of management of the disabled workers and in the extreme cases—the conflict with the law and its all consequences. That is why it is vital to make the strategy enclose the national policy and law making system in the area of employment of the disabled people and the programs of support where they exist. What needs to be emphasized is the fact that the effort and responsibility for securing these areas are on the hands of employers themselves. In this scope one shall have a look at the gained research result. The analysis showed that the lack of encompassing the policy of employment of the disabled persons in the strategy of gaining and managing the human capital took place in 65% of the instances of the studied companies. The question if the strategy of gaining and managing the human potential is ingrained into the policy of employment of the disabled people—42% of the respondents answered 'rather no' and 23% 'no'. At the same time the gained results prove that employers of the logistic companies in the great majority are characterized by the lack of readiness to employ the disabled people. In the prism of the results received there, it is indispensable to find out and indicate reasons of such course of action.

1. The issue of knowledge on the theme of tools of support in the array of employment of the persons with disabilities and also conditions which one is supposed to meet to get them

One is required to underscore the dependence which exists between the degree of readiness to employ the disabled persons and the level of the knowledge on the theme of tools of support in the range of employment of the persons with disabilities. It means that (in a reversed form) if an enterprise possesses such kind of the knowledge, then the readiness to employ the persons with disabilities is higher. Moreover the power of dependability between the variables indicated above is prominent, which is shown by the calculated coefficient of convergence of Czuprow's to 0.234. For convergence a remarkable power and direction of the relation between the knowledge on tools of support in the area of employment of the persons with disabilities and the extent of readiness to employ the persons with disabilities is unequivocally showed by the attained result of the correlation coefficient of Spearman's ranks which was 0.2672.

2. The problem of procedures in the range of receiving funds of support of employment for the persons with disabilities

As it is visible in the results of exploration, the major group of the firms employing the disabled people and possessing the entitlement to benefit from SoiD (https://docplayer.pl/3206536-Co-to-jest-sodir-system-obslugi-dofinansowan-i-refundacji.html 2019) did not take advantage of endowments to salaries or other subsidies in the result of the lack of the knowledge on the topic of procedures of their receiving. When the entrepreneurs who employed the persons with disabilities in 33% (summed up answer 'no' and 'rather no') did not have this data, the ones who did not employ responded so in 89% (60% answered 'no' and 29% 'rather no' so as to being knowledgeable).

3. The problem of the knowledge on the subject of legal privileges of the disabled persons

In the light of the possibility of benefiting from the range of the tools of support, employers are required to meet the defined duties. Among the other there are foremost the duty to abide legal privileges of the disabled workers which is connected with possessing the knowledge on them. Here the result of the study proved that the employers who do not provide jobs to the disabled persons in the great majority were not accustomed with the array of privileges for a disabled person as a worker. For the question: Does your company possess the knowledge concerning the legal privileges of the disabled persons as workers?—the employers in 37% indicated: 'rather no' and in 27% 'no'. A reverse tendency is present among the employers who do employ the persons with disabilities, who in 95% (the answer 'yes' got 80% and 'rather yes' 15%), but on the other hand what is troublesome is that 5% of the firms did not or rather not have such knowledge. One needs to mark that is the indispensable realm and one cannot diminish its prominence in relation to the benefits for employers, since first of all, meeting these requirements entitles to receive subsidies and moreover in extreme cases in the result of their lack of fulfillment may expose the company to serious sanctions. The study revealed that apart from the noneffective promotion and informative policy from the side of state institutions, the acquaintance of privileges of the disabled persons lays in the interest of the companies which have decided to implement into the strategy of management of the human capital employing the persons with disabilities.

4. The issue of promotion of employment of the disabled persons by public institutions

The efficiency of promotion of employment of the disabled persons by public institutions (which should take place among the others by indicating the economic benefits for the companies), was either by the employers providing jobs to the disabled persons or the ones not doing so assessed very strictly. In the eyes of 80% of the entrepreneurs not providing jobs to the disabled persons, the promotion of employment of these people was seen as not sufficient and according to 17% rather not sufficient. Even more strictly the entrepreneurs assessed the question of promotion of the human potential of the persons with divisibilities in the time of realization of the research. According to 93% of the investigated, the promotion of employment of the disabled people was regarded as insufficient (85% of the researched responded that it is insufficient and 17% that it is rather insufficient). In the prism of the appraisal received in this way, one awaits a brisk improvement in the range of the informative and promotional policy.

5. Problems of cooperation with non-government organizations in the range of employment of the people with disabilities

Taking into account the problems depicted above, a great solution may turn out to be the cooperation of the enterprises of the logistics companies with non-government organizations which often possess the essential and legal knowledge and also experience plus they can render skilled aid. Today, such organizations (due to benefiting from the EU funds) dispose very often of the array of not only essential support but also so as to financing apprenticeships, courses, trainings etc. However, the research has shown that that kind of help can be found among the enterprises employing the disabled persons (57%), but in reference to the ones which do not perform so, such a cooperation takes place in a minor scope (only 3% declared such cooperation). Moreover, the strength of dependence between the indicated variables is moderate, which is proved by the coefficient of convergence by Czuprow which amounted to 0.335. The moderate dependence of the power and its direction of the relation between the fact whether the company cooperates with some kind of organization of non-government origins which promotes the employment of the DP and the extent of readiness to provide jobs to them is demonstrated by the attained result of the correlation coefficient of Spearman's ranks which was 0.40453.

6. The issue of adjusting the architecture and the workplaces to the needs of a person with disability

To many issues in the area of benefiting from the human potential in the quantity and quality aspect in the firms of the logistic companies one needs to enclose the physical barriers connected with architecture of buildings and workplaces. The problematic is truly vital in the prism of the persons which are affected by disabilities in the motor organs, sight and hearing ones. In Poland there is approximately 3 million of the disabled persons with the impairment of the motor system, over 1300 thousand with the sight impairment and close to 700 thousand with the impairment of hearing (Czyrka 2013). In the structural picture for the production age, which is vital from the point of view of the labor market, the most common ailment in Poland is the one touching the motor organs 48.2% and then: cardiovascular system 32.9%, neurological damage 32.7%, sight impairment 20%, psychical ailments 14.6%, hearing impairment 6.2% and mental handicap 7.2% (GUS 2018). These numbers indicate a great role and at the same time a great responsibility or the people who decide on the shape of the architecture of the physical milieu and not only in companies but in all possible spheres of life of a person.

4 Conclusions

The conducted extrapolation proves that the extent of readiness of enterprises to employ the DPs is a factor which is decisive for benefiting from the usage of the human capital of the people with disabilities on logistics companies while the determinants of its amount are the factors of the internal and external origins.

As the research has shown, the primary and fundamental means in the aspect of using the human potential of the persons with disabilities of the logistics companies—is supporting and creating the approach of openness, understood as the process of breaking stereotypes, prejudices and apprehension connected with that, either speaking of the employers or the disabled themselves. One is required also to ponder on the implementation of a social campaign from the means of PFRON's on the central level, the so called media of the national range, with the usage of not only the form of standard commercials but also by the method of building up openness to employ the persons with disabilities via the so called' locating of the message' in serials, films and frameworks of a considerable audience. All the more so because of the mission of the State Fund for Rehabilitation of Disabled People (PFRON's) is to create conditions facilitating full participation in professional and social life for Persons with Disabilities. A good complementary solution seems to be the inclusion in this area the actions on the local area via booking and sharing funds for organizing conferences, courses and other initiatives which promote openness for employment of the disabled persons.

The study has proved that in reference to the readiness of enterprises to employ the disabled persons (a decisive factor) the issue of breaking stereotypes apart from its significance cannot be treated as the only, ultimate one since in the sphere of practical management it does not contribute anything. That is why the readiness of enterprises to profit from the human capital of the disabled persons shall be regarded not only in the categories of the attitude of openness for employment of the DPs but also in the practical consideration of the reaffirmed possibility:

- in the national policy and law making in the range of employment of the disabled persons especially in the programs of aid where they exist,
- in processes of management of human resources (recruitment and selection, the policy of professional development and trainings, the policy of appraisal, the policy of motivation, relations of interpersonal organizations).

In this context the process of the research reveled that:

Among the employers of the Among the employers of the logistics companies (even employing the disabled persons) there occurs a great shortage of the knowledge on the subject of the available tools of support for providing posts to the persons with disabilities as well a conditions of their receiving. (even employing the disabled persons) there occurs a great shortage of the knowledge on the subject of the available tools of support for providing posts to the persons with disabilities as well a conditions of their receiving. One shall claim that except for the fact of bringing to life the satisfactory array of tools of support of employment of the disabled persons, in the result of the lack of the properly conducted informative policy there takes place a considerable limitation of their role and function. First of all the encouraging function (the category of defined financial support for an employer). What needs to be emphasized is the matter that the statistically conducted proof process revealed that this factor has a huge meaning for the extent of readiness of enterprises to provide jobs to the DPs.

Among the employers of the Among the employers of the logistics companies (even employing the disabled persons) there occurs a great shortage of the knowledge on the subject of the available tools of support for providing posts to the persons with disabilities as well a conditions of their receiving. (even employing the disabled persons) there dominates the lack of acquaintance of the procedures in the range of acquiring aid funds for employment of the disabled people which constituted a barrier which hindered benefiting from these means. The first crucial cause of that fact created impediments in receiving the comprehensive and current knowledge in this scope. The second, fears of the complex process of procedures in the context of: the need to devote additional time and generating costs connected with that and also apprehension of extra audits in connection with profiting from the forms of aid. One shall mention that the core of this factor in the context of readiness of companies in order to provide posts to the DPs confirmed in the entire scope the statistic calculations. Bearing in mind the dimension of the influence of both of the factors above (the knowledge on the subject of the available tools of support and also conditions in order to receive them and the knowledge on the acquaintance of procedures in the area of aid means) on the readiness of enterprises of the logistics companies to profit from the human potential of the persons with disabilities and concurrently the fact of a breakthrough moment in the domain of law making regulating the question of using the aid scheme which is balancing the amount of endowments, starting from the 2nd quarter of 2014, to salaries on the open and the protected labor market, one shall recommend as a priority to run the campaign of promotion and information roots within the scope of employment of the disabled persons in media of the national range, underscoring the benefits for employers stemming from providing jobs to the people with disabilities which should simultaneously constitute a redirection of the formerly activated on-line central platform of employment of the disabled persons enclosing (providing):

- the collection of the indispensable date in the prism of employment of workers who are disabled, updated on a current basis and written in a simple language;
- the opportunity to conduct consulting which will be free of charge in the range of benefiting from the tools of support for employment of the disabled persons and procedures connected with that, including clearances and reporting in the traditional, on-line and call-in.

References

https://docplayer.pl/3206536-Co-to-jest-sodir-system-obslugi-dofinansowan-i-refundacji.html. Accessed 31 Oct 2019

According to the data of the Bureau of the Plenipotentiary of the Government towards the Affairs of the Disabled Persons in the study of the the Health of the People, GUS 2018

Commission Communication: "Europe 2020" (2010) A strategy for smart, sustainable and inclusive growth, European Commission, Brussels 3.3.2010, KOM, 2020. https://ec.europa.eu/eu2020/pdf/1_PL_ACT_part1_v1.pdf. Accessed 20 June 2019

Co-financing for the remuneration of disabled employees (2013) Proposed changes in 2014, Warsaw 2013. https://www.senat.gov.pl/gfx/senat/userfiles/_public/k8/komisje/2013/kbfp/opinie/127_20.pdf. Accessed 14 May 2018

Czyrka K (2013) Zarządzanie potencjałem ludzkim osób niepełnosprawnych w organizacjach, dysertacje doktorskie, Wyd. PWSZ w Gorzowie Wielkopolskim, Gorzów Wlkp, pp 1–307

European Health Institute Study (EHIS), GUS 2009

GUS BAEL–average annual data for 2014 for disabled people of economic working age 18–59/64. https://www.niepelnosprawni.gov.pl/index.php?c=page&id=80&print. Accessed 20 2019

Quarterly information on the labor market (2013) Monitoring of the labor market, Labor Department of GUS, Warsaw, 29 May 2013

Steczkowski J (1995) Metoda reprezentacyjna w badaniach zjawisk ekonomiczno-społecznych. Wydawnictwo naukowe PWN, Warszawa-Kraków, p 45

References

[illegible]

Application of the MAC Method for Risk Assessment During Handling of Loads

Adam Górny

Abstract Manual handling is a major category of occupational operations that entail a range of specific risks and other onerous consequences. These make it necessary to assess the implications of strains and consider the available improvement options. The scope and nature of such measures should ensure compliance with existing laws and be consistent with relevant assessment findings. The chapter describes an approach to such an assessment relying on the use of the Manual Handling Assessment Charts (MAC). The MAC method allows a good match with the specific nature of the risks and strains that arise in the course of manual load handling. The case of the job of the warehouse attendant has been used to assess strenuousness, identify areas in which strains exceed levels tolerable by worker bodies and suggest appropriate improvements. The use of a sequence of actions that is recommended in MAC guidelines has helped gather and compile the information necessary to identify the necessary modifications in the work environment and workflows.

Keywords Manual handling · Risk assessment · The MAC method · Improvement of work environment

1 Introduction

A substantial portion of the work performed in warehouses involves manual load handling. Such work often entails enormous risks resulting predominantly from muscoskeletal overexertion (Batish and Singh 2008; Bhattacharya 2014; Roman-Liu 2014; Bevan 2015). Some of the most common causes of strains include incorrect postures that are usually restrained during work and the handling of loads whose weight exceeds workers' capabilities. The result are excessive strains (Bhattacharya 2014; Bevan 2015) as well as excessively repetitive tasks (Enez and Nalbantoğlu 2019).

A. Górny (✉)
Faculty of Engineering Management, Poznan University of Technology, Institute of Safety and Quality Engineering, Prof. Rychlewski St. No 2 (room 357), 60-965 Poznan, Poland
e-mail: adam.gorny@put.poznan.pl

© The Author(s), under exclusive license to Springer Nature Switzerland AG 2020
P. Golinska-Dawson et al. (eds.), *Smart and Sustainable Supply Chain and Logistics – Trends, Challenges, Methods and Best Practices*, EcoProduction,
https://doi.org/10.1007/978-3-030-61947-3_19

Table 1 Characteristics of working conditions in transport activities in Poland in 2018 (Working conditions in 2018 2019)

Business activity section	Persons exposed to risk factors			Employed in conditions of excessive physical exertion	
	Total covered by survey	Employees with high risk			
		Total	Share among those covered	Total	Per 1,000 employees
Transportation and storage	577,669	29,342	5.1	11,744	20,3
Warehousing and support activities for transportation	137,119	13,839	10.1	7,736	53.5

The significance of the problem is indicated by the number of employees working in the conditions of hazards associated with carrying out transport. A summary of these data is presented in Table 1. Particular attention should be note to the section: Warehousing and support activities for transportation. This section has a 10.1% share of work performed during which significant (usually unacceptable) risk was diagnosed and 53.5% share of employees with excessive physical strain. When analyzing the data provided, it should be noted that this only covers the estimated part of employees covered by the diagnosis.

An additional problem is the large number of illness and accident events caused by onerousness and hazards, and consequently the benefits paid out because of it. In Poland in 2018 was over 260,000 cases, which was an infamous third position in the business activity section after manufacturing and mining.

The causes of the events are equally important. According to the reasons indicated by the Statistics Poland, the most common causes are (Working conditions in 2018 2019):

- The moved object is too heavy (too big weight), too large, bulky or difficult to maintain according with the verdict of the person employees managing,
- The object is unstable or its contents may move,
- The object is positioned so that it requires keeping or operating at a distance from the employee's torso or tends to bend or wrap around the employee's torso,
- The shape or structure of the object may cause injury to the employee, especially in the event of a collision,
- Moving the object can be done only by turning the torso,
- Performing the work requires tilting the employee's torso by more than 45° or performing unstable operations,
- The work place or its surroundings make impossible to move the object at a height ensuring safety or with the correct body position of the employees,
- The surface on which employees move is uneven, creates a danger when moving or is slippery in contact with the footing of employee footwear,

- The floor and work surface have different levels, which forces you to move objects at different heights,
- The floor or foot surface is unstable,
- The object limits the employee's field of view,
- Environmental conditions, in particular temperature, humidity and ventilation are not adapted to the nature of the work performed.

The above makes it necessary to assess risks, determine whether specific work tasks should be allowed and identify improvements necessary to eliminate risks or mitigate their effects. The need for risk elimination and mitigation is set forth in applicable laws. It is also recognized by employers who seek to minimize the strains that result from failures to ensure safe working conditions. The primary aim of the assessment is to contribute to the identification of solutions that will reduce the impact of existing risks on workers. The proposed technical and organizational solutions reflect the nature of non-conformities and the severity of their impacts on workers.

The occurrence of strains during manual handling and their effects alone does not oblige employers to employ countermeasures to mitigate their adverse consequences, which include accidents and diseases arising during the performance of work operations. The responsibility to take the measures follows from general safety criteria as well as manual handling requirements (Sado et al. 2019). Before proper improvement measures can be selected, an analysis of the nature of risks is in order that will allow one to identify and apply the most appropriate technical and organizational solutions (Górny 2017).

Both general manual handling guidelines and specific requirements regarding the modalities in which handling operations should be performed are laid down in the law. The fundamental instrument of the European law in this area, the so called framework directive that underpins more detailed recommendations, is Council Directive 89/391/EEC. More detailed modalities and methods of manual handling are provided for in Directive 90/269/EEC, which supports the framework directive. Together, the two directives define the responsibilities of employers and employees which, if adhered to, will ensure safety, lower the probability of the occurrence of risks and eliminate the risks that may potentially cause accidents (Bhattacharya 2014; Bevan 2015). What the directives do not describe are the specific ways to assess risks. It is therefore up to the employer to select the most appropriate assessment method that is best suited to the nature of the irregularities occurring at a particular workplace.

2 Assessment of Risks Associated with Manual Handling

General Guidelines for the Assessment of Existing Risks

The primary responsibility of employers is to keep their workers safe and appropriately deploy practicable safety solutions (Caroly et al. 2010). Such improvements need to reflect the circumstances at hand and be taken with their actual impact on safety in mind. The choice of tools and measures applied to improve working conditions should be based on an assessment of risks and options for modifying workstations and work tasks that are available in view of the current state of the art in technology. To that end, due account should be taken of:

- The nature of the strains in question, including the adequacy of the loads handled vis-à-vis worker capabilities, the possibility of holding loads in a firm and secure manner, the ways in which loads are gripped and handled and workers' proneness to injury during load handling,
- The acceptable level of physical exertion during task performance,
- Work environment specifications such as the room available to workers, surface types, the stability of foot support and physical parameters,
- The conditions in which work is performed such as the frequency and durations of spinal strains, the forms of rest, the available recovery periods, the pace of work, the extent of control over the pace of work given to workers, and the demand to handle excessive loads and/or carry them over excessive distances.

An essential aspect of risk assessment is the physical fitness of workers relative to the tasks they are expected to perform, the use of proper protective clothing and footwear and the completion of training in the performance of the required operations.

A workstation-specific risk assessment should be designed to offer answers to questions concerning the capability to complete relevant tasks (Bhattacharya 2014; Bevan 2015; Goode et al. 2019) such as:

- What risks occur at the workstation?
- Who is exposed to such risks?
- What types of injuries may result from the risks?
- What is the probability of the occurrence of circumstances that lead to risks in the work environment?
- What parts of the work environment need to be modified to reduce risks?
- What protections can be put in place to prevent the occurrence of circumstances that result in risks?
- What are the chances of injuries (accidents) being triggered by the circumstances that result in risks?
- How to protect people (workers) in circumstances that result in risks and in circumstances in which exposures to risks are unavoidable?

To be able to undertake measures aimed at eliminating irregularities, employers must design their workplaces to standards that will enable them to identify conditions that promote comfortable postures that are unlikely to result in excessive loads or strains (Pinder and Frost 2011; Salas et al. 2016).

How to Apply the MAC

The Manual Handling Assessment Charts (MAC) is a method used to examine and assess the risks that occur during manual handling. The tool is designed for identifying the factors that generate the biggest risks during the lifting, carrying and team handling of loads. The method is not used to assess pushing and pulling operations.

It is not a replacement for a full occupational risk assessment of jobs (Monnington et al. 2003; Pinder 2003; Pinder and Frost 2011).

The MAC requires users to go through a sequence of stages that help gather the information necessary for the assessment and identify the required changes. Such assessment stages are (the MAC tool):

- STAGE 1: Assessment of working conditions at workplace,
- STAGE 2: Consultations with workers and their representatives,
- STAGE 3: Ascertaining whether the work method and workflows in place are common for a given job,
- STAGE 4: Selection of an appropriate type of assessment of manual handling operations (lifting, carrying, team handling),
- STAGE 5: Completion of the score sheet,
- STAGE 6: Analysis of results,
- STAGE 7: Proposal of improvements in working conditions and of changes in the ways loads are handled.

The risk assessment involves assigning indicators and risk levels to assessment criteria, as per Table 2. Such risk levels will additionally provide an indication of how urgently improvements are needed.

The assignment of a particular operation to a specific color band helps identify the operations that belong to a specific task and that require particular attention. Where tasks involve multiple operations, total task scores help prioritize actions.

An assessment of lifting and carrying operations follows the guidelines provided in Tables 3 and 4 as relevant for each operation. Such an assessment relies on the identified risk levels with their assigned numerical scores of risk impact.

Once obtained, assessment scores are entered into the collective MAC score sheet. It is recommended that the tasks associated with particularly high risk levels be

Table 2. Risk levels (own work based on Monnington et al. 2003; Pinder and Frost 2011; The MAC tool 2020)

Color band	Risk level and risk description
G - green	Low risk Where needed, consider the impact of risks on vulnerable groups of workers (pregnant women, youngsters, etc.)
A - amber	Medium risk Closely examine the way tasks are performed
R - red	High risk Prompt action needed. A significant portion of the working population may be exposed to risk of injury
P - purple	Very high risk Operations may represent a serious risk of injury and should be examined closely, especially where the entire weight of a load is handled by a single individual

Table 3. Criteria applicable to the assessment of lifting operations (own work based on Monnington et al. 2003; Pinder and Frost 2011; The MAC tool 2020)

Assessment criterion	Description of assessment criterion	
A - Load weight / lifting frequency	Determines the number of lifts per hour and load weight in kg.	
	G/0	
	A/4	Dependent on the nature of the work
	R/6	
	P/10	50 kg or more
B - Hand distance from lower back	Determines the horizontal distance between the hands and lower back. It is recommended to assume the worst-case scenario	
	G/0	Close: upper arms upright, torso straight
	A/3	Medium: the upper part of the shoulders diagonally or torso forward
	R/6	Far: the upper part of the shoulders is skewed, and the trunk is tilted forward
C - Vertical lift zones	Determines the vertical position of the hands of the worker involved in load lifting. It is recommended to assume the worst-case scenario	
	G/0	Above the knees and/or above the elbow
	A/1	Below the knees and/or above the elbow
	R/3	At floor level or below and/or above the head
D - Torso twisting / sideways bending	Determines the position of worker's torso relative to hips and thighs	
	G/0	Torsion / lateral tilt small or none
	A/1	Torsion or lateral tilt
	R/2	Torsion and lateral tilt
E - Postural constraints	Determines constraints on worker's movements.	
	G/0	Uncomfortable
	A/1	Comfortable
	R/3	Very uncomfortable
F - Grip on the load	Determines load gripping comfort	
	G/0	Good: containers with well-designed, matching handles or holder
	A/1	Sufficient: container with incorrect handles or holders, fingers must be clamped at a 90-degree angle under the container
	R/2	Poor: poorly designed containers, loose parts, irregular objects, bulky or difficult to grasp
G - Floor surface condition	Determines floor surface condition in the load handling area.	
	G/0	Dry in good condition
	A/1	Dry in poor condition or uneven
	R/2	Dirty, wet, uneven or unstable
H - Other environmental factors	Determines environmental factors. Particular attention should be paid to air temperature, lighting and strong air movements.	
	G/0	No risk factors
	A/1	One risk factor
	R/2	Two risk factors

Table 4. Criteria applicable to the assessment of carrying operations (own work based on Monnington et al. 2003; Pinder and Frost 2011; The MAC tool 2020)

Assessment criterion	Description of assessment criterion	
A - Load weight/ lifting frequency	Notes the weight of the load carried by worker and carry frequency. The assessment is based on the relationship between load weight and carry frequency.	
	G/0	Dependent on the nature of the work
	A/4	
	R/6	
	P/10	50 kg or more
B - Hand distance from lower back	Determines the horizontal distance between the hands and the lower back of the worker carrying loads.	
	G/0	Close: upper arms upright, torso straight
	A/3	Medium: the upper part of the shoulders diagonally or torso forward
	R/6	Far: the upper part of the shoulders is skewed, and the trunk is tilted forward
C - Asymmetrical torso or load	Determines the position of the load relative to the torso.	
	G/0	Load symmetrically, in the front, in both hands
	A/1	Load asymmetrically or sideways, in two hands
	R/2	Two-handed side-load transfer
D - Postural constraints	Determines freedom to carry loads seen as the absence of restrictions on movement.	
	G/0	Uncomfortable
	A/1	Comfortable
	R/3	Very uncomfortable
E - Grip on the load	Determines load gripping comfort and the possibility of the load falling out of the worker's hands.	
	G/0	Good: containers with well-designed, matching handles or holders
	A/1	Sufficient: container with incorrect handles or holders, fingers must be clamped at a 90-degree angle under the container
	R/2	Poor: poorly designed containers, loose parts, irregular objects, bulky or difficult to grasp
F - Floor surface condition	Notes floor surface condition (dry, uneven, worn, debris, damp, inclined, unstable).	
	G/0	Dry in good condition
	A/1	Dry in poor condition or uneven
	R/3	Dirty, wet, uneven or unstable
G - Other environmental factors	Determines environmental factors, the most noteworthy of which are air temperature, lighting and strong air movements.	
	G/0	No risk factors
	A/1	One risk factor
	R/2	Two risk factors
H - Carry distance	Notes the distance to which the load is carried. The greater the distance, the greater the risk associated with the carry distance.	
	G/0	2 m—4 m
	A/1	4 m—10 m
	R/3	10 m or more
I - Obstacles on route	Notes obstacles along the carrying route.	
	G/0	No risk factors
	A/1	One risk factor
	R/2	Two risk factors

indicated. For added score sheet clarity, the letters used should reflect the color associated with identified risk levels.

3 An Example of Using the MAC Method in Workplace Risk Assessment

Risk Factor Identification

The manual handling strains were assessed for the case of a warehouse attendant in a wholesale electrical supplies establishment. The company's core business is the wholesale and retail sale of cables, cords and other fitting items.

The warehouse attendant works on a three-person team in a two-shift system. His job description includes typical operations involved in the receiving of goods and their preparation for shipping as well as their placement on shelves and in a storage yard. In performing his work, the warehouse attendant uses technical tools that are adequate for its nature.

Some of the tasks performed by the warehouse attendant require manual load handling. The work is done in keeping with the available job manuals and using knowledge learned in his training. Observations have shown that when handling materials, the worker frequently assumes wrong body postures that place excessive strains on his muscoskeletal system.

During risk factor identification, NIOSH checklists were used to identify ergonomic risk factors in jobs characterized by a high likelihood of developing muscoskeletal disorders and locomotor diseases (Harward 2004; Górny 2020). Two checklists were used that are helpful in assessing strains in warehouse attendants:

- List 5A: General checklist of ergonomic risk factors,
- List 5F: Checklist for manual load handling.

An assessment based on affirmative answers relied on direct observations at workstation and interviews with the warehouse attendant. Table 5 provides sample questions and answers that help identify the risk factors associated with manual handling operations.

Risk Assessment

For the assessment to be appropriate, it is essential to demonstrate in advance that no counter indications exist to the performance of specific work.

The individual assessed is a 32-year-old man who has been employed as a warehouse attendant for the last 4 years. The worker is in good health and very fit. His work involves carrying and lifting loads. The nature of the operations he conducts is described and the relevant assessment scores are provided in Table 6. The score sheet additionally offers reasons why specific scores were assigned to individual risk levels. The worker is not involved in any team handling.

Table 5 Risks arising during manual handling

Question asked to determine the presence of risks or strains	Does the risk/strain exist?
List 5A, Part A.1: Manual load handling	
A.1.1 Does the worker lift tools, parts and/or other loads?	✓
A.1.2 Does the worker lower tools, parts and/or other loads?	✓
A.1.4 Does the worker bend torso when handling tools, parts and/or other loads?	✓
List 5A, Part A.2: Energy expended	
A.2.1 Do the tools and parts weigh more than 5 kg?	✓
A.2.4 Is the lifting or lowering of loads the worker's primary occupational task?	✓
A.2.5 Is walking and/or carrying of loads by the worker a key work factor?	✓
A.2.7 Is pushing and/or pulling of loads by the worker a key work factor?	✓
List 5F	
F.2 Are materials carried over minimal distances?	✓[a]
F.3 Is the distance between worker body and the item handled minimal?	✓[a]
F.4.3 Are the routes and surfaces along and on which workers move dry and clean?	✓[a]
F.6 Are the objects handled fitted with handles?	✓[a]
F.9 Is there sufficient room for maneuvering?	✓[a]
F.11 Are work surfaces adjustable to ensure their height is a best fits the type of manual work being performed?	✓[a]
F.12.1 Can the worker avoid handling loads below the height of hand knuckles and above shoulder height?	✓[a]
F.12.4 Can the worker avoid torso twisting when handling loads?	✓[a]
F.13 Is there help available to workers lifting heavy loads and/or assuming unnatural postures?	✓[a]
F.14.1 Are risks associated with highly repetitive tasks eliminated by e.g. rotating workers?	✓[a]
F.15 Is the need to use force in pulling or pushing minimized or eliminated?	✓[a]
F.17 Is there an injury prevention program in place that helps workers use auxiliary equipment?	✓[a]

[a]Negative answers point to the presence of the risk factor

An observation of the performance of the work performed by the warehouse attendant revealed a number of irregularities reflected in the final assessment score. These pertain in particular to factors designated with the letter R (red). Verification (i.e. further observation, in-depth examination) is also required for operations marked with the letter A (amber). The numerical score should be seen as a general assessment

Table 6. MAC score sheet for the job of warehouse attendant

Task description:

Carrying: The warehouse attendant manually carries cable coils from the winding site in the warehouse to pallets on racks. The coils are approximately 50 to 70 cm in diameter and weigh approximately 20 kg each. The workbench top on which the cables are placed is 105 cm high. The bench is 7 m away from the rack on which the coils are placed. Each carry operation last ca. 30 seconds. An average of two such operation are performed every hour.

Lifting: The warehouse attendant manually moves boxes containing electrotechnical materials with a pallet truck from receiving to appropriate storage locations. During the operation, he lifts the boxes from the pallet truck and places them on a rack shelf at shoulder height. An average of fifteen lifting operation are performed by the worker every hour.

Are there indications to suggest that the task at hand is a very high risk? (check if applicable):

☐ accidents have occurred during load handling

☐ the task is known to usually require substantial force or to be a very high risk

☐ the workers performing the task display signs of fatigue (accelerated heavy breathing, redness of face, sweating)

☐ other indications, if any ....

Factor assessed	Color band (G, A, R or P) and score		Reason assessment is needed
	Lifting	Carrying	
Load weight/frequency of lifting/carrying	G/0	A/4	The average weight of the coil carried is 20 kg, the carrying operation is performed twice every hour. The loads handled weigh approximately 5 kg and are lifted an average of 15 times per hour.
Distance between arms and lower back	A/3	A/3	The upper arms of the warehouse attendant are angled away from the torso. During the lifting of boxes, the worker's upper arms are angled away from the torso.
Vertical lifting zone	A/1	–	Boxes are placed on shelves at shoulder height, i.e. above the worker's elbows.
Torso twisting/ sideways bending. Load and torso asymmetry (during lifting)	A/1	R/2	The warehouse attendant carries coils with one hand by holding them on one side of his torso; the picking up and putting down of boxes requires torso twists.
Postural constraints	G/0	A/1	The carrying takes place in a highly confined space. During the lifting, the worker has much freedom of movement.
Factor assessed	Color band (G, A, R or P) and score		Reason assessment is needed
	Lifting	Carrying	
Load grip	A/1	R/2	Coil dimensions prevent a comfortable grip. When lifting boxes, the worker has no means of gripping the load firmly.
Floor surface condition	R/2	R/2	The warehouse floor is wet and covered with debris.
Distance carried	–	A/1	The carrying distance is 7 m.
Obstacles on route (concerns carrying only)	–	A/2	The warehouse is disorderly, unsecured power cords used for machines (e.g. winder) are strewn on the floor.
Other environmental factors	G/0	G/0	There are no other environmental factors that would reduce work comfort.
Score	8	17	

that cannot be associated with any specific tasks. In the case in question, one can assume that particular risks to worker health arise during carrying operations.

The assessment results advised the determination of the scope of technical and organizational solutions to be recommended to adequately address the existing problems and their causes.

Improvement Measures

Workstation observations and an occupational risk assessment helped identify improvements affecting work methods.

The suggested technical and organizational solutions reflect the nature of risks and strains and the severity of their impacts on workers. Workstations were designed to best fit the postures assumed by workers during the work process and the forces exerted during manual handling. It is also crucial to account for static loads resulting from the need to maintain a restrained body posture, the highly repetitive nature of the work and the substantial amounts of energy expended in performing operations.

The recommended technical solutions designed to mitigate risks in the performance of tasks are as follows:

- Improvements in the manner in which cable coils are carried from the winding site to storage racks—the improvements rely on the use of containers with properly designed handles to carry cable coils in both hands in front of the torso; one way to do away with carrying the coils altogether is to move them on manual pallet trucks or in wheeled containers,
- The use of stands for storing cable spool axles horizontally and eliminating the need to carry them from the storage site to the cable winder,
- Improvements in the way boxes are lifted from manual pallet trucks and placed on storage shelves at shoulder height. The related risks can be mitigated by using work platforms, which are particularly helpful for shorter workers who are otherwise forced to collect items from shelves located above their shoulder height and place items on them,
- Improvement in the way boxes are lifted from manual pallet trucks and placed on warehouse shelves at the height of more than 6 cm above warehouse floor; strap handles on carboard boxes will help eliminate the existing risks associated with the performance of such work.

The organizational solutions are designed to modify the work environment to best fit worker needs and raise worker awareness of the existing risks. The suggested organizational measures involve:

- Changing the location of the work bench used for winding cable coils, thus shortening the distance over which workers carry the load,
- Placing a fence around cable coil storage to separate it from warehouse travel routes,
- Changing the storage location used for fixing axles to provide workers with freedom of movement during their handling,

- Designating a storage zone as close to receiving as possible to reduce worker exposure to untoward external environment factors,
- Setting aside a location for the storage of heavy and rarely used items on the lowest rack shelves while keeping the most frequently used items between knee and elbow height,
- Making work breaks longer or establishing an additional break,
- Providing the workers with footwear with non-slip soles, steel toes and cushioned heels,
- Ensuring that the weight of the loads handled reflects worker capabilities.

Regardless of the above, it is essential to make safety improvements. These include regular checks of warehouse floors for cleanliness, as well as worker training and physical examinations of workers with a view to preventing muscoskeletal disorders.

Quite commonly, even tasks viewed as light may result in excessive strains on the muscular system. Such strains typically result from incorrect working technique (Pandya and Desai 2019; Chiasson et al. 2015). Hence, even minor modifications of the method and course of work and in the work environment itself may be well-advised (Sadłowska-Wrzesińska et al. 2016).

Additional benefits can be derived from strict adherence to technical standards. Technical documents may be referred to for information on the latest technical advances and best organizational practices in workplace design to ensure proper protection of worker health and safety.

The working conditions in place must facilitate injury-free performance. It is crucial to enshrine the principles of work performance in in-house rules and ensure such principles are well respected.

4 Conclusions

As manual handling is associated with the risk of injury, employers are often exposed to significant financial losses. The nature and severity of such risks have compelled the legislator to adopt laws that offer guidelines on ways to assess whether specific work can be performed and that identify the inherent risks.

The use of the MAC method for risk assessment helps identify the nature of problems and the improvement potential. The solutions suggested in the follow-up to such assessments aim to eliminate excessive strains on workers, ensure effective improvements and raise efficiency. The end result is safer manual handling and mitigation of the adverse financial consequences of incidents.

References

Batish A, Singh TP (2008) MHAC-an assessment tool for analysing manual material handling tasks. Int J Occup Safety Ergon 14(2):223–235

Bevan S (2015) Economic impact of musculoskeletal disorders (MSDs) on work in Europe. Best Pract Res Clin Rheumatol 29(3):356–373

Bhattacharya A (2014) Costs of occupational musculoskeletal disorders (MSDs) in the United States. Int J Ind Ergon 44(3):448–454

Caroly S, Coutarel F, Landry A, Mary-Cheray I (2010) Sustainable MSD prevention: management for continuous improvement between prevention and production. Ergonomic intervention in two assembly line companies. Appl Ergon 41(4):591–599

Chiasson ME, Imbeau D, Major J, Aubry K, Delisle A (2015) Influence of musculoskeletal pain on workers' ergonomic risk-factor assessments. Appl Ergon 49:1–7

Council Directive 89/391/EEC of 12 June 1989 on the introduction of measures to encourage improvements in the safety and health of workers at work; OJ L 183, 29.6.1989, pp 1–8, as amended

Council Directive 90/269/EEC of 29 May 1990 on the minimum health and safety requirements for the manual handling of loads where there is a risk particularly of back injury to workers (fourth individual Directive within the meaning of Article 16(1) of Directive 89/391/EEC); OJ L 156, 21.6.1990, pp 9–13, as amended

Enez K, Nalbantoğlu SS (2019) Comparison of ergonomic risk assessment outputs from OWAS and REBA in forestry timber harvesting. Int J Ind Ergon 70:51–57

Goode N, Newnam S, Salmon PM (2019) Musculoskeletal disorders in the workplace: development of a systems thinking-based prototype classification scheme to better understand the risks. Saf Sci 120:146–156

Górny A (2017) The use of working environment factors as criteria in assessing the capacity to carry out processes. In: MATEC web of conferences, vol 94, p 06011

Górny A (2020) The ergonomics of work conditions as force element of the OHS management. In: Murata A, Goossens RHM (eds) Advances in social and occupational ergonomics, advances in intelligent systems and computing, vol 970. Springer Nature Switzerland, pp 184–194

Harward J (2004) NIOSH workers health chartbook. In; NIOSH workers health chartbook. NIOSH Publishers, Washington, DC, pp 58–192

Manual handling assessment charts (the MAC tool), Health and Safety Executive, https://www.hse.gov.uk. Accessed 20 Feb 2020

Monnington SC, Quarrie CJ, Pinder ARJ, Morris LA (2003) Development of manual handling assessment charts (MAC) for health and safety inspectors. In; McCabe PT (ed) Contemporary ergonomics, taylor & francis, London, New York, pp 2–7

Pandya MG, Desai DA (2019) Systematic review on various risk assessment techniques of musculoskeletal disorder. Ind Eng J 12(6):1–7

Pinder ADJ, Frost G (2011) Prospective evaluation of the Manual handling Assessment Charts. Occup Environ Med 68(Suppl. 1):A73–A73

Pinder ADJ (2003) Benchmarking of health and safety inspectors' manual handling assessment charts (MAC). In: McCabe PT (ed) Contemporary ergonomics, taylor & francis, London, New York, pp 14–19

Roman-Liu D (2014) Comparison of concepts in easy-to-use methods for MSD risk assessment. Appl Ergon 45(3):420–427

Sadłowska-Wrzesińska J, Górny A, Mościcka-Teske A (2016) The outcomes of shift working the context of psychosocial functioning-sex aspects. In: Arezes PM et al (eds) Occupational safety and hygiene IV. CRC Press, Taylor and Francis Group, Boca Raton, pp 197–203

Sado F, Yap HJ, Ghazilla RAR, Ahmad N (2019) Design and control of a wearable lower-body exoskeleton for squatting and walking assistance in manual handling works. Mechatronics 63:102272

Salas EA, Vi P, Reider VL, Moore AE (2016) Factors affecting the risk of developing lower back musculoskeletal disorders (MSDs) in experienced and inexperienced rodworkers. Appl Ergon 52:62–68

Working conditions in 2018 (2019), Statistics Poland, Statistical Office in Gdańsk, Warszawa, Gdańsk

Modeling for Human Resources Management by Data Mining, Analytics and Artificial Intelligence in the Logistics Departments

Magdalena Graczyk-Kucharska, Maciej Szafrański, Selma Gütmen, Marek Goliński, Małgorzata Spychała, Gerhard-Wilhelm Weber, Zbigniew Włodarczak, Semih Kuter, and Ayşe Özmen

Abstract As a result of the environment changes, all Logistic Departments need to cover gaps between their needs and possibilities in regard to resources potential available in market and customer requirements. In addition to material re-sources, there are also intangible resources like knowledge, attitude and skills (part of HR). A pioneering mathematic-supported study was done with respect to technical skills of students from IT Departments at technical high schools. We achieve a mathematical model representing possible contributions of students into jobs through professional skills, subject to soft skills, common skills and other socio-economic variables in time. A general aim is to explore the effects between variables, the structure, stability and sensitivity of the model. Thus, the needs in Logistics are addressed through decision aid, educational improvements, programs and measurements. We take a genuine lead to networking and modelling side of HRM by modern Data Mining, Analytics and AI. Herewith, human and educational factors are addressed in Logistics, eventually for a best balance between job offers and demands. The resulting models are compared by the help of statistical performance criteria, they are discussed, interpreted, evaluated, and economic as well as educational implications are derived.

Keywords Professional skills · Labor market · Supply chain management · Logistics · HRM · MARS · ANN

M. Graczyk-Kucharska (✉) · M. Szafrański · S. Gütmen · M. Goliński · M. Spychała · G.-W. Weber · Z. Włodarczak · A. Özmen
Poznan University of Technology, Poznań ul. Jacka Rychlewskiego 2, 60-965 Poznań, Polska
e-mail: magdalena.graczyk-kucharska@put.poznan.pl

S. Kuter
Çankırı Karatekin University, Cankiri, Turkey

© The Author(s), under exclusive license to Springer Nature Switzerland AG 2020
P. Golinska-Dawson et al. (eds.), *Smart and Sustainable Supply Chain and Logistics – Trends, Challenges, Methods and Best Practices*, EcoProduction,
https://doi.org/10.1007/978-3-030-61947-3_20

1 Introduction

Resources encompass everything that adds value to a product or service in the process of design, manufacture, delivery or use. Usually, one speaks of simple or complex resources, physical resources, financial resources, personal resources, information resources, tangible and intangible resources and internal and external resources in reference to a specific entity. Given that the environment shapes ways of functioning on the market, decision-makers should take advantage of that impact, capitalizing on existing opportunities (potential, resources). Resources which are a source of competitive advantage can be characterized in the operational sense by dividing them into:

- tangible resources,
- intangible resources,
- capabilities—which enable effective use of organizational resources to generate value (Prahalad and Hamel 1990; Hall 1993; Hamel and Prahalad 1994; Mintzberg et al. 1998; Eisenhardt and Martin 2000).

Competence is often portrayed as a combination of components, such as knowledge, skills, abilities, capabilities and resources (Derwik and Hellström 2017) and in the times of knowledge management (source) it is the key value which supports competitive advantage building. In scientific literature, more and more emphasis is being put on competencies in the area of logistics, including supply chain management (Derwik et al. 2016), transport (Bazaras et al. 2016) or Logistic Enterprises on Service Innovation (Yang and Lu 2017). In the age of the globalization of the economy, the modern logistics industry has increasingly greater potential for fast and better development. In order to ensure such action, it is necessary to access resources, including skills that ensure and support planning, organization, realization and control of the company's processes in order to meet its organizational objectives. Although the issue is touched upon in many scientific articles, it is still not exhausted (Derwik et al. 2016), particularly in the context of mathematical modelling of the availability of resources in the context of competence needs in the area of logistics on the labour market.

It is the purpose of the article to present a mathematical model representing possible contributions of students into jobs through professional skills, subject to soft skills, common skills and other socio-economic variables in time. A general aim is to explore the effects between variables, the structure, stability and sensitivity of the model. The first part of the article presents a synthetic literature review in reference to competencies in the logistics sector. It showcases research methods, i.e., the Multivariate Adaptive Regression Splines (MARS) and Artificial Neural Networks. The research part is characterized by a description of the data structure as well as the presentation and comparison of the models. The last part focuses on conclusions and outlook.

2 The Importance of Competencies in Logistics

The dynamic of changes in the economy is very intense, with it being particularly visible in the area of logistics where far-fetched market transformations have appeared in the practice of logistics and supply chain management (Derwik et al. 2016). High competitiveness, complexity and uncertainty in the area of logistics forced companies to intensify active cooperation between human resources involved in the supply chain (Mendoza-Fong et al. 2020). The analysis of employer expectations concluded unambiguously that given such change dynamics, attention must be drawn to current employee competencies as well as the expected competencies as that guarantees success in the future (Palšaitis et al. 2017). Based on literature analysis in the area of logistics (in any part of the world) both technical as well as social competencies and a theoretical background are needed and practice-based experience is required (Sangka et al. 2019; Lawrence et al. 2019; Takele 2019; Szafrański et al. 2019).

Regardless of the logistics specialization and the employment level, the scope of the most important competencies also shows much repeatability in the analysed literature sources. The key competencies which are sought in the area of logistics most often include "ability to manage under the pressure of work", "crisis management skills", "interpersonal skills", "ability to use sources of information", and "ability to effectively manage and motivate teamwork" (Spychała et al. 2017; Wróbel-Lachowska et al. 2017; Ding et al. 2019).

The issues of demand for human resources in logistics and ways of looking for proper employees for the company are of key importance to the companies' success. The matters connected with high-quality talent and employee attraction are closely connected with HR management and due to the significant demand for employees in logistics and its global reach, recruitment processes are more and more often based on the Internet and social networks (Kiessling and Harvey 2014; Spychała et al. 2019).

The specifics of the logistics sector include the significant importance of professional experience as a precondition and an employability factor. It shows that logistics professions require a certain set of skills, therefore they demand trained and educated personnel. That confirms the expectations of professional logistics associations that predict demand for highly skilled staff in their sector. Job offers in logistics are available to people on all education levels, combine competencies from different sectors, careers can develop in a very wide area and there are more and more opportunities for the employment of women (Rudy 2012; Pearson 2015).

Human resources are an important element in the rivalry of businesses, they are correlated and equally important as infrastructure and legal regulations (Melliana et al. 2019). Particularly in the logistics sector, one can see big influence of choices of proper employee competencies on the effectiveness and efficiency of activities and thus competitiveness of the company, which is particularly visible in the era of Industry 4.0 (Melliana et al. 2019; Graczyk-Kucharska et al. 2018; Goliński and Miądowicz 2019).

Competence acquisition in the logistics sector is a continuous process. With respect to high-school technical professional education and university education, literature indicates that apart from acquiring basic skills and competencies, students need to be opened to continuing training at the workplace. Schools and universities should develop close cooperation with logistics companies and offer postgraduate studies or courses in order to continuously train logistic personnel to adjust to the changing needs of the sector (Kotzab et al. 2018). When the employability of school and university graduates is a key factor that affects the development of the entire economy, professional improvement should be subject to continuous development. Many educational solutions are today based on remote internet training and simulations (Butin 2006; Saifudin et al. 2015). The use of simulation methods has been popularized for years particularly for promising sectors in which teaching rare and innovative competencies might involve significant costs (Thai 2012).

3 Methodology

3.1 Artificial Neural Network (ANN)

Inspired by the biological nervous system, artificial neural network technology is being used to solve a wide variety of complex scientific, engineering, and business problems (Haykin and Network 2004). An ANN makes it possible to create complex non-linear models relatively easily, by "*learning*" from the presented examples (Fausett 1994). A neural network consists of a large number of simple processing elements that are variously called neurons or nodes. Each neuron is connected to other neurons by means of direct communication links, each with an associated weight. The weights represent information being used by the net to solve a problem. The neural network usually has two or more layers of neurons in order to process non-linear signals (Haykin and Network 2004). A neuron receives multiple inputs from different sources, and has a single output. There are various functions used for activation (Vanneschi and Castelli 2019).

One of the problems that may occur during neural network training is called overfitting. The error on the training set is driven to a very small value, but when new data is presented to the network the error is large. One of the frequently used methods for improving network generalization is to use an adequate-size network, which is just large enough to provide an adequate fit. The larger a network is, the more complex are the functions that the network can create which may cause overfitting. If we use a small enough network, it will not have enough power to over fit the data (Demuth and Beale 1998). It is difficult to know beforehand how large a network should be for a specific application (Fletcher and Goss 1993).

3.2 Multivariate Adaptive Regression Splines (MARS)

Multivariate Adaptive Regression Splines (MARS) is an innovative and flexible modelling tool that automates the building of accurate predictive models for continuous and binary dependent variables. It is a non-parametric regression method in which no assumption is made regarding the functional relationship between dependent and independent variables. Instead, MARS builds this relationship from a set of coefficients and basic functions, which in turn are heavily influenced by the regression of the data. The operating method involves partitioning the area of entry into regions, each with its own regression equation (Hill and Lewicki 2006). The MARS method creates a regression equation for each region. MARS regression model is defined as

$$Y = c_0 + \sum\nolimits_{n=1}^{N} c_n B_n(\boldsymbol{X}) + \varepsilon \tag{1}$$

where

B_n: n-th basis function from the set of basic functions,
c_n: coefficient of n-th basis function,
c_0: constant "absolute" coefficient,
ε: random noise term (Friedman 1991).

Its non-parametric nature, its transparency in terms of the relevant variables, and its adaptability to the data give MARS great potential as a multi-spectral classifier (Molina et al. 1994).

The basic advantages of the MARS method compared to other methods is the fact that its use does not require any specific and restrictive special assumptions about the functional connections between dependent and independent variables. An important feature of MARS is the nature of the variables it operates on—they can have a nonparametric form and a multidimensional range. When developing a database for calculations at the input to the analysis, no list of assumptions about the type of dependencies between independent and dependent variables is required. Thanks to these MARS features, the method is very useful in applying to inferences using large amounts of various data.

MARS method is often interchangeably used with methods using an Artificial Neural Network, imitating a biological neural network (Merolla et al. 2014). Both techniques are successfully employed in social research and technical sciences because as accurate tools they allow the identification of many, even very small dependencies, and ultimately the development of an over-fit model.

4 Data Preparation

The analysis was based on the data collected in the project *Time of Professionals—Wielkopolskie professional education*. The collected data come from the years 2013–2015 and concern a comparison of skills sought on the labour market with skills held by potential job candidates. Skills of every student were not compared with the skills required in every offer; instead, the data were linked with one another if the student was interested in a given offer.

The other data selection criterion, apart from time, was the limitation of the area of interest to skills connected with the profession of the logistics technician. The skills were attributed according to the curriculum for teaching the profession of the logistics technician which was in effect at the time. In the description of the profession, the curriculum is divided into three categories: professional (Z), common for the educational area (O) and common for all the professions (W). Every individual skill which is an element of a logistics employee's skills is attributed to exactly one of the aforementioned categories. The categories were selected as the aggregating criterion as part of which indicators $X_O, X_Z, Y_W, X_{2O}, X_{2Z}, Y_{2W}$ were computed.

Due to the method chosen for the analyses of the dependencies, data which contained incomplete information about the student regarding their gender, date of birth, district of residence and about the offer regarding the work commencement date and the district where the offer maker is seated were additionally removed from the set of data which met the aforementioned criteria of selection.

As a consequence, the authors arrived at 1436 records containing *student_id* and *offer_id* identifier pairs together with the properties and indicators which describe them. The data used in the analysis were aggregated in the table composed of 21 data columns including, e.g., gender, date of profile creation, common skills, general skills.

Due to the requirements of the method chosen for the data analysis, all source data were recoded to the numerical format according to the following rule:

- data type: a conversion involved transforming the data into the number of days which passed from 1999-12-31 to the specified date,
- logical type: converted from 0–1 to 1–2,
- text type: a numerical identifier was chosen for every value.

5 Models

The main aim is to obtain a model which shows the relations between dependent variable and independent variables. Firstly, as pre-processing part, the collected data has been arranged and predictor and response variables have been decided (cf. Sect. 3). Later on, the dataset of 1436 real-life observations was divided into 2 groups by random sampling method as %70 of the whole dataset for training with 1035 cases, and %30 for testing with 401 cases. Herewith, while MARS will be applied as a fundamental model, ANN will be particularly employed to compare with the MARS results. In this work, results will be shown only for the train dataset. All results including the test performance, will be presented during further works.

5.1 MARS Model

In MARS methodology, in order to get an optimal model, we started with determining a maximum number of basis functions, a number of interactions, a penalty and a threshold value in statistical software Statistica 13.3. At the beginning of MARS' forward step, maximum number of BFs, M_{max}, has been preassigned as 90, while the maximal degree of interactions is 2. After pruning in the MARS' backward stage, the number of BFs got down to 54, and the number of terms has been decided by MARS as 33. We concluded both MARS' forward and backward steps by addressing *Generalized Cross-Validation* (GCV) as introduced in Sect. 3. In consequence, the best MARS model with its BFs is shown as follows:

$$
\begin{aligned}
y_Z = &-8.7247 - 7,7882 \cdot \max\{0, X_7 - 5.156\} - 7.9166 \cdot \max\{0, 5.156 - X_7\} + \\
&1.7764 \cdot \max\{0, X_6 - 5.157\} + 1.7933 \cdot \max\{0, 5.157 - X_6\} + 3.1182 \cdot \max\{0, X_4 - 5.280\} \\
&+ 6.2939 \cdot \max\{0, 5.280 - X_4\} - 1.3311 \cdot \max\{0, X_6 - 5.248\} \\
&+ 1.84 \cdot \max\{0, 5.28 - X_4\} \cdot \max\{0, X_O - 1.2\} - 1.6050 \cdot \max\{0, X_2 + 1.274\} \\
&+ 2.156 \cdot \max\{0, X_4 - 5.28\} \cdot \max\{0, X_5 - 5.622\} \\
&+ 1.191 \cdot \max\{0, 5.28 - X_4\} \cdot \max\{0, X_7 - 5.156\} \\
&- 1.7899 \cdot \max\{0, 5.28 - X_4\} \cdot \max\{0, X_7 - 5.191\} \\
&- 2.024 \cdot \max\{0, X_5 - 5.714\} \cdot \max\{0, X_6 - 5.248\} \\
&- 4.38 \cdot \max\{0, X_6 - 5.248\} \cdot \max\{0, 1.3 - X_W\} \\
&+ 1.166 \cdot \max\{0, X_2 + 1.52\} - 7.9212 \cdot \max\{0, -1.274 - X_2\} \cdot \max\{0, 5.172 - X_7\} \\
&+ 1.4416 \cdot \max\{0, -1.274 - X_2\} \cdot \max\{0, 5.539 - X_6\} \\
&+ 1.6691 \cdot \max\{0, -1.274 - X_2\} \cdot \max\{0, X_4 - 5.233\} \\
&- 5.5611 \cdot \max\{0, -1.274 - X_2\} \cdot \max\{0, 5.233 - X_4\} \\
&- 1.259 \cdot \max\{0, X_1 - 1\} \cdot \max\{0, X_6 - 5.157\} + 1.2833 \cdot \max\{0, X_1 - 1\} \cdot \max\{0, X_7 - 5.156\} \\
&- 7.39 \cdot \max\{0, X_{13} - 1\} - 1.615 \cdot \max\{0, 5.156 - X_7\} \cdot \max\{0, X_W - 6.6\} \\
&- 2.3787 \cdot \max\{0, 5.28 - X_4\} \cdot \max\{0, X_W - 9.4\} \\
&+ 1.1838 \cdot \max\{0, 5.28 - X_4\} \cdot \max\{0, X_W - 6.1\} \\
&- 2.583 \cdot \max\{0, X_5 - 5.568\} \cdot \max\{0, X_7 - 5.156\} \\
&+ 3.5333 \cdot \max\{0, X_3 - 5.516\} \cdot \max\{0, X_W - 0\} + \\
&4.4346 \cdot \max\{0, X_2 + 7.69\} \cdot \max\{0, 9.4 - X_W\} - 4.548 \cdot \max\{0, X_2 + 1.52\} \cdot \\
&\max\{0, X_W - 1.3\} + \varepsilon.
\end{aligned}
$$

In our MARS model, 10 variables seem important, or centrally involved statistically, among 15 input variables. These affecting variables are: student's gender (X_1), student's birthday (X_2), student's profile creation date (X_3), job offer visible from

(X_4), job offer visible to (X_5), job offer date of work start (X_6), job offer date of creation (X_7), whether job offer and student are in the same country (X_{13}), student's total evaluation value of common skills (X_O), and student's total evaluation value of general skills (X_W). Hence, it is astonishing that 6 of 10 predictors are related to *time* and in our MARS model we have all time variables. This shows that time is particularly important for our model and applications in human resource management of the logistics sector.

5.2 *Artificial Neural Network Design*

The ANN models were generated in MATLAB. A multilayer feedforward ANN with one hidden layer was employed in the study. The network had 15 nodes in the input layer and 1 node in the output layer, i.e., the number of nodes in the input and output layers were set equal to the number of predictor and response variables, respectively. Gradient-based Levenberg-Marquardt (LM) back-propagation was used as network training algorithm. Hyperbolic tangent was used as transfer function in both input and output layers. Gradient-based Levenberg-Marquardt (LM) back-propagation was selected as network training algorithm. The number of neurons in the hidden layer was decided through a trial-and-error approach by varying the number of neurons from the set $\{7, 10, 13, 16, 19, 22, 25, 27, 29, 31\}$. Related information about ANNs can be found in (Haykin and Network 2004; Fausett 1994).

Since an ANN comes from a computational family of methods, it shows the results just as an input-output automaton. In ANN methodology, because of the implicit and hidden of finding y, it does not provide an analytic understanding about which predictors play an affecting role, and with which sensitivities, as it is possible by our MARS model explicitly. Hence, for ANN techniques, we cannot assess and quantify the important variables and their contributions in our setting from logistics and education.

6 Comparison of the Models

For the comparison of the two methods' performances, the results obtained from the training dataset were evaluated in this work. Before comparing an optimal MARS model with best ANN system, we decided about our model and systems among other possible cases, respectively. In order to do that, we addressed the *first order* performance criteria: Absolute Average Error (AAE), Root Mean Square Error (RMSE), and then continue with checking the *second order* performance criteria: Multiple Coefficient of Determination (Adjusted R^2) and Correlation Coefficient (R). The comparison has shown accuracy and stability performance within the scope of statistical evaluation. Through these criteria, we measured the predictive capability of the models.

7 Professional Skills Training

Table 1 shows the comparison result of MARS with two selected best ANNs called ANN_1 and ANN_2. In the table, both ANN systems represent different numbers of nodes: 25 and 10, respectively. In comparison, MARS has different performances with respect to different ANNs. In the first case, while MARS does not perform well enough compare to ANN_1, in the second case, MARS works better than ANN_2. The two approaches show that MARS plays a very competitive role when comparing with ANN for the matched areas of Logistics and Education. Overall, in our study, MARS has proven to be equally well against its ANN counterpart.

In Fig. 1, the *histogram* demonstrates that residuals, as a sample of the noise terms in the model, accumulated around mean 0. Mostly, the squared residuals have

Table 1 Results of statistical performance criteria based on training dataset for MARS and ANN_1

	Professional skills training		
	MARS	ANN_1	ANN_2
AAE	17.7119	14.5708	19.2966
RMSE	25.8319	23.1142	29.1537
Adjusted R^2	0.6206	0.6836	0.5240
R	0.7910	0.8359	0.7345

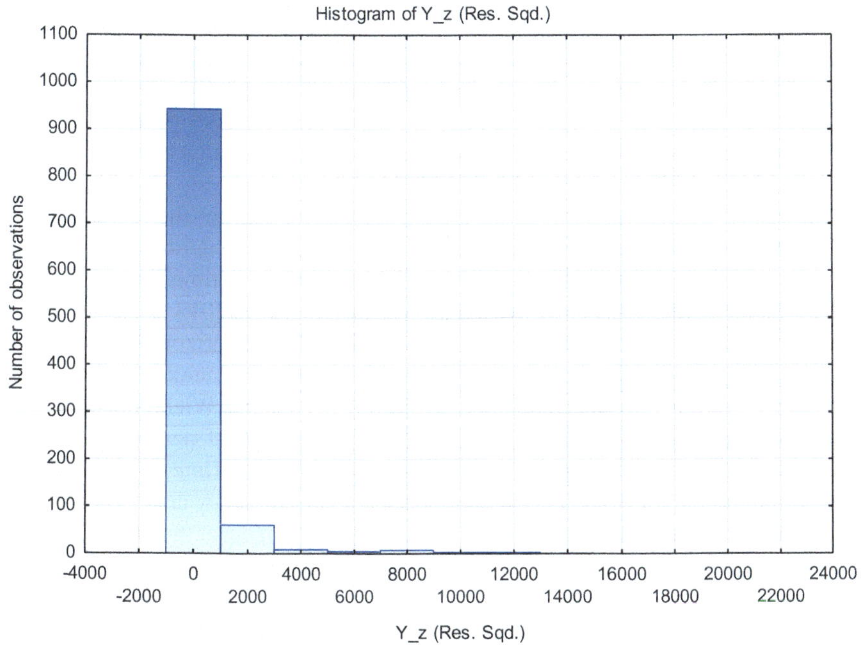

Fig. 1 The squared residuals related to our MARS model, based on training dataset

their value close to 0. (For closer statistical discussions we refer to (Aster et al. 2018). Our good performance criteria's results also imply the characteristic of our dataset in terms of complexity and accuracy properties in despite of our dataset is inhomogeneous and nonconvex. We also recall that MARS, differently from Linear model, does not need a particular distribution assumption.

8 Conclusion and Outlook

The application of MARS in comparison with ANN delivered extremely promising results with respect to the analysis and creation of decision-supporting models, taking into account significant variables in order to match professional competencies in the area of logistics with the needs of the labor market. The analyses made it possible to study the key variables out of the 21 analyzed ones, with as many as 6 out of 10 of them referring to time. The conclusions in that area are unambiguous. The works connected with the spatiotemporal analysis in the context of competencies need to be continued and improved. The aspect is important from the point of view of improving the competencies of candidates and adjustment to the needs of employers in the logistics sector. It might also be important to public education institutions which should flexibly respond to the needs of the labor market and the employers' demand for specific competencies in the era of continuous technological changes as well as social, ecological and health-related challenges.

In further studies, there are many challenges connected with the continuation of the works. Further activities should include spatiotemporal analyses, taking into account other macroeconomic factors such as unemployment, the gross domestic product, level of employment in the sector, number of logistics companies or their location. The location and availability of competency resources might also be significant in the context of mobility of human resources.

There are certain limitations to the study results. The data for the research were collected in the period of 2013–2015 and thus they might differ significantly from the current needs of the labor market. It is the case because, over that time, unemployment in the Wielkopolskie province decreased from 6.8% reported in 2013 to 3.1% in 2020. Therefore, the studies should also encompass the current time horizon and up-to-date indices and companies' demand for competency needs of young candidates. The other important limitation is the fact that the studies were carried out in Wielkopolska which is one of 16 provinces in Poland. Wielkopolska is a region of nearly 3.5 million inhabitants and, despite the intentional choice in the project *Time of Professional BIS—professional Wielkopolska*, the samples are so big that they cover nearly 28% of all students learning the profession of the logistics technician at the time. The third and the last limitation is the difficulty in acquiring data for analyses in the context of intangible resources, in this case, competencies. The analyzed data are based on self-evaluation of competencies and the level of their mastering. However, it is worth emphasizing that the analyzed competency profiles of candidates learning to be logistics employees were most often created under teacher supervision and thanks to being able to talk to the instructor, it was easier for the students to understand separate skills and evaluate them correctly.

The results of the studies confirm that it is worth continuing the works on statistical analyses possibilities in the context of human resources modelling on the labor market. The developed MARS model generates promising results regarding its use in further studies. Comparing the performance criteria, the MARS model provides comparable and, depending on the ANN model, also better results, a fact which makes us inclined to broaden the data for the analyzed factors and develop increasingly more accurate models describing the needs in the logistics sector. Further works connected with data and competence models on the labor market concern applying MARS (Kuter et al. 2018) and ANN, and further data mining methods: RMARS (Robust MARS) (Özmen and Weber 2014), CMARS (Conic MARS) (Kuter et al. 2015), RCMARS (Özmen et al. 2014), to represent and understand *P*, *O* and *G*, comparing all models and systems by statistical performance criteria, error diagrams and sensitivity analyses.

Acknowledgements We would like to thank the Regional Government of the Wielkopolskie Province—The Partner and Leading Partner in projects supporting the activities for the use of data for the purposes of improving education processes for the needs of a knowledge-based economy, including projects such as *The Wielkopolska System for Monitoring and Forecasting*, *Time of Professionals—Wielkopolska vocational education*, *Time of Professional BIS—professional Wielkopolska* which are funded from WRPO programs.

References

Aster R, Borchers B, Thurber C (2018) parameter estimation and inverse problems. 3rd Edn, Elsevier

Bazaras D, Čižiūnienė K, Palšaitis R, Kabashkin I (2016) Competence and capacity-building requirements in transport and logistics market. Transp Telecommun J 17(1):1–8

Butin DW (2006) The limits of service-learning in higher education. Rev Higher Educ 29(4):473–498

Demuth H, Beale M (1998) Neural network toolbox for using Mathlab—user's guide. MathWorks Inc, 24 Prime Park Way, Natick, MA

Derwik P, Hellström D, Karlsson S (2016) Manager competences in logistics and supply chain practice. J Busin Res 69(11):4820–4825

Derwik P, Hellström D (2017) Competence in supply chain management: a systematic review. Supply Chain Manag: Int J 22(2):200–218

Ding JF, Kuo JF, Tai WH (2019) Using fuzzy AHP method to evaluate key competency and capabilities of selecting middle managers for global shipping logistics service providers. Pomorstvo 33(1):3–10

Eisenhardt KM, Martin JA (2000) dynamic capabilities: what are they? Strategic Manag J 21:1105–1121

Fausett L (1994) Fundamentals of neural networks. Prentice-Hall, Englewood Cliffs, NJ

Fletcher D, Goss E (1993) Forecasting with Neural Networks: an application using bankruptcy data. Inf Manag 24:159–167

Friedman JH (1991) Multivariate adaptive regression splines. Ann Statist 19(1):1–67

Goliński M, Miądowicz M (2019) management of employee competencies in resource management of an enterprise. In: ECKM 2019 20th European Conference on Knowledge Management Vol 2. Academic Conferences and publishing limited, 405–414

Graczyk-Kucharska M, Szafranski M, GolinskiM, Spychala M, Borsekova K (2018) Model of competency management in the network of production enterprises in industry 4.0—Assumptions. In: Advances in manufacturing. Springer, Cham, 195–204

Hall PA (1993) Policy paradigms, social learning, and the state: the case of economic policymaking in Britain. Comparative politics, 275-296

Hamel G, Prahalad CK (1994) Competing for the future. Harvard Busin Rev 72(4):122–128

Haykin S, Network N (2004) A comprehensive foundation. Neural Netw 2:41

Hill T, Lewicki P (2006) Statistics: methods and applications. a comprehensive reference for science, Industry and Data Minining. StatSoft; Tulsa, OK, USA

Kiessling T, Harvey M (2014) Human resource management issues associated with the globalization of supply chain management and logistics. Int J Phys Distrib Logist Manag 44(8/9)

Kotzab H, Teller C, Bourlakis M, Wünsche S (2018) Key competences of logistics and SCM professionals—the lifelong learning perspective. Supply Chain Manag Int J 23(1):50–64

Kuter SZ, Akyurek Z, Weber G-W (2018) Retrieval of fractional snow covered area from MODIS data by multivariate adaptive regression splines. Remote Sens of Environ 205:236–252

Kuter S, Weber G-W, Akyürek Z, Özmen A (2015) Inversion of top of atmospheric reflectance values by conic multivariate adaptive regression splines. Inverse Problems Sci Eng (IPSE) 23(4):651–669

Lawrence JM, Hossain NUI, Nagahi M, Jaradat R (2019) Impact of a cloud-based applied supply chain network simulation tool on developing systems thinking skills of undergraduate students (Conference Paper). Proceedings of the International Conference on Industrial Engineering and Operations Management, 4th North American IEOM Conference. IEOM 201. Holiday Inn Toronto International Airport Toronto. Canada, 23-25.10. 878–889

Melliana Sinulingga S, Nasution H, Matondang N (2019) Impact competence of human resources and infrastructure in logistic performance improvement. IOP Conference Series: Materials Science and Engineering, 1st International Conference on Industrial and Manufacturing Engineering, ICI and ME 2018, Medan City North Sumatera, Indonesia 505(1)

Melliana, Sinulingga S, Nasution H, Matondang N (2019a) Competence model of human resources, infrastructure, and regulation in improving logistics performance. Int J Civil Eng Technol 10(1):2577–2586

Mendoza-Fong JR, García-Alcaraz JL, Marmolejo-Saucedo JA, Díaz-Reza JR (2020) Impact of managers and human resources on the supply chain performance. Techniques tools and methodologies applied to global supply chain ecosystems. Springer, Cham, pp 3–23

Merolla PA, Arthur JV, Alvarez-Icaza R, Cassidy AS, Sawada J, Akopyan F, Jackson BF, Imam N, Guo Ch, Nakamura Y, Brezzo B, Vo I, Esser SK, Appuswamy R, Taba B, Amir A, Flickner MD, RiskW P, Manohar R, Modha DS (2014) A million spiking-neuron integrated circuit with a scalable communication network and interface. Science 345(6197):668–673

Mintzberg H, Ahlstrand B, Lampel J (1998) Strategy safari: a guided tour through the wilds of strategic management. Prentice-Hall, New York

Molina R, De la Blanca NP, Taylor CC (1994) Modern statistical techniques. Machine learning, neural and statistical classification, 29–49

Özmen A, Weber GW (2014) RMARS: robustification of multivariate adaptive regression spline under polyhedral uncertainty. J Comput Appl Math 259:914–924

Özmen A, Batmaz I, Weber G-W (2014) Precipitation modeling by polyhedral RCMARS and comparison with MARS and CMARS. Environ Model Assess 19(82):425–435

Palšaitis R, Čižiūnienė K, Vaičiūtė K (2017) Improvement of warehouse operations management by considering competencies of human resources. Procedia Eng 187:604–613

Pearson M (2015) Pearson on Excellence: Exception management is becoming the rule [WWW Document]. http://www.logisticsmgmt.com/article/pearson_on_excellence_exception_management_is_becoming_the_rule. Accessed 28 Feb 2020

Prahalad CK, Hamel G (1990) The core competence of the corporation. Harvard Bus Rev 68(3):79–91

Rudy P (2012) Ten reasons you should consider a career in logistics [WWW Document]. http://www.supplychaindigital.com/top10/2512/Ten-Reasons-You-Should-Consider-a-Career-in-Logistics Accessed10 April 2012

Saifudin AM, Zainuddin N, Bahaudin AY, Zalazilah MH, Jamaludin R (2015) Enriching students' experience in logistics and transportation through simulation. Int J Econ Financ Issues 5(1S):343–348

Sangka BK, Rahman S, Yadlapalli A, Jie F (2019) Managerial competencies of 3PL providers a comparative analysis of Indonesian firms and multinational companies. Int J Logist Manag 30(4):1054–1077

Spychała M, Goliński M, Szafrański M, Graczyk-Kucharska M (2019) Competency models as modern tools in the recruitment process of employees. In: European conference on intangibles and intellectual capital. Academic Conferences International Limited, 282–291

Spychała M, Szafrański M, Graczyk-Kucharska M, Goliński M (2017) The method of designing reference models of workstations. In: European conference on knowledge management. Academic Conferences International Limited, 930–939

Szafrański M, Goliński M, Graczyk-Kucharska M, Spychała M (2019) Cooperation of education and enterprises in improving professional competences-analysis of needs. Int Scient-Techn Conf Manuf. Springer, Cham, pp 155–168

Takele TB (2019) The relevance of coordinated regional trade logistics for the implementation of regional free trade area of Africa. J Transp Supply Chain Manag 13(1):1–11

Thai VV (2012) Competency requirements for professionals in logistics and supply chain management. Int J Logist Res Appl 15(2):109–126

Wrobel-Lachowska M, Wisniewski Z, Polak-Sopinska A, Lachowski R (2017) ICT in logistics as a challenge for mature workers. Knowledge management role in information society. In: International conference on applied human factors and ergonomics. Springer, Cham, 171–178

Yang B, Lu J (2017, February) Empirical research on the factors affecting core competence evolution of logistics enterprises based on service innovation. In: 2017 international conference on humanities science, management and education technology (HSMET 2017). Atlantis Press, 1205–1209

Vanneschi L, Castelli M (2019) Multilayer perceptrons. Encyclopedia of Bioinformatics Comput Biol Refer Module Life Sci 1:612–620

Modeling and Optimization of the Supply Chain and Logistics Operations

Classification of Trends and Supply Chains Development Directions

Katarzyna Grzybowska and Agnieszka Stachowiak

Abstract The base of knowledge related to supply chain (SC) research is dynamically expanding. The growing trend of publishing scientific papers on the subject is observable, which indicates that the supply chain concept is in the centre of attention of scientists and researchers. It can be assumed that the tendency will continue. This proves the expansive nature of research in the analysed area. The goal of the chapter to present the classify trends and development directions of supply chains. The goal is of cognitive and conceptual character and the research implemented benefits from text mining method. Research is necessary from academic perspective, not only to organize the knowledge but also to identify future research potential. To achieve the research goal, an original methodology was designed by selecting research methods and tools to answer comprehensively to the research questions. The importance of research results for the development of supply chain, management and quality science is indisputable. Research to be conducted strives to prove that heterogeneity, multi-facetedness and multi-directionality of research on supply chain is a difficult but definable phenomenon.

Keywords Supply chain management · Green supply chain · Sustainable supply chain · Flexible supply chain · Service supply chain

1 Introduction

The supply chains are in a state of transformation. The better we understand the future needs, the better modern and future supply chain will function.

Supply chains are inherently complex, dynamic (Surana et al. 2005) and dispersed and open systems (Grzybowska and Hoffa 2015). Collaboration across SC partners

K. Grzybowska (✉) · A. Stachowiak
Poznan University of Technology, Poznan, Poland
e-mail: katarzyna.grzybowska@put.poznan.pl

A. Stachowiak
e-mail: agnieszka.stachowiak@put.poznan.pl

© The Author(s), under exclusive license to Springer Nature Switzerland AG 2020
P. Golinska-Dawson et al. (eds.), *Smart and Sustainable Supply Chain and Logistics – Trends, Challenges, Methods and Best Practices*, EcoProduction,
https://doi.org/10.1007/978-3-030-61947-3_21

has to be smart, innovative to form value-creating networks. So, we see a different kinds of supply chain, different trends of supply chain.

It can be observed that the amount of knowledge related to research on supply chains is significantly spreading. The growing trend of publishing scientific papers can be observed, which indicates that research on the concept of supply chains is still in the spotlight of scientists and researchers (Hoffa-Dąbrowska et al. 2020). The supply chains are in a state of transformation. The transformation of the supply chain of today to the supply chain of the future is an enormous task. The better we understand the future needs, the better smart and sustainable supply chain will function.

The publication aims at presenting the classify trends and development directions of supply chains. Research to be conducted strives to prove that heterogeneity, multi-facetedness and multi-directionality of research on supply chain is a difficult but definable phenomenon.

The structure of the work is as follows: Sect. 2 presents the literature review. The subsequent part of the article discussed the research methods. Section 4 focuses on and comments results of the research– presents the main trends. The paper is finished with a summary.

2 Supply Chain—Literature Review

Nowadays, changes are faster and more unpredictable. Supply chains must respond quickly to challenges and opportunities in the business world. This makes supply chains evolve and adapt. Global competition and rapidly changing expectations and requirements of the final customer force major changes in the style and configuration of business organizations and supply chains. The sequential and centralized system is not flexible enough to meet these expectations. Industry 4.0 and Circular Economy together have motivated business organizations to evolve towards effective and prompt sustainable supply chain management (Ballouki et al. 2017; Hernández et al. 2014; Lv 2017).

It can be observed that the amount of knowledge related to research on supply chains is significantly spreading (Fig. 1). The growing trend of publishing scientific papers can be observed, which indicates that research on the concept of supply chains is still in the spotlight of scientists and researchers.

With access to the global market, the supply chain is becoming increasingly important. Interest in the supply chain concept results from the significant commitment of theorists and practitioners to the integrated flow of goods from the supplier to the final customer in an increasingly turbulent, unpredictable and global market. It was in the 1980s that organizational solutions and the overall concept of supply chain management became an alternative to the traditional (usually transactional) way of seeing relationships and doing business between cooperating suppliers and customers. Since then, this concept has developed significantly in both the theory

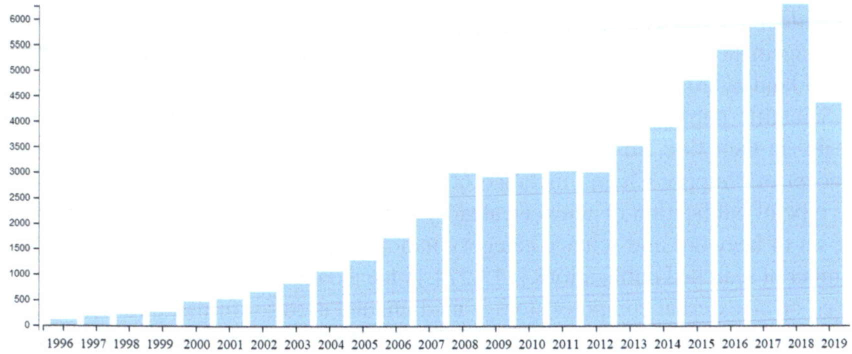

Fig. 1 Number of publications in the area of supply chain research (Hoffa-Dąbrowska et al. 2020)

and practice of logistics management. As a consequence, we also have much more knowledge about organization, functioning and supply chain management (Mentzer et al. 2001; Nagurney 2006).

3 Research Methods

In order to accomplish with the objective of investigating classify trends and development directions of supply chains used an initial search for relevant publications using a literature review. Have been used Web of Science and Scopus databases. This is in line with the recommendations regarding the number of databases used (Green et al. 2006).

The literature review was carried out in three steps, in accordance with guidelines (Strozzi et al. 2017, 2–3):

1. identifying the scope of the analysis,
2. identifying and locating keywords, type of documents, language, databases,
3. selection, assessment and synthesis of the existing set of completed, reviewed and registered in the analysed base of scientific papers, developed by researchers, scientists and practitioners.

The method of bibliometric analysis adopted was the Systematic Literature Network Analysis (SLNA) method introduced by Colicchia and Strozzi (2012), which combines Literature Review and analysis and visualization of a bibliographic network. Adopting this approach allows the identification of trends of key issues that affect the development of knowledge in a given field in a more scientific and objective way than descriptive reviews, which are based on subjective criteria for the selection of works and the classification of research input.

As a result of literature review, a set of keywords has been identified that were used to collect metadata from Web of Science and Scopus databases. As a result of this action, a set of concepts was identified. The basic term "Supply Chain" (SC) has been identified in the collection. The set of concepts has been narrowed down to the "Management" science category. Research and creation of a local base was carried out in January 2020.

Scientific publications from 1989–2019 were searched with the term 'Supply Chain' in their title, abstract or keywords. The initial search for the defined basic concept of "Supply Chain" included 60,652 references, not including identifying the type of publication or science category. Then the search was narrowed down to the set of keyword and science category indicated earlier (management). The total number of searched publications is 13,274. A local database was created, which was saved on an external computer storage medium and used for further analysis.

Further refinement of search results was narrowed down to selected types of scientific publications. For analysis were selected publications that were subject to the review process: (1) articles from scientific journals, (2) articles as part of conference proceedings and (3) chapters in edited books. Duplicates have also been removed.

Subsequent search attempts yielded a total of 13,148 scientific publications (Fig. 2).

Statistics show that for the publication of scientific articles published by scientific journals, two of them contributed to the publication of 1,730 identified scientific articles. This represents 13.16% of all published papers of this type. The dominant magazine is "European Journal of Operational Research". As part of the magazine, a total of 1,089 searched scientific publications were published. It is a prestigious scientific journal about operations management and operational research. It is worth noting that "European Journal of Operational Research" as the most-published journal from the studied area is located in the area of operations and operational research, not supply chain management. This suggests that there is a strong connection between these two disciplines.

For the surveyed population of 13,148 identified scientific publications, statistics show that researchers from the United States (3,679) and China (3,470) dominate in the analyzed set of scientific papers. It is a leadership that significantly contributes

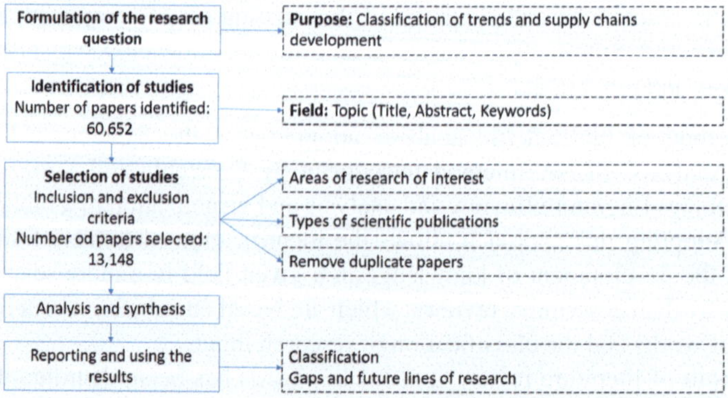

Fig. 2 Summary of the Systematic Literature Review. Developed by the authors

to the development of the studied area. They published a total of 7,149 scientific papers. This gives a rate of 54.37%. England comes in third with 1,202 publications.

4 Identification of Main Trends

As a result of the above actions, a total of 23,701 keywords were specified that describe the scientific papers identified in the local database over the years 1989–2019. Among them was a set of keywords that take into account the minimum occurrence of keywords. Their total number is 965 expressions. On their basis, a semantic map was developed for the examined period (Fig. 3).

The collection of scientific publications from the analyzed period can be divided into seven clusters (Fig. 3). The collection of scientific publications from the analyzed period can be divided into seven clusters (Fig. 3).

Cluster 1 (red area) consists of 296 keywords and relates, among others, to aspects of Behavioral Supply Management. The BSM is define as the study of how judgment in supply management decision-making deviates from the assumptions of homo economicus. This cluster concern issues related to integration and relationships in supply chains, too. Concern issues related to integration and relationships in supply chains. Cluster 1 also includes problems related to: (1) supply chain collaboration, (2)

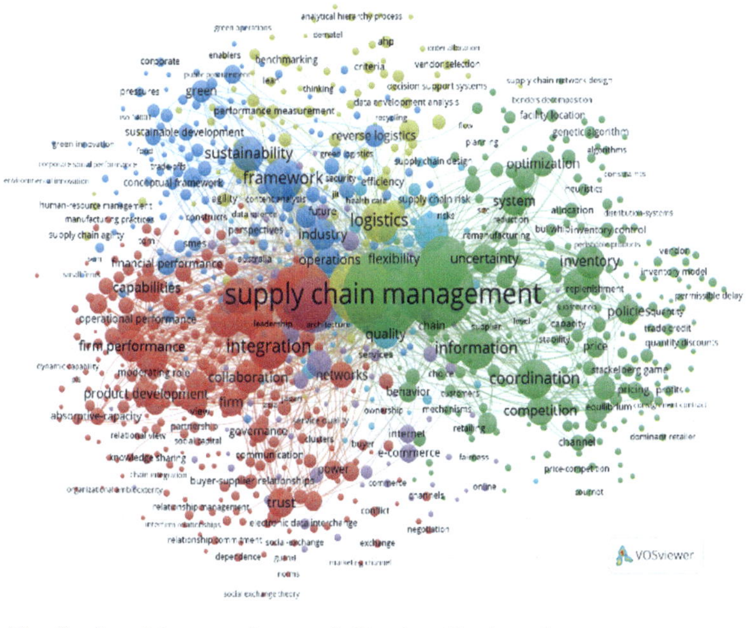

Fig. 3 Visualization of the semantic network. Developed by the authors

supply chain flexibility, (3) supply chain integration, (4) supply chain performance, (5) supply chain relationships, (6) supply chain structure.

Two more clusters concern the greening and social sustainability of supply chains. Cluster 2 (green area) grouped keywords that relate to three supply chain development trends: Closed-loop Supply Chain and Reverse Supply Chain. These two trends are universally recognised as two environmentally friendly practices that could help in greening conventional supply chains. This area also touches on problems related to (1) supply chain contracts, (2) supply chain coordination, (3) supply chain finance, (4) supply chain network, (5) supply chain planning. This cluster consists of 258 keywords.

In the next cluster 3 (blue area) 157 keywords were grouped. This area puts emphasis on topics related to responsible, sustainable and ecological problems of the modern business world. Green capability is characterized in terms of product recovery, supply chain integration, and environmentally. Global competition and rapidly changing expectations and requirements of the final customer force major changes in the style and configuration of supply chains. The following development trends can be identified here: Green Supply Chain, Sustainable Supply Chain, Food Supply Chain and Global Supply Chain. In cluster 3 a clear problem area was identified: supply chain sustainability.

The yellow area (cluster 4) includes 113 keywords. Two other emerging supply chain development trends have been identified here, i.e. Humanitarian Supply Chain and Service Supply Chain. Problems have also been reported: (1) supply chain agility, (2) supply chain strategy.

Cluster 5 (purple area) concerns new technologies. Along with the development of artificial intelligence and intelligent agent technology, topics related to intelligent solutions are taken up. The following problems that supply chains have identified here are: (1) supply chain innovation, (2) supply chain processes, (3) supply chain visibility.

The last of the identified clusters (sea area), cluster 6, consists of 62 scientific concepts. Issues of disruption, risk and reliability are becoming particularly important in supply chains. Risk, disruptions and uncertainty are becoming the main factors limiting the achievement of a high level of supply chain efficiency and disturbing the current functioning of cooperating companies in the supply chains. They concern a group of problems: (1) supply chain complexity, (2) supply chain design, (3) supply chain disruption, (4) supply chain resilience, (5) supply chain risk, (6) supply chain security.

The next part discusses the key development trends of modern supply chains.

4.1 Closed-loop Supply Chain

The Closed-loop Supply Chain is the concept based on integration of primary flows, between supplier and customer, and reverse flows, between customers and vendors. Closed-loop supply chains (CLSC) are supply chain networks that "include the returns processes and the manufacturer has the intent of capturing additional value and further integrating all supply chain activities" (Guide et al. 2003). The definitions refer also to Close-loop Supply Chain Management which is design, control, and operation of a system to maximize value creation over the entire life cycle of a product with dynamic recovery of value from different types and volumes of returns over time. Closed-loop Supply Chain is the most comprehensive and holistic trend in research on supply chains, covering the forward and backward flows of material, benefiting from reverse logistics concept (Hosseini-Motlagh et al. 2020).

4.2 *Reverse Supply Chain*

Reverse Supply Chain is complementary to supply chain in its traditional meaning, these chains integrated constitute Closed-loop Supply Chain (Frei et al. 2020). Hence, the Reverse Supply Chain is responsible for flows initiated by customers and eventually expected by the vendors Reverse Supply Chain benefits from reverse logistics which is the process is the process of planning, implementing and controlling the efficient and effective inbound flow and storage of secondary goods and related information for the purpose of recovering value or proper disposal (Stock 1992). The concept is one of the oldest development concepts for supply chains and evolved over time in works by (Stock 1998) and (Rogers and Tibben-Lembke 1999) who greatly popularized the idea not only among academics (in Web of Science Core collection there is over 3000 publications on "reverse logistics" indexed since 1992), but also among business. The dissemination of Reverse Supply Chains and reverse logistics in business was greatly supported by the list of expected benefits from Reverse Supply chain implementation which include pro-longing product lifecycles, decreased lead time in production, improved image of company and many others.

4.3 *Green Supply Chain*

The Green Supply Chain idea focuses on ecological aspects of supply chain organization and management (Abdi et al. 2020). The key issue in Green Supply Chain concept is to reduce environmental pollution generated on the supply chain scale by implementation of green practices in business operations (Beamon 1999). Practices, such as green purchasing, cooperation with customers, eco-design, and investment recovery, are designed to positively impact environmental performance, moreover there are models proving their direct and positive impact on operational performance (Green et al. 2012). There is some ambiguity between Green Supply Chain and Sustainable Supply Chain since they emerged in the same time. The ambiguity may be clearly explained, as the first concept is to be driven by ecological and economic factors while the latter with ecological, economic and factors. Nevertheless, many researchers use both terms interchangeably. Considering time frame, Green Supply

Chain is the concept older and more often considered as the research topic (in Web of Science Core collection there is over 1600 publications on "green supply chain" indexed since 1997 and their number is growing dynamically after 2010).

4.4 Sustainable Supply Chain

The trend recognized as Sustainable Supply Chain refers to the idea of sustainability and its implementation to supply chain structure (Mardani et al. 2020). The generic term for sustainability is sustainable development which is defined as "a development that meets the needs of the present without compromising the ability of future generations to meet their own needs"(WCED 1987). Since the term is general it requires operationalization which is usually provided by implementing the so-called triple bottom line approach. The approach encompasses three dimensions: economic, social and environmental (Elkington 2002) the combination and balance of which is supposed to result in equilibrium in economy. The Sustainable Supply Chain requires a broadened approach to the Supply Chain (Grzybowska 2012). In the case of sustainable or "green" SCM, sup-ply chain members are encouraged to fulfil customers' needs concerning ecological or social products (Zhu and Sarkis 2004). Sustainable development and triple bottom line ideas were originally defined in world economy context. Nevertheless, it is now used also in the context of business units, especially Supply Chains as the often operate on global scale and hence can have significant effect on societies, economies and eco-systems. Sustainable supply chain is not only the trend recognized in research (in Web of Science Core collection there is over 1200 publications on "sustainable supply chain" indexed since 2000, the number of works published is dynamically increasing after 2010), but also practical approach promoted by organizations (i.e. SCF—Sustainable Supply Chain Foundation) and authorities.

4.5 Food Supply Chain

Food supply chain management is developing as a research discipline spanning local, regional, national and international arenas and it has progressed from a series of shorter, independent transactions to more collaborative relationships between producers, processors, manufacturers and retailers (Bourlakis 2004). Many key trends observable for supply chains are also evident in food supply chains, including mentioned in previous sections trend on sustainability but also implementation of information technology for supply chain coordination and management (Mondragon et al. 2020). Food supply chain is the area of interest important for academics, nevertheless it is an issue of great practical importance.

4.6 Global Supply Chain

The general trend recognized in the economy is globalization. It affects all areas of business activity (and does not limit to business activities, affecting science, education, leisure activities and entertainment). Global Supply Chain can be simple defined as supply chain operating in global market, while Global Supply Chain Management is defined as the distribution of goods and services throughout a trans-national companies' global network to maximize profit and minimize waste (Bhatnagar 2012). The scale is crucial and determines the trend and size and scope of problems dealt with. Since global supply-chain management is the general trend dealing with numerous specific problems, it is operationalized in six dimensions namely: logistics management, competitor orientation, customer orientation, supply-chain coordination, supply management, and operations management (Hult 2003). Due to general formulation of the trend its representation in literature is fuzzy as researchers tend to focus on specific problems.

4.7 Humanitarian Supply Chain

Humanitarian supply chain is the flow of relief aid and the related information between the beneficiaries affected by disaster and the donors so as to minimize human suffering and death (Cozzolino et al. 2012). Since there were 6,637 natural disasters between 1974 and 2003 worldwide, with more than 5.1 billion affected people, more than 182 million homeless, more than 2 million deaths, and with a reported damage of $1.38 trillion USD (Center for Research on the Epidemiology of the Disasters (CRED)) (Ergun et al. 2009) such solution becomes not only a trend but also the necessity. In 2005 alone, over 180,000 deaths and $200 billion USD economic losses have occurred due to disasters according to the Disaster Resource Network Humanitarian Relief Initiative (HRI). The September 11 attacks (2001), tsunami in South Asia (2004), Hurricane Katrina (2005), earthquakes in Pakistan (2005) and Java (2006), COVID19 situation and many more till now make humanitarian aspect of supply more and more important.

4.8 Service Supply Chain

World's economy is facing the shift from economy based on manufacturing to economy based on services. And as goods move in supply chain structures so do the services coordinated with material flows, completing them and adding value to materials as they flow. Due to the nature of services, SCM in services is more complicated than in manufacturing. Service Supply Chain can be defined as a network of suppliers, service providers, consumers and other supporting units that performs the function of transaction of resources required to produce service followed by transformation of these resources into supporting and core services and finally delivery these services to customers (Baltacioglu et al. 2007). Major elements of Service Supply Chain Management which differentiate it from manufacturing include bidirectional supply chains, perishability and simultaneous management of both capacity and demand (Shahin 2010). It has been pointed out that SSCM could provide considerable benefits to service organizations, including flexibility in delivery, dynamic scheduling and process orientation. The Service Supply Chain seems to be an interesting stressing the importance of logistics services.

4.9 Problems and Issues Identified in Literature on Supply Chains

The list of problems and issues identified in literature on supply chain with systematic literature review includes 22 elements linked with clusters presented in the previous section. The problems listed are of a different nature and scope. However, they can be grouped according to their character and origin. With such approach, the first group includes problems arising from and connected to supply chain's operation, so the character of the problems is endogenous, while the second includes problems arising from interaction with environment or environment itself, so the character of problems is exogenous.

One of the endogenous problems is **supply chain structure**, as it refers to number and types of enterprises involved in supply chain operation, as well as links created between them. Supply chain structure can be referred to as **supply chain network** (Abdi et al. 2020). The definition of the structure and of the principles of its operation depends on many factors, including product design, technology, market etc., which makes it a complex decision. The process of definition of supply chain structure is **supply chain design** (Tirkolaee et al. 2020). Supply chain design results in developing general principles for supply chain constitution and operation. The more elements in a supply chain, the more complex it is. **Complexity of supply chain** can be analyzed in two dimensions, namely the number of elements and the diversity of elements, both resulting in numerous challenges for supply chain management (Piya et al. 2019). Moreover, supply chain structure, design and complexity have a strong impact on supply chain performance and characteristics.

Considering the above mentioned, **supply chain performance** is also endogenous problem, greatly influenced by supply chain structure, complexity (Zhan and Tan 2020) and **supply chain processes** (Zimon et al. 2019). The problem of supply chain performance is of great complexity as it refers not only to performance of individual enterprises that cooperate within supply chain, but also to overall performance of the entire supply chain. The latter approach requires considering numerous trade-offs and synergies within the structure and processes. Performance of the supply chains is expected to improve with integration level. However, it can be compromised by supply chain complexity.

Hence, **supply chain integration** is also an endogenous issue. Integration is operationalized by the number and range of links between supply chain elements (enterprises) and is generally perceived as positive phenomena (Durach and Machuca 2018). Thanks to integration material and information flows in supply chains are faster, more efficient and economic. Moreover, research prove that supply chain integration mitigates Bullwhip effect (Zhang and Zhang 2020).

The problem closely linked to supply chain integration and performance is **supply chain coordination**. The coordination of actions means combined operation of the subjects, which: (1) is directed towards achievement of set, common, mutually compliant targets, (2) encompasses systematizing, ordering and approving process and various components of the system, (3) takes place in the agreed time and (4) exerts some influence on behavior of the cooperating entities (Grzybowska and Hoffa 2015). The problem of supply chain coordination was first modeled and analyzed in (Kumar 1992). The problem is also of endogenous character as aims at improving supply chain performance by aligning the plans and the objectives of individual enterprises. Hence, it focuses on supply chain operation. Aligning of plans and objectives could lead to integration by inventory management and ordering decisions taken in distributed inter-company settings but striving for the same goal.

Goal definition is the part of strategy. **Supply chain strategy** is an endogenous problems as strategies are often resource-based and strive to exploit potential of business entities (von Falkenhausen et al. 2019). Strategy includes goals not only in ongoing perspective but also for the ones to be achieved in the future. According to some authors, strategy is mostly about planning for the future. **Supply chain planning** is another endogenous supply chain problem identified in the literature (Lejarza and Baldea 2019). Supply chains case give lots of opportunities when strategy is concerned, as potential of supply chain cane be increased easier than potential of single business (by including new elements in the structure). Moreover, supply chains can benefit from shared resources and synergy resulting from integration and coordination of their structure. The resources considered include human, technical and financial ones. **Supply chain finances** are the problem that should be interpreted on one hand as an advantage of complex structures, as potentially they have access to larger financial resources, on the other it could be disadvantage due to problems with money distribution (Jia et al. 2020). Undoubtedly, financial resources support goal realization when managed well. The goals most often mention in supply chain context are **supply chain flexibility** (Novais et al. 2019), **supply chain agility** (Shamout 2020), and **supply chain resilience** (Sawyerr and Harrison 2020). The supply chain should be characterized by durability, denoting the ability to resist rapid change and the ability to survive (Grzybowska and Hoffa-Dąbrowska 2017). Many definitions of organizational flexibility may be found though it is commonly accepted that organizational flexibility stands for the organization's capability to adopt to changes taking place in its environment (Buła and Bernard 2011). The definition of agility is quite similar, as the terms are sometimes confused. Agility can be defined as the capability of a company to rapidly change or adapt in response to changes in the market. When searching for differences, one of the most important is that flexibility is adaptive approach, while agility tends to be pro-active one. Considering resilience, it is defined as the ability of an organization to anticipate, prepare for, respond and adapt to incremental change and sudden disruptions in order to survive and prosper. Resilience is thus the ability of a system to return to its original or desired form after a disturbance has occurred (Konecka et al. 2018). Organizational resilience is supposed to provide **supply chain security** on numerous levels, starting from security of the entire supply chain structure, through security of supply chain components and security of individual employees (Shamsi et al. 2019).

All these goals included in plans are, as mentioned, supported by integration and coordination of supply chain. However, they can be compromized by diruptions in supply chain environment. **Supply chain disruption** should be perceived as risk source and exogenous factor as it results from environment characteristics (Lawson et al. 2019). Contemporary market environment is often described as volatile, turbulent or even chaotic due to scope and pace of changes it undergoes. The changes make management difficult inreasing **supply chain risk** (Xu et al. 2020). Hence, implementation of well-known, recognized and established management methods inefficient or even impossible. Hence the interest in the problem has both academic (as a challenge to existing knowledge on management) and practical (as a challenge to practitioners in their everyday activities) perspectives.

The disruptions can be dealt with thanks to innovative approach to management. **Supply chain innovation** (Chen et al. 2020) is also the problem identified in systematic literature review which generally corresponds with the trends recognized in many areas, including industry (Industry 4.0 solutions). Industry 4.0 focuses on nine technologies: autonomous robots, system integration, the Internet of Things (IoT), simulation, additive manufacturing, cloud computing, augmented reality, big data, and cybersecurity (Kaur et al. 2020). The innovative solution from industry affect supply chains, yet logistics solutions, namely Logistics 4.0 concept, empowers contemporary supply chains equipping it with data necessary for making decisions and methods and tools supporting information and material flows, even in turbulent environment and dynamically changing conditions. Supply chain innovations are perceived as exogenous problems as they are usually produced by research institutes and/or academia and implemented to supply chains. Thanks to real time communication that benefits from contemporary technology **supply chain collaboration** (Shou et al. 2020) can be increased via new forms of **supply chain contract** (Min et al. 2020), together with (21) supply chain relationships with environment (Davis et al. 2019) and (22) supply chain visibility (Cousins et al. 2019). These problems are also addressed by literature and recognized in hereby paper.

5 Conclusion

The factor that inspires academics, stimulates research and leads to generating new knowledge is undoubtedly the changeability of the economy. Both, enterprises and SCs must respond very quickly to the challenges and opportunities of the business world, adjusting their structure and processes to requirements of customers by introducing all kinds of innovations.

The base of knowledge related to supply chain research is dynamically expanding. The growing trend of publishing scientific papers on the subject is observable, which indicates that the supply chain concept is in the center of attention of scientists and researchers. Identification of these trends seems to be important from academic perspective, giving inspiration do scientists and research, as well as to business people. Contemporary research stress the importance of advanced ICT solutions in supply chains, cyber-physical systems, intelligent and autonomous solutions. Considering continuous development of knowledge generated in the field, in-depth research to systematize and rationalize it is required. Analysis of the process of knowledge development and its dynamics to reveal and characterize the evolution and identify potential development trends seems necessary from academic perspective, not only to organize the knowledge but also to identify future research potential. The research conducted allowed to identify not only contemporary trends in research on supply chains, but also problems dealt with. It can be used as the basis for deepened research on relations and interactions between trends and problems identified.

Acknowledgments Support for this research has been provided by **0812-SBAD-4170**.

References

Abdi A, Abdi A, Akbarpour N (2020) Innovative approaches to design and address green supply chain network with simultaneous pick-up and split deliver. J Cleaner Prod 250, In Press

Ballouki I, Douimi M, Ouzizi L (2017) Decision support tool for supply chain configuration considering new product re-design: an agent-based approach. J Adv Manuf Syst 16(4):291–315

Baltacioglu T, Baltacioglu A, Kaplan E, Kaplan D, Oznur M, Oznur Y, Kaplan Y (2007) A new framework for service supply chains. Serv Indust J 27:105–124

Beamon BM (1999) Designing the green supply chain. Logist Inform Manag 12(4):332–342

Bhatnagar K (2012) Customer-oriented global supply chains: concepts for effective management. Hershey, Pennsylvania, Information Science Reference

Bourlakis W (2004) Food supply chains management. Blackwell, Oxford

Buła P, Ziębicki B (2011) Organizational flexibility as a challenge of contemporary management. Determinants and methods of measurement. Acta Commercii 11:171–180

Chen Y, Duan L, Zhang W (2020) Effect of user involvement in supply chain cloud innovation: a game theoretical model and analysis. J Global Inform Manag 28(1):23–38

Colicchia C, Strozzi F (2012) Supply Chain risk management: a new methodology for a systematic literature review. Supply Chain Manage: Int J 17(4):403–418

Cousins PD, Lawson B, Petersen KJ (2019) Investigating green supply chain management practices and performance The moderating roles of supply chain ecocentricity and traceability. Int J Oper Product Manag 39(5):767–786

Cozzolino A, Rossi S, Conforti A (2012) Agile and lean principles in the humanitarian supply chain: the case of the united nations world food programme. J Humanitarian Logist Supply Chain Manag 2(1):16–33

Davis DF, Davis-Sramek B, Golicic SL (2019) Constrained choice in supply chain relationships: the effects of regulatory institutions. Int J Logist Manag 30(4):1101–1123

Durach CF, Machuca JAD (2018) A matter of perspective—the role of interpersonal relationships in supply chain risk management. Int J Oper Product Manag 38(10):1866–1887

Elkington J (2002) Cannibals with forks: the triple bottom line of 21st century business. Capstone, Oxford

Ergun O, Karakus G, Keskinocak P, Swann J, Villarreal M (2009) Humanitarian supply chain management – an overview. Models Algorithms Optim Logistics

Frei R, Jack L, Krzyzaniak S (2020) Sustainable reverse supply chains and circular economy in multichannel retail returns, Business Strategy and the Environment, In Press

Green K, Morton B, New S (1998) Green purchasing and supply policies: do they improve companies' environmental performance? Supply Chain Manag: Int J 3(2):89–95

Green BN, Johnson CD, Adams A (2006) Writing narrative literature reviews for peer-reviewed journals: secrets of the trade. J Chiropractic Med 5(3):101–117

Green KW Jr, Zelbst PJ, Meacham J, Bhadauria VS (2012) Green supply chain management practices: impact on performance. Supply Chain Manag: Int J 17(3):290–305

Grzybowska K (2012) Supply chain sustainability—analysing the enablers. In: Golinska P, Romano CA (Eds), environmental issues in supply chain management—new trends and applications, Springer, 25–40

Grzybowska K, Hoffa P (2015) Approving with application of an electronic bulletin board, as a mechanism of coordination of actions in complex systems. In: Omatu S, Malluhi QM, Rodríguez Gonzalez S, Bocewicz G, Bucciarelli E, Giulioni G, Iqba F (Eds.), Distributed Computing and Artificial Intelligence, 12th International Conference Advances in Intelligent Systems and Computing, 373, 357–365

Grzybowska K, Hoffa-Dąbrowska P (2017) Analysis of the survival of complex systems with an actions coordination mechanism, In: Bajo J et al (eds) Highlights of Practical Applications of Cyber-Physical Multi-Agent Systems. Communications in Computer and Information Science, Springer International Publishing, 722, 197–208

Guide VDR, Jayaraman V, Linton JD (2003) Building contingency planning for closed-loop supply chains with product recovery. J Oper Manag 21:259–279

Hoffa-Dąbrowska P, Grzybowska K, JPE, de Souza (2020) Sustainable Supply Chain: a content analysis based on published case studies, in press

Hernández JE, Lyons AC, Poler R, Mula J, Goncalves R (2014) A reference architecture for the collaborative planning modelling process in multi-tier Supply Chain networks: a Zachman-based approach. Product Plann Control 25(13–14):1118–1134

Hosseini-Motlagh S, Ebrahimi S, Zirakpourdehkordi R (2020) Coordination of dual-function acquisition price and corporate social responsibility in a sustainable closed-loop supply chain. J Cleaner Product 251, In Press

Hult GTM (2003) An integration of thoughts on knowledge management. Decis Sci 34(2):189

Jia F, Zhang T, Chen L (2020) Sustainable supply chain Finance: Towards a research agenda. J Cleaner Product 243, In Press

Kaur R, Awasthi A, Grzybowska K (2020) Evaluation of key skills supporting industry 4.0—a review of literature and practice. In Grzybowska K, Awasthi A, Sawhney R (eds), Sustainable logistics and production in industry 4.0, EcoProduction (Environmental Issues in Logistics and Manufacturing). Springer, Cham, 19–29

Konecka S, Łupicka A, Grzybowska K (2018) Disruptions in supply chains with special reference to road transport processes. Innovation Management and Education Excellence Through Vision 2020, I–XI, 3841–3849

Kumar A (1992) Supply contracts and manufacturing decisions. Carnegie Mellon University, Pittsburgh, PA, Graduate School of Industrial Administration, p 15213

Lawson B, Potter A, Pil FK (2019) Supply chain disruptions: the influence of industry and geography on firm reaction speed. Int J Oper Product Manag 39(9/10):1076–1098

Lejarza F, Baldea M (2019) Closed-loop optimal operational planning of supply chains with fast product quality dynamics. Comput Chem Eng 132, In Press

Lv Q (2017) Supply Chain coordination game model based on inventory information sharing. J Interdisciplinary Math 20(1):35–46

Mardani A, Kannan D, Hooker RE (2020) Evaluation of green and sustainable supply chain management using structural equation modelling: a systematic review of the state of the art literature and recommendations for future research. J Cleaner Product 249, In Press

Mentzer JT, DeWitt W, Keebler JS, Min S, Nix NW, Smith CD, Zacharia ZG (2001) Defining supply chain management. J Busin Logist 22(2):1–25

Min J, Zhou YW, Zhao J (2010) An inventory model for deteriorating items under stock-dependent demand and two-level trade credit. Appl Math Model 34(11):3273–3285

Nagurney A (2006) Supply chain network economics: dynamics of prices, flows and profits. Edward Elgar Publishing, Cheltenham, England

Novais L, Maqueira JM, Ortiz-Bas A (2019) A systematic literature review of cloud computing use in supply chain integration. Comput Ind Eng 129:296–314

Piya S, Shamsuzzoha A, Khadem M (2020) An approach for analysing supply chain complexity drivers through interpretive structural modelling. Int J Logistics Res Appl 23(4):311–336

Rogers DS, Tibben-Lembke R (2001) An examination of reverse logistics practices. J Bus Logistics 22(2):129–148

Sawyerr E, Harrison C (2020) Developing resilient supply chains: lessons from high-reliability organisations. Supply Chain Manage: Int J 25(1):77–100

Shahin A (2010) SSCM: service supply chain management. Int J Logistics Syst Manage 6(1):60–75

Shamsi J, Urban AS, Imran M, Trizio LD, Manna L (2019) Metal halide perovskite nanocrystals: synthesis, post-synthesis modifications, and their optical properties. Chem Rev 119:3296

Stock JR (1992) Reverse logistics, Oak Brook, IL: Council of Logistics Management

Stock JR (1998) Development and implementation of reverse logistics programs. In: annual conference proceedings, Oak Brook, IL: Council of Logistics Management, 579–586

Strozzi F, Colicchia C, Creazza A, Noè C (2017) Literature review on the "Smart Factory" concept using bibliometric tools. Int J Prod Res 55(22):6572–6591

Surana A, Kumara S, Greaves M, Raghavan UN (2005) Supply-chain networks: a complex adaptive systems perspective. Int J Prod Res 43:20:4235–4265

Tirkolaee EB, Mardani A, Dashtian Z, Soltani M, Weber GW (2020) A novel hybrid method using fuzzy decision making and multi-objective programming for sustainable-reliable supplier selection in two-echelon supply chain design. J Clean Prod 250:119517

von Falkenhausen, Fleischmann M, Bode C (2019) How to find the right supply chain strategy? An analysis of contingency variables. Decis Sci 504:726–755

Xu S, Zhang X, Feng L, Yang W (2020) Disruption risks in supply chain management: a literature review based on bibliometric analysis. Int J Prod Res https://doi.org/10.1080/00207543.2020.1717011

Zhan Y, Tan KH (2020) An analytic infrastructure for harvesting big data to enhance supply chain performance. Eur J Operat Res

Zhang S, Zhang M (2020) Mitigation of bullwhip effect in closed-loop supply chain based on fuzzy robust control approach. Complexity, 1085870

Zhu Q, Sarkis J (2004) Relationships between operational practices and performance among early adopters of green supply chain management practices in Chinese manufacturing enterprises. J Oper Manage 22:265–289

Zimon D, Tyan J, Sroufe R (2019) Implementing sustainable supply chain management: reactive, cooperative, and dynamic models. Sustain 11:7227

Framework of the Model of Dissemination and Absorption of Logistics 4.0 Solutions—Causal Loop Dynamics of Relations Between Academia and Business

Agnieszka Stachowiak, Joanna Oleśków-Szłapka, Przemysław Niewiadomski, and Natalia Pawlak

Abstract The changes in contemporary economy are widely recognized and dealt with advanced tools and methods supporting companies facing dynamics and competitiveness of market environment. The list of tools and methods is long and usually they are jointly referred to as Industry 4.0. The general idea of Industry 4.0 is to benefit from contemporary technologies to increase efficiency and competitiveness of company, however, when referring specifically to material and information flows the Logistics 4.0 term is used. Logistics 4.0 concepts benefits from advanced communication protocols, cooperation between humans and machines, which requires knowledge and skills that are provided by higher education institutions. According to authors, knowledge gap filling is a dynamic phenomenon that can be interpreted and analysed with system dynamics approach. The dynamics of knowledge gap filling could be described as goal-seeking dynamics. The loop dynamics shows that when considering the expectations and taking corrective actions, the knowledge gap can be easily filled in a given time period. The goal of the chapter is to present the original model of dissemination and absorption of Logistics 4.0 solutions developed based on literature review and authors' experience.

Keywords Logistics 4.0 · Knowledge absorption · Causalloops

A. Stachowiak (✉) · J. Oleśków-Szłapka · N. Pawlak
Poznan University of Technology, J. Rychlewskiego 2, 60-965 Poznan, Poland
e-mail: agnieszka.stachowiak@put.poznan.pl

J. Oleśków-Szłapka
e-mail: joanna.oleskow-szlapka@put.poznan.pl

N. Pawlak
e-mail: natalia.pawlak@put.poznan.pl

P. Niewiadomski
University of Zielona Gora, Gora, Poland
e-mail: pniewiadomski@uz.zgora.pl

© The Author(s), under exclusive license to Springer Nature Switzerland AG 2020
P. Golinska-Dawson et al. (eds.), *Smart and Sustainable Supply Chain and Logistics – Trends, Challenges, Methods and Best Practices*, EcoProduction,
https://doi.org/10.1007/978-3-030-61947-3_22

1 Introduction

As the world and economy changes, so change the needs and requirements on knowledge and skills of workers. Knowledge and skills of workers have to be consistent with the content of work and since the content changes continuously, workers have the solid background within their specialization but also ability to adapt and learn. Flexibility is a personal feature, which can be practiced and developed while solid knowledge background needs to be provided by education institution. Nevertheless, taking into account the operation mode of education system the process of educating and knowledge gap closing or minimizing is disturbed by delays: in identification of divergences, as they are recognized after the students graduate, hence the feedback opportunities are limited. Moreover, there are some delays connected with designing and then implementing corrective actions. Thus, the gap filling dynamics is more likely to be recognized as oscillatory dynamics. The delays result in increased difficulties, time and money consumption in knowledge gap filling process. To deal with them, identification of knowledge needs and corrective actions design and implementation need to be the processes carried on not only regular bases but also concurrent to providing education.

Knowledge requirement in contemporary economy is determined by processes realized and technologies implemented. The list of technologies, tools and methods that contemporary companies benefit from includes Internet of Things, cloud computing, Big Data processing an many more. These technologies are referred to (among others) as pillars of Industry 4.0 (Rüßmann 2015). The general idea of Industry 4.0 is to benefit from contemporary technologies to increase efficiency and competitiveness of company. Since implementation of production processes is dependent on logistics, feeding production and delivering final products to customers, contemporary solutions for logistics required. Hence, when referring specifically to material and information flows the Logistics 4.0 term is used. Logistics 4.0 concepts benefits from advanced communication protocols, cooperation between humans and machines, which makes it demanding in terms of knowledge.

The following paper aims to introduce original model on knowledge on Logistics 4.0 absorption based on causal loops dynamics.

The chapter is structured as follows: after introduction, the terms and definitions of Logistics 4.0 are introduced in Theoretical background section, research methodology, including knowledge absorption idea and causal loops dynamics is explained in Material and method section, the original model is presented and discussed in Model introduction section. The final section is Summary.

2 Theoretical Background

According to the Council of Supply Chain Management Professionals (previously the Council of Logistics Management), logistics is the process of planning, implementing and controlling procedures for the efficient and effective transportation and storage of goods including services and related information from the point of origin to the point of consumption for the purpose of conforming to customer requirements and includes inbound, outbound, internal and external movement. Since the definition is quite general, it is not challenged. However, considering characteristics of contemporary market, change of methods and tools than logistics benefits from seems unavoidable due to contemporary material and information flows' complexity. Globalization is recognized as the determinant of most of the changes in the contemporary economy, and one of its consequences is development of a global supply chain as a solution enabling companies and customers benefiting from global market.

The Logistics Trend Radar report, developed by the DHL Company in 2014, was one of the first publications that linked Industry 4.0 with logistics. It drew attention to the benefits emerging from using modern technologies within the field of Industry 4.0 as part of the provision of logistic services. Since production is closely linked to Logistics through material and information flows feeding the processes, the paradigm shift caused by Industry 4.0 brought about a fundamental change in the entire logistics delivery process, shifting it towards service-oriented logistics on demand (Bubner 2014). According to the report, the Internet of Things will support the introduction of intelligent machinery and devices that enable the control of modern logistics processes to the logistics market (Bubner 2014), contributing to development of Logistics 4.0 idea. More recent work on logistics' (especially inboud/internal logistics') role is presented in (Kostrzewski et al. 2020).

Logistics 4.0 is defined as "A strategic technological direction that integrates different types of technologies to increase both the efficiency and effectiveness of the supply chain, shifting the focus of the organizations to value chains, maximizing the value delivered to the consumers as well as the customers by raising the levels of competitiveness. In literature on Industry 4.0, the term used to refer to Logistics processes performed at excellent level is Smart Logistics (Barreto et al. 2017). Smart Logistic is defined as the „logistics system, which can improve the flexibility, the adaptation to the market changes and bring the company closer to customer needs" (Strandhagen et al. 2017). Disregarding small terminological differences between Smart Logistics and Logistics 4.0, they both are designed to support individualized and technologically advanced production processes performer at the global scale with corresponding (in terms of complexity and technological level) logistics processes (Szymańska et al. 2017).

This is achieved by increasing the levels of transparency and decentralization among different parties through digitalization" (Amr et al. 2019). Hence, the definition of Logistics 4.0 combines two aspects: processual (supply chain processes are a subject of the Logistics 4.0 actions) and technical (tools and technologies that support internal processes in the supply chains) (Szymańska et al. 2017). The tools and technologies are mostly within IT range as digitization creates plenty of advantages for the supply chain. Considering scope and intensity of logistics operation in global markets, where companies manage millions of daily shipments, large amounts of data is obtained via orders transcripts, smart low battery consuming sensors, GPS, RFID tags, weather forecasts or even social media. These solutions are usually associated with transport processes. Warehousing processes can also benefit from advanced technical solutions thanks to improvement by combining automation, the Internet of things, drones, 3D printing and innovative applications, becoming smarter, more connected, automated and robotized. The list of existing solutions that support Logistics 4.0 includes the technologies: internet of things, cyber-physical systems, Big Data, cloud computing, mobile-based systems, social media-based systems and further technologies (Winkelhaus and Grosse 2019).

The expected benefits include access to data and ability to process them, which not only improves logistics services provided, but also consequently results in increased customer service level. Moreover, companies reduced complexity, increased reliability, predictability and thus minimized risks, reduced errors, reduced transport cost, creating new business areas and thus turnover potential, increased innovation capability, increased agility and flexibility, e.g. in case of new market requirements (Strandhagen et al. 2017). As a result, Logistics 4.0 plays the same role in managing supply chains as Industry 4.0 for modern manufacturing factories. However, to benefit from Logistics 4.0 companies need to adjust their operations and management to requirements of contemporary technologies by improving their flexibility and openness through (Heistermann et al. 2018):

- decentral self-organization through ad hoc interconnection;
- virtual ad hoc organizations, value added chains;
- autonomous self-organizing logistics and production entities;
- expansion through "up-numbering" (modularization);
- flexible deployment of employees (availability calendars, specialist knowledge catalogue).

Meeting the requirements listed requires deepened knowledge in management, while implementing technical solutions associated with Logistics 4.0 requires technical, IT and engineering knowledge.

3 Materials and Methods

The methodology implemented to develop the model of dissemination and absorption of Logistics 4.0 solutions is based on two pillars: knowledge absorption phenomena and causal loops dynamics.

Knowledge is an extremely important resource at many levels: economy/society (knowledge contributes to well-being of societies by developing innovations), organizations (organizational knowledge contributes to creating competitive advantage) and individual (personal knowledge increases potential of a person, opens professional opportunities and contributes to higher level of well-being). The process of knowledge absorption is consisting of a number of different sub-processes, such as recognizing the value of external knowledge, acquiring external knowledge, transforming or assimilating it, and exploiting external knowledge (Zerwas 2014). Considering the complexity of the term, many theoreticians in this area structure absorptive capacity dividing it into dimensions and components (Burcharth et al. 2015) where dimensions include acquisition, assimilation, transformation and exploitation of knowledge (Zerwas 2014), as listed and defined below (Ferreras-Méndez et al. 2015):

- Ability to acquire knowledge (Aac): understood as the ability to locate and identify sources, and also its valuation and acquisition as a basis for operational activities.
- Ability to assimilate knowledge (Aas): understood as the ability to assimilate it, consisting of ordering, correct analysis, interpretation, and understanding.
- Ability to transform knowledge (Atr): understood as the ability to combine new knowledge with knowledge already possessed, through adding or eliminating specific components of knowledge or by interpreting knowledge in a new innovative way.
- Ability to use knowledge (Aus): understood as the ability to incorporate knowledge into operational activities and improve processes and competences.

Based on the structure of the concept, knowledge absorption requires an active approach, openness to interaction with business and scientific environments, and knowledge flows and diffusion within the company itself. Moreover, the findings from the study by Ferreras-Méndez et al. (2015, Kostopoulos et al. (2011) show that firms' involvement in innovation collaborations with various external parties enriches their knowledge base (Gebauer et al. 2012) and develops a better ability to assimilate and exploit external knowledge (Camison and Fores 2010) to facilitate the understanding of new concepts and effectively implement innovation (Pai and Chang 2013).

Knowledge absorption is intensified thanks to knowledge diffusion. Knowledge diffusion is the responsibility of educators at all the levels of educational system, starting from primary level, through secondary and vocation education, to professional and academic level as well (Biesta 2006).

To describe mechanism of knowledge diffusion and absorption, causal loops dynamics is implemented. Since knowledge absorption is a process, implementing dynamic analysis based on causal seems to be appropriate.

Causal loops visualize interrelations between variables in systems. Thanks to visualization, understanding the situation and consequences of changes of variables is easier. Causal loop diagram consist of nodes representing variables, and links representing connections and relations (Sterman 2000). The links can be marked (S) for positive relations meaning that changes in variables go in the same direction or (o) for negative relations, meaning that changes in variables go in opposite directions. The changes in variables tend to be cyclical—causal diagrams take the form of loops. The loops can be reinforcing (R) or balancing (B). The nature of the loop is based on the feedback it is based on (Lopez-Caamal et al. 2014). Positive feedback is a process that occurs in a feedback loop which exacerbates the effects of a small disturbance, which means that the effects of a perturbation on a system include an increase in the magnitude of the perturbation, while negative feedback occurs when some function of the output of a system, process, or mechanism is fed back in a manner that tends to reduce the fluctuations in the output, whether caused by changes in the input or by other disturbances (Ford 2010). Hence, considering the nature of the feedback, the reinforcing loop is a cycle in which the effect of a variation in any variable propagates through the loop and returns to the variable reinforcing the initial deviation. A balancing loop is the cycle in which the effect of a variation in any variable propagates through the loop and returns to the variable a deviation opposite to the initial one (Sterman 2000).

4 Research Background

The first phenomenon referred to in the research is knowledge diffusion. Knowledge on Logistics 4.0 dissemination is performed mostly at academic level as the term and solution is relatively new. Growing number of publications on the subject proves researchers' interest in Logistics 4.0 and Smart Logistics. The trend is presented in Fig. 1.

Knowledge is also disseminated by providing educational services. In the scope of logistics, they are offered at vocational level, by HEIs at first cycle courses and second cycle (MSc), as well as on numerous courses and training sessions.

The second phenomenon into consideration is knowledge absorption. For the research purposes, knowledge absorption was analysed at organizational and individual level. Considering companies, Polish companies were analysed. The research was on how they absorb knowledge on Logistics 4.0 by implementing solutions it encompasses. It relies on secondary sources of data from Smart Industry reports, statistical data and literature analysis. Fundamental sources of data are Smart Industry reports commissioned by the Ministry of Development and Siemens: editions from 2016, 2017 and 2018. Report from 2016 comprises companies employing 250 people

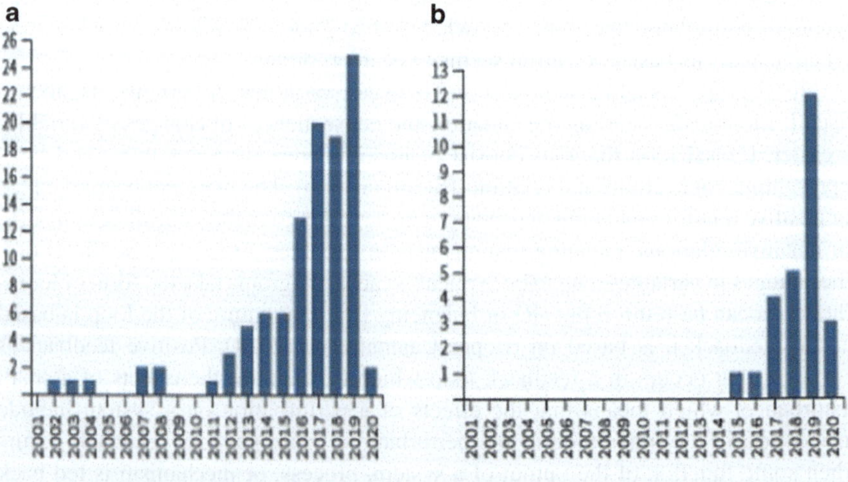

Fig. 1 Number of publications on **a** "Logistics 4.0" topic and **b** "Smart Logistics" topics (*Source* own elaboration work based on WOS data: February 2020)

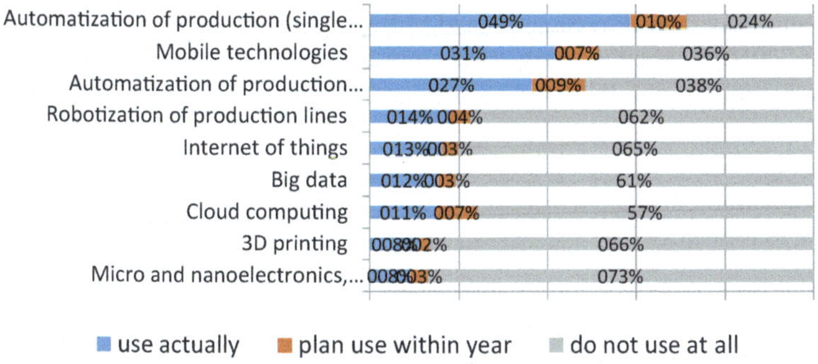

Fig. 2 Technologies that use SMEs in Poland (*Source* own elaboration work based on Smart Industry Report 2017)

and more, while report 2017 and 2018 were conducted on a sample of enterprises from micro, small and medium groups employing no more than 250 people. The sample from 2018 includes micro, small and medium enterprises in the following sizes: N = 60 micro companies (1–9 employees); N = 90—small companies (10–49 employees); N = 50—medium companies (50–250 employees). According to the study, "Smart Industry Poland 2018" Smart Industry is based on three pillars (Siemens 2018):

- digitizing information that allows to create a more effective value chain and more efficiently manage production processes at all levels,
- flexible and intelligent production technologies,
- modern communication using the technologies and capabilities of contemporary networks between market participants, systems and end users.

The pillars refer to manufacturing but also to logistics. The diagnosis of the innovativeness of the Polish economy allows to state that the Polish industry (with few exceptions) is in fact at the third stage of development, which is even before the Smart Industry stage.

Polish industrial companies associate smart production most of all with operational efficiency perceived from many different perspectives. Such an approach can be found in the conclusions presented in Siemens's Smart Industry reports. The most common answers to the question on how the respondents understand the concept under analysis focused on (Siemens 2018):

- efficiency in terms of production management, including the use of smart equipment and new technologies,
- production cost optimization possibilities,
- flexible response to customers' needs, with processes optimized in accordance with the customers' orders,
- modern communication that also includes sharing information directly with product consumers.

The study conducted by Siemens and Millward Brown on Smart Industry in 2016 (Siemens 2016) showed that only 25% representatives of Polish production companies were familiar with the concept of Smart Industry. At the same time, a significantly higher number of people declared that their organizations used technologies and solutions characteristic for smart factories. Elements of Smart Industry used and planned for implementation within surveyed companies comprised solutions (also referring to logistics):

- for the digitization of information enabling optimal and effective production management at all its levels—73.2%
- aimed at improving communication with the use of modern technologies and networks inside and outside the company—66%
- enabling the use of flexible and intelligent technologies in production, which can be influenced on a regular basis and quickly respond to the changing expectations of target recipients—58.8%.

According to Global Entrepreneurship Monitor (GEM) Poland has been included in the efficiency economies with the aspirations to join the group of innovation-oriented countries in recent years.

The results of the survey show that the technologies that give the greatest competitive advantage are the automation and robotization of production. The most commonly used solution currently supporting innovativeness is automation with the use of individual machines—in 48.6% of companies have already been implemented, and 10.4% of enterprises have plans for the next year. In turn, the robotization of the entire production line already was implemented in 14.3%, and implementation is planned in 3.6% of companies (see Fig. 3.).

Smart Industry 2018 Report concludes that 60% of entrepreneurs have not heard of the Industry 4.0 concept. However, this is not synonymous with the non-use of

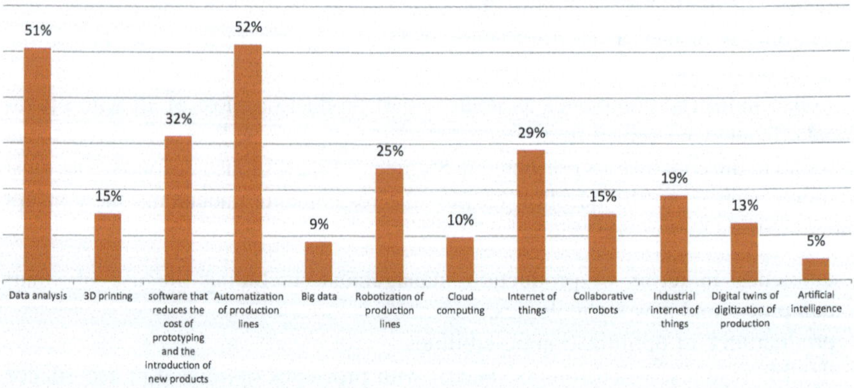

Fig. 3 Quantity of SMEs in Poland that use Smart Industry Technologies (*Source* own elaboration work based on Smart Industry 2018 report

modern technologies by these companies. It has already been proved by findings from previous reports 2016 and 2017 (Siemens 2017).

The most important barriers limiting the adaptation of technology are related to the lack of financial resources for such activity (almost 65% of responses), lack of time (62%) and lack of qualified staff (53%). Among external factors decelerating the innovation process with the use of technological solutions, the respondents indicated bureaucracy (over 78% of answers), difficulties in acquiring competent employees (71.5%) and lack of support from public authorities (71%).

Automation of production lines (52% of indications), data analysis (51%) and software supporting the reduction of prototyping costs of new products (32%) are the most frequently used technologies by SMEs (see Fig. 2).

At the same time, these are also technologies (including also robotization of production lines used by 25% of respondents) with the greatest impact on building competitive advantages and profits of the surveyed enterprises.

The survey Smart Industry 2018 was conducted on a nationwide sample of 200 companies from the industrial or production sector with the number of employees up to 249 employees, conducting production activity in Poland, i.e. having a production plant or plants operating in Poland. The analysis of the results takes into account the specificity of the heavy and light industry sector.

The least popular solution used in business practice is currently artificial intelligence, which was indicated only by 5% of respondents. At the same time, the relative low percentage of responses to Big Data and Cloud Computing solutions is puzzling.

Automatization of production lines, data analysis, software reducing cost of prototyping, robotization of production lines, Industrial Internet of things, collaborating robots, cloud computing and big data are use in practice in majority by medium sized companies. Whereas Internet of things and Artificial intelligence is used by medium and small enterprises in a similar range. Mainly micro companies use one-component-3D printing and more than 26% micro companies indicated that use this tool in the practice. SMEs assessed within survey considered Artificial Intelligence the most difficult to implement (45% of surveyed companies).

Hence, it can be stated that companies are striving for new solutions implementation, and consequently knowledge absorption.

Knowledge absorption on individual level was also briefly characterized based on secondary data. According to research by Polish Ministry of Entrepreneurship and Siemens (Siemens 2019) over 50% of engineers, working in industry think that educational offer is not well adjusted to requirements of innovative industry (see Fig. 4.). They lack knowledge and understand that the knowledge they need was not provided to them during educational processes.

The observation is recognized by academics and result in research initiatives undertaken to adjust the offer to requirements. The example of these initiatives are, among other University of the Future project (https://universitiesofthefuture.eu) and IE3: Industrial Engineering and Management of European Higher Education (https://ie3.eu/the-project/). The conclusion arising from the background research is the following: knowledge on Logistics 4.0 is disseminated and absorbed in the economy; however, the processes are not as efficient as they could be. The authors believe that it

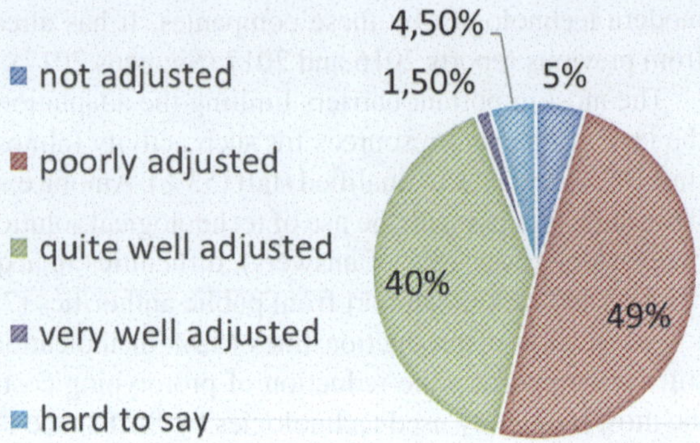

Fig. 4 Managers' perception of adjustment of educational offer to requirements of innovative industry (*Source* own elaboration work based on Smart Industry Report, 2019)

is caused by delays in both processes, dissemination and validation. The mechanism is to be illustrated by the original model to be presented in the next section.

5 Model Assumptions

As discussed above, individuals and companies challenged by a lack of knowledge resources, which is especially important because knowledge is a prerequisite for innovation and firms can generate long-term competitive advantages through innovation (Zerwas 2014). The stakeholders involved directly in knowledge dissemination and absorption are characterized in the Table 1.

As it is presented in the Table 1, the task of developing innovations is performer by both, academia and business, since academia can support companies when developing innovations with scientific expertise. There are programs (by local authorities, governments, UE etc.) financing such, cooperation and it can also take the form of industrial projects founded by companies themselves.

The lack of knowledge on Logistics 4.0 perceived at companies and individual level is being eliminated by educational offer of the system (including academia).

Table 1 Stakeholders of knowledge diffusion and absorption process

Stakeholder	Characteristics	Tasks
Companies	Industrial companies, implementing innovation, striving to increase their competitiveness, adjusting to market requirements	Searching for innovation, implementing innovations, developing solutions (innovations)
Academia	HEI and research centres generating knowledge	Conducting research on advanced solutions, developing innovations, providing educational offer for students, teaching

(*Source* own elaboration)

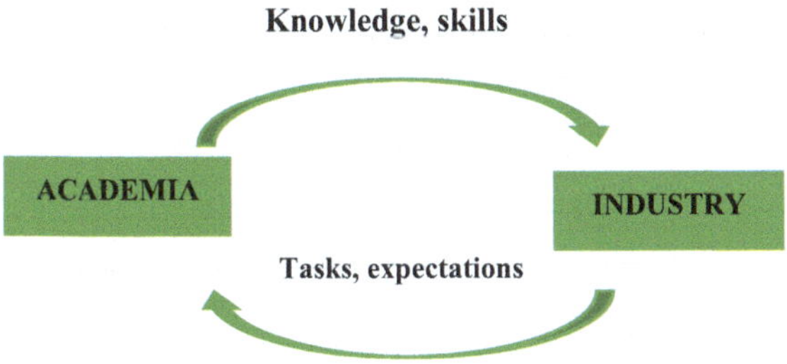

Fig. 5 Direct knowledge gap filling (*Source* own elaboration)

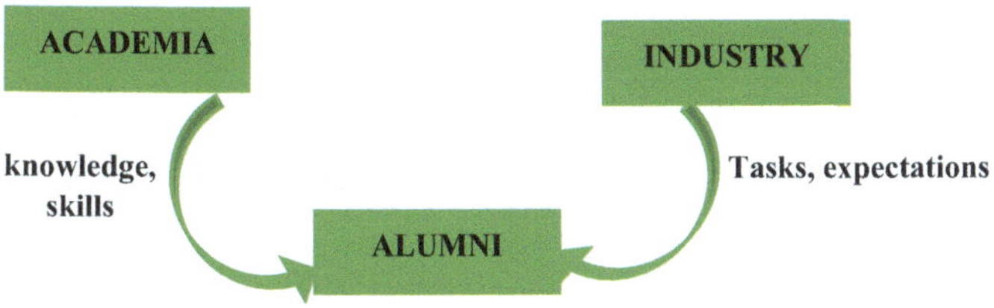

Fig. 6 Indirect knowledge gap filling (*Source* own elaboration)

The knowledge gap can be filled in directly or indirectly, as presented in the Figs. 5 and 6 respectively.

The direct knowledge gap filling is driven by requirements of companies needing knowledge and innovation implementation. It includes industrial projects and expertise by universities and its nature is determined by clear and comprehensive identification of needs. In this case, companies willing to implement or develop innovative solutions cooperate with academia on market basis, feeding it with data necessary to develop a solution. Product of such cooperation is customized to specific needs and requirements of companies. If the gap between expectations and products is recognized, it is immediately filled in. Knowledge flow is initiated by specified need; hence, it is pulled from both academia and industry.

The indirect knowledge gap filling is through the University alumni, equipped with knowledge and working for industry.

The alumni are the intermediary with accumulated knowledge they were provided with and they are supposed to use in their work. The nature of the phenomena is more complex. Students during their courses are taught and trained; knowledge flow is pushed by academia. The expectations of industry and tasks to be performed are communicated to alumni/employees and this is the moment when the gap is recognized. Recognized gap needs to be filled in and usually it is done by internal training, mentoring and courses by companies or self-learning by alumni/employees. The dimensions of the gap are to be the subject of the further research, followed by defining and providing tools to fill in the gap. According to authors, knowledge gap filling is a dynamic phenomenon that can be interpreted and analysed with system

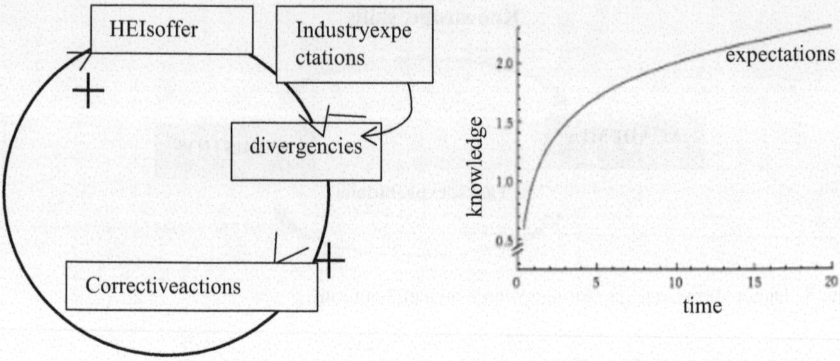

Fig. 7 Dynamics of gap filling: goal-seeking dynamic (*Source* own elaboration)

dynamics approach. The dynamics of knowledge gap filling could be presented with the model depicted in Fig. 7.

The loop dynamics is based on archetypes of system behaviour. The archetype was selected basing on the analysis of phenomena. The model shows that when considering the expectations and taking corrective actions, the knowledge gap can be easily filled in a given time period. Nevertheless, taking into account the operation mode of education system the process of knowledge gap closing or minimizing is disturbed by delays: in identification of divergences, as they are recognized after the students graduate, hence the feedback opportunities are limited. Moreover, there are some delays connected with designing and then implementing corrective actions. Thus, the Fig. 8 more likely reflects the gap filling dynamics.

The delays make the knowledge gap filling more difficult, and time- and cost-consuming. To deal with them, identification of knowledge needs and corrective

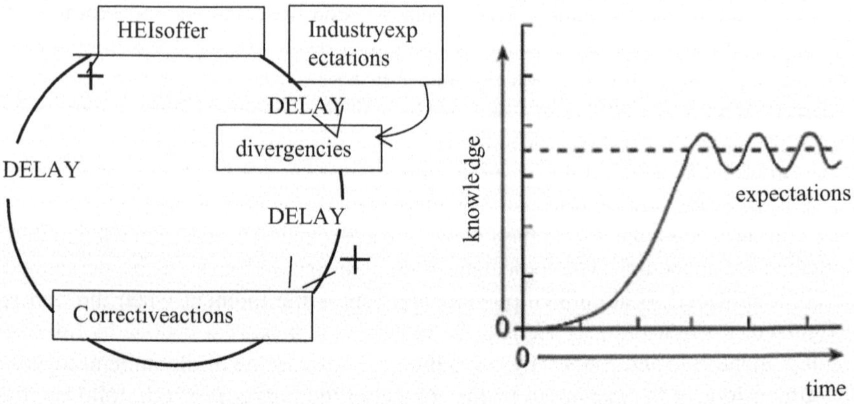

Fig. 8 Dynamics of gap filling: oscillatory dynamics (*Source* own elaboration)

actions design and implementation need to be the processes carriedon not only regular bases but also concurrent to providing education.

6 Conclusions

Contemporary economy generates numerous challenges for companies. Customers' requirements are changing immanently and these changes on one-hand follow technological changes, on the other induce them. The range and scope of changes in technology and management results in developing many methods and tools of different character but grouped under a flag of Industry 4.0. The general idea of Industry 4.0 is to benefit from implementation of contemporary technologies to improve efficiency and competitiveness, however, when referring specifically to material and information flows the Logistics 4.0 term is used. Logistics 4.0 concepts benefits from advanced communication protocols, cooperation between humans and machines, which requires knowledge and skills that are provided by higher education institutions. According to authors, knowledge gap filling is a dynamic phenomenon that can be interpreted and analysed with system dynamics approach. The dynamics of knowledge gap filling could be described as goal-seeking dynamics. The loop dynamics shows that when considering the expectations and taking corrective actions, the knowledge gap can be easily filled in a given time period. The goal of the paper is to present the original model of dissemination and absorption of Logistics 4.0 solutions developed based on literature review and authors' experience. The model presented shows the mechanism of knowledge diffusion and absorption. It show the mechanisms of processes and stresses the delays role. The greater the delay, the greater consequences—gaps in knowledge acquired and produced (disseminated). Knowing the mechanisms of the phenomenon enables it management. All the stakeholders involved namely academia, industry, alumni and student can potentially benefit from the research results:

a. Academia by understanding the importance of adjusting its educational offer to needs and requirements of contemporary industry, which will make its educational offer more attractive to students and future students and as results will improve its competitive position.
b. Industry by understanding the importance of acquiring employees equipped with relevant knowledge, which will probably contribute to developing and implementing high quality organizational solutions and innovations, increasing performance and competitiveness of companies.
c. Alumni by understanding the importance of knowledge, which is making them attractive for future employers and ready for job market challenges.
d. Students by stimulating their need for knowledge, as they will see it is useful and in-line with business and industrial practices.

Further research is in progress. It encompasses simulation experiments on system behaviour to identify the most important parameters of diffusion and absorption process, as well as critical sizes of delays and their influence on the size of the gaps.

References

Barreto L, Amaral A, Pereira T (2017) Industry 4.0 implications in logistics: an overview. Proceedia Manuf 13:1245–1252

Biesta G (2006) What's the point of lifelong learning if lifelong learning has no point? on the democratic deficit of policies for lifelong learning. Eur Educat Res J 5(3–4):169–180

Bubner NN, Bubner R,Helbig MJ (2014) Logistics trend radar, delivering insight today. Creating value tomorrow!, Pub. DHL Customer Solutions & Innovation, Troisdorf

Burcharth ALLA, Lettl C, Ulhøi JP (2015) Extending organizational antecedents of absorptive capacity: organizational characteristics that encourage experimentation. Technol Forecast Soc Change 90:269–284

Camison C, Fores B (2010) Knowledge absorptive capacity: New insights for its conceptualization and measurement. J Bus Res 63:707715

Ezzat AM, Kassem S (2019) Logistics 4.0: definition and historical background, novel intelligent and leading emerging sciences conference (NILES), Giza, Egypt, 2019, pp 46–49

Ferreras-Méndez JL, Newell S, Fernández-Mesa A, Alegre J (2015) Depth and breadth of external knowledge search and performance: The mediating role of absorptive capacity. Indust Market Manag 47:86–97

Ford A (2010) Modeling the environment. Island Press

Gebauer H, Worch H, Truffer B (2012) Absorptive Capacity, learning processes and combinative capabilities as determinants of strategic innovation. Eur Manag J 30:57–73

Heistermann F, van Hompel M, Mallée T (2018) Digitisation in logistics, BVL International www.bvl.de/en

Kostopoulos K, Papalexandris A, Papachroni M, Ioannou G (2011) Absorptive capacity, innovation, and financial performance. J Bus Res 64:1335–1343

Kostrzewski M, Varjan P, Gnap J (2020) Solutions dedicated to internal logistics 4.0. In: Grzybowska K, Awasthi A, Sawhney R (eds) Sustainable logistics and production in Industry 4.0. EcoProduction (Environmental Issues in Logistics and Manufacturing). Springer, Cham

Lopez-Caamal F, Middleton HR, Huber H (2014) Equilibria and stability of a class of positive feedback loops. J Mathemat Biol 68(3):609–645

Pai F-Y, Chang H-F (2013) The effects of knowledge sharing and absorption on organizational innovation performance—a dynamic capabilities perspective. Interdiscip J Inform Knowl Manag 8:83–97

Rüßmann M et al (2015) Industry 4.0: the future of productivity and growth in manufacturing industries. Boston Consulting Group (BCG), 1–14

Siemens, Raport Smart Industry Polska 2016, https://publikacje.siemens-info.com/pdf/123/Raport%20Smart%20Industry%20Polska%202016.pdf

Siemens, Raport Smart Industry Polska 2017, https://publikacje.siemens-info.com/webreader/00131-000758-raport-smart-industry-polska-2017/index.html

Siemens, Raport Smart Industry Polska 2018, https://publikacje.siemen-s-info.com/webreader/00175-001613-raport-smart-industry-polska-2018/index.html

Siemens, Raport Smart Industry Polska 2019, https://publikacje.siemens-info.com/pdf/594/Raport%20Smart%20Industry%20Polska%202019.pdf

Sterman JD (2000) Business dynamics: systems thinking and modeling for a complex world. McGraw Hill/Irwin

Strandhagen JO, Vallandingham LR, Fragapane G et al (2017) Logistics 4.0 and emerging sustainable business models. Adv Manuf 5:359

Szymańska O, Cyplik P, Adamczak M (2017) Logistics 4.0 A new paradigm or a set of solutions. Res Logist Product 7(4):299–310, Poznan University of Technology Publishing House, Poznan

Winkelhaus S, Grosse EH (2019) Logistics 4.0: a systematic review towards a new logistics system. Int J Product Res 58(1):18–43

Zerwas D (2014) The impact, meaning and challenges of knowledge absorption. In: Organizational Culture and Absorptive Capacity. Springer Gabler, Wiesbaden

The Supply Planning in the Distribution Network Using the Dynamic Programming Concept

Fertsch Marek

Abstract The chapter presents concept of method of supply planning in the distribution network using the dynamic programming. The main goal of the chapter is to present the results of the preliminary phase of research on supply planning methods in the distribution networks. The method presented in the chapter uses the dynamic programming for purpose of supply planning in the distribution network. A distribution network structure model has been developed. Algorithm for supply planning has also been developed. The chapter ends with a presentation of a plan for further research.

Keywords Distribution network · Supply planning · Dynamic programming

1 Introduction

Supply planning in distribution networks is based on two methods: the first is the so-called "traditional" method of operating a market. The second is the method of distribution requirement planning (DRP I).

The basis of the traditional method of operation of the market is supply planning, which assumes an evenly distribution of supply in the planning horizon at a constant delivery cycle. The size of supply for a given assortment item is determined on the basis of the maximum inventory norm, actual inventory, determined at the time of supply planning and the sales forecast to the next delivery. The described planning method works well in conditions of continuity and stability of demand. It is usually used in a situation of decentralized delivery planning, when ordering decisions are shifted to the level of the last link in the network (market). When this method is used, it is possible to use partial IT support to support individual activities in the planning process. A uniform standard of such support has not yet been developed. The author did not find any sources in the literature to identify the origin or author of

F. Marek (✉)
Faculty of Management Engineering, Poznan University of Technology, Jacka Rychlewskiego 2, 60-965 Poznań, Poland
e-mail: marek.fertsch@put.poznan.pl

© The Author(s), under exclusive license to Springer Nature Switzerland AG 2020
P. Golinska-Dawson et al. (eds.), *Smart and Sustainable Supply Chain and Logistics – Trends, Challenges, Methods and Best Practices*, EcoProduction,
https://doi.org/10.1007/978-3-030-61947-3_23

the traditional method of the market operation. This state prompts him to hypothesize that it arose a long time ago as a result of practical experience.

The method of distribution requirements planning is an adaptation for the needs of distribution the method material requirements planning (MRP I) originally developed for the production needs. The basis of its operation is a description of the structure of the distribution network. In this description, each node of the network (market or intermediate warehouse) is assigned a node from which deliveries of a specific assortment item take place. Each assortment item has a fixed delivery cycle and delivery quantity (alternatively, instead of a fixed delivery quantity the method of determining its size may be specified). Delivery plans are built for individual assortment item and then decomposed into supply plans for individual network nodes. The DRPI method works well in conditions of continuity and stability of demand. Is usually used in the situation of centralized supply planning, when the decisions on supply planning are moved to the level of network management. This method has a proven IT support standard. This is the so-called ERP system equipped with the DRPI module. Distribution requirements planning method, established in the 70 s of the last century, it is described in detail in the literature.

A characteristic feature of both methods discussed above is that they work well in conditions of continuity and stability of demand. In distribution networks, such conditions are enforced by actions in the areas of shaping the sale offer and determining the rules of its availability. The flexibility of operating and responding to changes in demand, which is necessary in practical conditions, is achieved through the interventions of planners.

2 Delivery Plan as an Object of Optimization—An Attempt to Systematize

Both of previously described supply planning methods are heuristic. They are widely used in practice. Their use allows for the development of satisfactory (but not optimal) plans. Under real conditions, their use does not ensure the development of optimal plans or identify how far the satisfactory plans are distant from the optimal plan. This situation has prompted many researchers to attempts to formulate the model of optimal supply planning. Many models of optimal supply planning and a large number of publications were created. There are so many of them that an exhaustive critical discussion of them all probably exceeds the capabilities of one researcher. One of the reasons for these difficulties is that the problem of optimal supply planning occurs in many areas of business practice. In addition to the distribution network, it occurs in the power industry as a problem of electricity distribution (Kersting 2012), in water supply (Bhave and Gupta 2006), in city logistics as a problem of the availability of public transport (Mason et al. 2017). A characteristic feature of each of the above-mentioned sub-areas is the occurrence of a number of unique restrictions resulting from the specificity of a given sub-area. In such a complex situation, the area of problems considered in this article will be limited to the sphere of physical distribution of material goods, characteristic of the operation of sales networks. The analysis of the state of knowledge in this area will be limited to an attempt to build a classification of supply planning methods used in of physical distribution of material goods found, in the literature.

The first of the approaches to optimal supply planning identified in the literature is the search optimum understood as minimal total cost network operations. The most widely presented example of such an approach in the literature is the minimum-cost flow problem—MCFP (Ahuja et al. 1993). MCFP is an optimization problem to find the cheapest possible way of sending a certain amount of flow through a flow network. Many models based on other criteria have also been developed For optimal supply planning such criteria are:

- minimum cost of delivery (Sifaleras 2013)
- minimizing the total journey time (Merline et al. 2012)
- minimizing transport work (Bravo and Vidal 2013)
- minimization of resource size (Ming-Fung et al. 2015)
- minimizing negative impact on the environment (Golinska et al. 2020)
- multi-criteria optimization (Bouvy and Lukas 2007).

The second approach identified in the literature for supply planning is an approach that can be described as integrated. It is based on the assumption that supply planning should be integrated with demand planning. The inspiration for an integrated approach is an article published in 2015 in Sloan Management Review (Tate et al. 2015). Following the literature on the subject and materials available on the Internet, one can get the impression that this publication has slightly interested the scientific sphere, but it has stimulated broad activity in the sphere of practice. The network has dozens of sources posted by consulting and training companies as well as software providers offering their services in the field of supply and demand integration.

Another of the approaches to supply planning identified in the literature is the approach that can be described as holistic (global). It does not consider the problem of optimal supply planning as a separate process, but connects it with the environment in which this process takes place (supply chain) or with other processes (selection of partners in the chain) (Mak and Su 2007).

Another approach identified in the literature for optimal delivery planning is an approach that can be described as tool-based. It does not focus on the optimality of the solution received but emphasizes the tools supporting the process of supply planning. Analyzing the materials available on the network, it can be seen that the tools offered are limited to ERP systems, sometimes supplemented with additional functionalities (advanced planning). Materials posted on the web (offers, system rankings) come from software suppliers or consulting companies.

3 Method of Supply Planning in the Distribution Network

The method presented in this article tries to combine two of the approaches briefly discussed above. The approach of searching for the optimal solution has been modified to strive to approach such a solution The criterion adopted in the method is to minimize total transport costs. However, the method does not guarantee finding the optimal solution. In turn, the tool approach drew inspiration to create an IT tool that will cooperate and support solutions used in practice (traditional market operation method, DRP I).

A distribution network structure model has been developed. This model can be implemented to describe the structure of the network using the Bill of Materials module in any ERP class system.

Model of distribution network structure:

The structure of the distribution network for which delivery planning takes place consists of the following elements:

a set of locations from which deliveries are made is given

$$D = \{d_i\}, \ i = 1, \ldots, n \tag{1}$$

where: n—the number of locations from which deliveries take place

- to each location d_i is assigned the amount of a given assortment item a_{ij} that a given location can provide in the planning horizon

$$A = \{a_{ij}\}, \ j = 1, \ldots, m, \ i = 1, \ldots, n \tag{2}$$

where: m—the number of items in the assortment that are subject to delivery

- a set of locations to which deliveries should be planned is given

$$S = \{s_k\}, \ k = 1, \ldots, p \tag{3}$$

where: p—the number of locations to which deliveries should be planned

$$p > n \tag{4}$$

The number of locations to which deliveries should be planned is greater than the number of locations from which deliveries take place

- for each location s_k, the quantity of a given assortment item z_{jk} is assigned which should be delivered to a given location in the considered planning horizon

$$Z = \{z_{jk}\}, \ j = 1, \ldots, m, \ k = 1, \ldots, p \tag{5}$$

- for any j—assortment item, the sum of the quantities that can be delivered from all n-locations from which deliveries take place is greater than the sum of the quantities to be delivered to all k—locations to which deliveries should be planned

$$\sum_{i} a_{ij} > \sum_{k} z_{jk} \tag{6}$$

- quantity of any j—assortment item quantity a_{ij} that can be delivered by any location d_i is smaller than the sum of the quantities to be delivered to all k—locations to which deliveries should be planned

$$a_{ij} < \sum_{k} z_{jk} \tag{7}$$

- any k—location to which deliveries should be planned can be supplied from any i—location from which deliveries take place
- each pair of locations consisting of any i—location from which deliveries take place and any k—location to which deliveries should be planned is assigned the parameter l_{ik} which specifies the distance between the two locations

$$L = \{l_{ik}\} \tag{8}$$

- the supply of transport services is higher than the demand
- Transport costs depend on the distance traveled and the size of the transport. They are calculated as the product of the distance, transport volume and rate per unit of transported load.
- The criterion and optimization model were formulated on the basis of the developed network structure model. They have the form:

  Criterion:

$$\sum_{I,k} l_{ik} x_{ik} \text{—min} \tag{9}$$

where: x_{ik}—volume of deliveries from i—location to k—location.
Restrictions: for any j—product item

$$\sum_{Ik} x_{ik} = \sum_{jk} z_{jk} \tag{10}$$

$$a_{ij} < \sum_{k} x_{jk} \tag{11}$$

The algorithm of the method operation was developed. It is based on dynamic programming (Fig. 1).

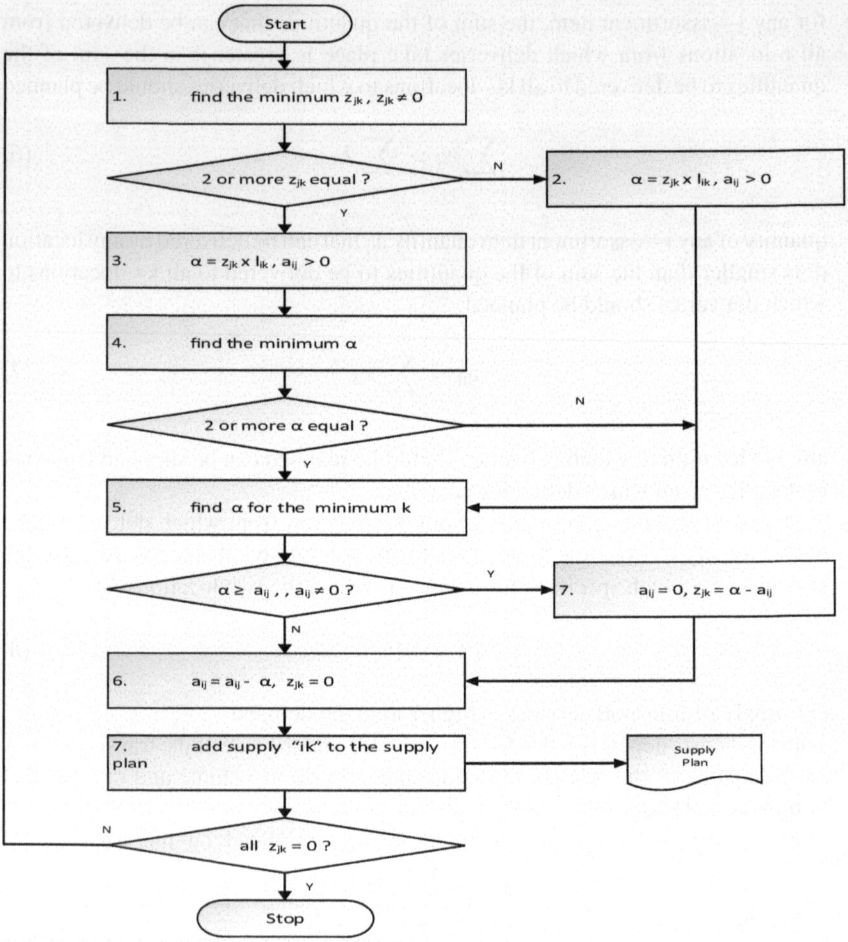

Fig. 1 Algorithm for supply planning (*Source* own work)

4 Conclusions

The method of dynamic supply planning in the distribution network presented in this article combines elements of two identified approaches to supply planning—searching for the optimal solution and a tool approach. It can complement and support the methods used in practice (traditional market operation method, DRP I). It uses the data generated by these methods as input in the distribution network model. The method does not find an optimal solution supply plan. Enables you to improve pre-solutions generated by other methods at selected stage of planning.

The further research program assumes:

- identification based on the literature of the reference method providing an optimal solution to the problem of developing an optimal delivery plan.
- development of a simple simulator, modeling the behavior of commonly used in practice delivery planning methods in cooperation with the developed method,
- conducting a numerical experiment and comparing the results obtained using the simulator and the reference method.

Acknowledgements The research was financed from the statutory research on topic no. 11/140/DSPB/4163 of the Faculty of Engineering Management at the Poznan University of Technology.

References

Ahuja RK, Magnanti TL, Orlin JB (1993) Network flows. theory, algorithms, and applications. Prentice Hall

Bhave PR, Gupta R (2006) Analysis of water distribution networks. Alpha Science International

Bouvy C, Lukas K (2007) Multicriterial optimization of communal energy supply concepts. Energy Conver Manag 48(11):2827–2835. https://doi.org/10.1016/j.enconman.2007.06.046

Bravo JJ, Vidal CJ (2013) Freight transportation function in supply chain optimization models: a critical review of recent trends. Expert Syst Appl 40(17):6742–6757 https://doi.org/10.1016/j.Esse.2013.06.06

Golinska P, Wang L, Hui M (2020) Research on joint emission reduction in supply chain based on carbon footprint of the product. J Cleaner Product 121086. https://doi.org/10.1016/j.jclepro.2020.121086

Kersting WH (2012) Distribution system modeling and analysis. CRP Press Taylor & Francis Group, Boca Raton, FL

Mak KL, Su W (2007) Partner selection and production-distribution planning for optimal supply chain formation, Proceedings of the World Congress on Engineering 2007 Vol II WCE 2007, July 2–4, 2007, London, U.K

Mason R, Trentini A, Lehuédé F, Malhéné N, Péton O et al (2017) Optimization of a city logistics transportation system with mixed passengers and goods. EURO J Transp Logist 6(1):81–109. Springer

Merline JV, Rithaa W, Abrahamb A (2012) Total time minimization of fuzzy transportation problem. J Intell Fuzzy Syst 23:93–99. https://doi.org/10.3233/IFS-2012-0498

Ming-Fung FS, Ming L, Simaan MA, Resource supply-demand matching scheduling approach for construction workface planning. J Constr Eng Manag 142(1). https://doi.org/10.1061/(ASCE)CO.1943-7.862.0001027

Sifaleras A (2013) Minimum cost network flows: problems, algorithms, and software. Yugoslav J Oper Res 23(1):3–17. https://doi.org/10.2298/YJOR121120001S

Tate WL, Mollenkopf D, Stank T, Lago da Silva A (2015) Integrating supply and demand. MIT Sloan Management Review

Dual Channel Warehouse Sizing Under a Piecewise Linear Warehousing Cost Structure

Fabian Alexander Torres Cárdenas and Carlos Eduardo Díaz Bohórquez

Abstract Since the recent growth of e-commerce, manufacturers are adapting traditional warehouses to dual-channel warehouses, which has a unique structure that is divided into two zones: the first one dedicated to fulfill online channel orders and the second dedicated to load and fulfill orders from the traditional retail channel. However, adding a new sales channel creates new challenges in terms of inventory management due to differences in order size and fulfillment times. In addition to the large uncertainty of demand in the short and long term, it makes planning for storage capacity and allocation complex. Shorter delivery times and the trend of urbanization have led to dual-channel warehouses often near cities where access to land is difficult and expensive. Manufacturers must find ways to maximize capacity utilization at their specific facilities or new facilities. The objective of this research is to establish an integrated inventory management and sizing model for the dual-channel warehouse, where the total size of the warehouse is determined, as well as the inventory policy for each of the channels. A linear storage cost structure by parts is assumed, a difference from typical economical lot size models, the storage cost structure examined in this case is not the simple unit rate type but rather a more realistic function of the warehouse space to be acquired. A heuristic method for solving the problem was formulated and validated using numerical examples according to the problem.

Keywords Dual-channel warehouse · Inventory management · Sizing warehouse · Online fulfilment · Probabilistic demand

---

F. A. Torres Cárdenas (✉) · C. E. Díaz Bohórquez
Universidad Industrial de Santander, Bucaramanga, Colombia
e-mail: Fabian.torres1@correo.uis.edu.co

C. E. Díaz Bohórquez
e-mail: cediazbo@uis.edu.co

© The Author(s), under exclusive license to Springer Nature Switzerland AG 2020
P. Golinska-Dawson et al. (eds.), *Smart and Sustainable Supply Chain and Logistics – Trends, Challenges, Methods and Best Practices*, EcoProduction,
https://doi.org/10.1007/978-3-030-61947-3_24

1 Introduction

Electronic commerce has shown significant growth in the last decade; according to the transfronterizo e-commerce study, the value of sales for 2019 is estimated at $3.4 trillion and will move $900 billion in international purchasing expenses by 2020. More than 2,140 million is expected by 2021 people buy goods and services online worldwide, compared to 1.66 billion global digital buyers in 2016 (Statista 2014). This is due largely to manufacturers operating mainly in the traditional retail channel are adding an online sales channel. These channels differ in opportunity costs from lost sales and therefore require different levels of service. That should be added at the general inventory level and a company fulfilment policy (Agatz et al. 2008). On the other hand, uncertainty in demand in the short and long term makes planning and allocating capacity complex. The scientific community has focused primarily on workforce planning, largely neglecting infrastructure and physical capacity issues such as storage area and equipment (Hübner et al. 2015). Therefore, warehouses or distribution centers must be prepared to fulfill orders from retail stores and online shoppers.

The conventional warehouse designed for order delivery from physical stores makes online order fulfillment complex, so the adaptation of the traditional warehouse to the dual-channel warehouse is gaining popularity, which has a unique structure that is divided into two zones: one to fulfill online channel orders and the other to store and fulfill traditional channel retail store orders, the advantages of this structure include reducing the cost of facilities through the construction of a warehouse integrated, reduced warehouse space and inventory required for both channels, increased coordination capacity and flexibility to fulfill online and offline orders, and higher service levels (Alawneh et al. 2017). However, with shorter delivery times, the urbanization trend has led to dual-channel warehouses often being located near cities where access to land is difficult and expensive, in which manufacturers seek ways to maximize the use of capacity in your existing facilities or the planning of a new facility in terms of storage size and the size of the storage areas for the traditional and online channel (Kembro et al. 2018).

Sizing is closely related to inventory policies, which are based on forecasting and demand analysis. This step should consider related costs and estimates of needs above and below, since an estimate of the excessive size of space needs could generate wasted space, while a smaller one can generate overcrowded conditions (Pires and Amorim 2015). Although, in the previous works they take into account the capacity of the warehouse, this is represented as a space constraint, they assume the size of the warehouse as a known parameter where the cost of storage is part of the cost of inventory maintenance and is generally assumed which is proportional to the amount of inventory stored (Allgor et al. 2004; Xu 2005; Alawneh et al. 2017). However, warehouse space in real environments is often only available in a few fixed sizes, for example, if a multi-store warehouse were to be built, the total available space would depend on the number of floor levels required or in the case In a leased warehouse, the leasing space units are generally located in discrete parts of the gross floor area (Goh et al. 2001). The main objective is to establish an integrated inventory management and sizing model for the dual-channel warehouse in which the total size of the warehouse is determined, as well as the inventory policy for each of the channels. The linear piecewise storage cost structure provides a more realistic cost value of the warehouse space to be purchased. Since the problem is a constrained nonlinear programming model, the bisection algorithm is applied to solve the dual Lagrange problem in which the capacity constraint is equal to zero to achieve a balance between purchased space and inventory level of each of the channels to minimize the cost of inventory and the cost of storage in the warehouse. The method was validated by numerical examples according to the problem.

The rest of this article is organized as follows. Section 2 presents the literature review. Section 3 presents the description of the problem and the notation. In Sect. 4 the integrated mathematical model. Section 5 exposes the methodology used. Section 6 modified bisection method. Section 7 presents and discusses numerical results and sensitivity analysis, respectively. Finally, the conclusion, discussion and future works are discussed in Sect. 8.

2 Literature Review

The distribution channel plays a very important role in the management of the supply chain. Initially, manufacturers produce the products and ship them to retailers, then customers visit retail stores to purchase the product based on their needs. This distribution system is called the traditional channel that still exists in business (Karim and Xu 2018). However, with the expansion of electronic commerce, new distribution systems have been created (Clemons et al. 2000) an example of these is the dual-channel supply chain, in which manufacturers in addition to selling in the traditional channel has added a direct online channel (Chiang and Monahan 2005).The direct online channel allows manufacturers to directly interact with customers to offer their products and services without intermediaries that guarantee a greater profit margin and minimize costs (Dumrongsiri et al. 2008; Cattani et al. 2006). This allows to collect information about customers, to understand their preferences, and also to produce personalized products that adapt to their needs (Batarfi et al. 2016).

Two research approaches were observed in dual-channel supply chain inventory management investigations; the first approach is coordination between channels and the second approach is competition between them. Chiang and Monahan (2005) proposed a dual channel supply chain coordination model, inventory is held in the manufacturer's warehouse, to meet both direct channel demand online and demand from a retail store; each channel receives independent requests; however, there is a probability of going to the other channel if any of them is out of stock. The authors develop an inventory control strategy, where they calculate the operating cost of the system by determining inventory holdings and lost sales costs, and compare the dual channel strategy with that of a single channel obtaining better results when using the dual channel strategy. Geng and Mallik (2007) studied the competition between a manufacturer-owned direct online channel and a retail channel. Considering that the manufacturer has limited production capacity. Defined that, in one channel you find out-of-stock products, a fraction of dissatisfied customers visit the other channel. They developed a Stackelberg approach, which is that one of the two channels is the leader, in this case the manufacturer and the other is the follower, the retailer. The leader has two decisions: the inventory level and the quantity of dispatch in each order to the retailer and based on these decisions the retailer decides his inventory level. They demonstrated that the manufacturer had the possibility of denying the inventory to the retailer even when its capacity was sufficient, obtaining greater benefit from this decision.

On the other hand, (Teimoury et al. 2008) developed a mathematical model for the dual channel supply chain between a manufacturer with a direct online channel and a retailer in which they define that when considering the costs of sales lost in a way Independent; that is, each channel manages its own cost, this can have a significant effect on the inventory level. Then (Takahashi et al. 2011) studied a dual channel supply chain in which they consider the decision of production and delivery of products and proposed an inventory control strategy, with the aim of reducing costs inventory maintenance, lost sales costs as well as production and delivery costs.

Warehouse management has been a fundamental aspect of managing the dual-channel supply chain since these have undergone major changes in recent decades, influencing mainly the design and operations of the warehouse. The complexity of warehouse operations has been multiplied by storing a large variety of SKUs in small quantities (De Koster et al. 2017). The small size of the transaction differs with the size of the store's replenishment order and requires pick-and-pack order fulfillment operations that are labor intensive (high variable cost) or automation (high capital investment) (Agatz et al. 2008; Hübner et al. 2015; Kembro et al. 2018). Therefore, the adaptation of operations and design to the particular context is receiving more attention in storage theory.

Two common strategies for warehouses in the dual channel business environment are decentralized and centralized policies. A company with a decentralized warehousing policy establishes an electronic fulfillment warehouse, where the online channel order is selected, packaged, and shipped independently to the primary warehouse, which means having separate inventory teams and business operations teams (Bendoly 2004; Alawneh et al. 2017; Ishfaq and Bajwa 2019).

Among the major studies related to decentralized policy highlights made by Allgor et al. (2004) and (Xu 2005). Allgor et al. (2004) investigated the setup of different inventory models in a compliance warehouse, which was divided into two areas: collection area, where items are stored individually in containers, and a reserve area, where items are stored in bulk on pallets bearing note the space constraint. This to minimize the cost of maintenance and labor, with a policy of periodic inventory review with stochastic demand. For its part, the study by (Xu 2005) considered delays in sending and preorders information. While the option of separating channels provides versatility in order fulfillment online, this option requires additional operating capital, as well as a higher level of inventory in the system, due to the need to maintain separate security stocks at various facilities in the distribution network (Ishfaq and Bajwa 2019).

On the other hand, the centralized policy is based on using a centralized warehouse, that is, an integrated warehouse or several warehouses grouped in the same location, to serve both direct online channel orders and traditional channel orders. Seifert et al. (2006) developed mathematical models for the decentralized and centralized dual channel supply chain. They showed that when the chain is centralized, the cost savings are significant, in addition to benefiting the retailer and customers. Yao et al. (2009) studied three different inventory strategies: (1) centralized, (2) Stackelberg's approach in which the retailer is the leader, and (3) a strategy in which the operation of direct channel online is outsourced to a third party logistics provider (Third Party Logistic, 3PL), in a dual channel supply chain comprising a manufacturer and a retailer, in which it defined the inventory level for both direct channel online from manufacturer to retailer, in order to maximize the expected profit.

The first study of the dual-channel warehouse was carried out by Alawneh et al. (2017) based on inventory management in the dual-channel supply chain, in which a manufacturer uses the warehouse that is divided into two areas with different levels inventory: one low-density area for online order fulfillment (Stage 1), where all products must be accessed, individually stored in containers, and the other high-density area to fulfill larger orders for Retail stores (Stage 2) that are stored on pallet and replenish low density area to fulfill online orders. They developed a multi-item inventory model under a continuous inventory review (Q, R) policy in which they consider the warehouse capacity constraint, the uncertainty in demand and the delivery time in which the inventory level is determined. for each of the channels and the replacement of each one of them. They highlight for future research the importance of determining the size of each of the areas (Stages 1 and 2) and how it affects total costs.

3 Problem Description and Notation

3.1 Problem Description

It is considered a dual channel warehouse owned by a manufacturer under a decentralized policy which satisfies the demand of the traditional channel of retail stores and the direct online channel of the manufacturer. Orders placed by the manufacturer reach the reserve area (Step 2) from where it is delivered to retail stores and exclusive collection area for online order fulfillment (Stage 1); An important difference between these channels is the order delivery time, for stores they tend to be longer because they are scheduled in advance and for online orders that have a shorter time due to the speed required to fulfill them. Due to the importance of having an integrated policy to define the inventory level of each of the channels and an adequate warehouse size to meet the demand of both retail stores and the direct online channel, an integrated inventory management model is proposed. and sizing for the double-channel warehouse in which the total size of the warehouse is established, as well as that of each of the storage areas, the inventory management policy, taking as a reference that the demand is uncertain, as well as the time delivery. Difference from previous studies this study takes the storage cost structure behaves as a piecewise linear function based study (Goh et al. 2001) for sizing stores. In which \gamma: storage space required by a maintenance unit is the space, *bn*: Annualized fixed cost to acquire an amount of space *sn* and *cn*: Variable cost of storage per unit of item in *sn*. Thus, integrating the inventory management and sizing model.

3.2 Notations and Assumptions

The following notation and assumptions were used for this investigation

i	Item index
j	Stage index, where j = 1 for warehouse area dedicated to satisfying online demand (online picking area), and j = 2 for warehouse area dedicated to satisfying both retail and dedicated online area demands
n	Sizes of warehouse index
L_{ij}	Length of lead time for item i in stage j (random variable)
D_{ij}	Expected annual demand for item i in stage j
h_{ij}	Holding cost per unit time for item i at stage j
B_{ij}	Backorder cost per unit for item i at stage j
A_{ij}	Ordering cost per order for item i at stage j
x_{ij}	Demand during lead time (DDLT, random variable) for item i in stage j
$f(x_{ij})$	Probability density function of lead-time demand for item i at stage j
μ_{ij}	Mean of item i demand in stage j
σ_{ij}	Standard deviation of item i demand in stage j
γ_{ij}	Storage space required by a stock keeping unit in stage j
α	Minimum required probability that total order quantities will be within warehouse space
S_n	Discrete choices of warehouse sizes available
b_n	The annualized fixed cost of acquiring S_n
C_n	Variable storage cost per unit item en el size warehouse S_n
θ	Lagrange multiplier for the warehouse space constraint
Q_{i2}	Order quantity for item i in Stage 2
Q_{i1}	Order quantity for item i in Stage 1
R_{i2}	Reorder point when new order is placed for item i in Stage 2
R_{i1}	Reorder point when new order is placed for item i in Stage 1

(continued)

(continued)

i	Item index
C	total inventory cost and warehousing costs: Please note that the first paragraph of a section or subsection is not indented. The first paragraphs that follows a table, figure, equation etc. does not have an indent, either.

The assumptions used in this research:

- The demand rate per unit time is a random variable and the item lead times are constant.
- The inventory policy used is a continuous review (Q, R).
- Backorder conditions occur when a request cannot be met from the existing provisions with a prescribed penalty fee.
- Each stage has its own reorder point value obtained from the sum of DDLT expectation values and safety stock.
- For the case of multi-SKUs stored in the same warehouse, we assume that each type of SKU is assigned some dedicated storage space. Consequently, the total warehouse space required is the sum of the ordering sizes for all the SKUs stored.

4 Mathematical Model

Objective: Min the total expect cost (the combined inventory and warehousing costs)

$$\begin{aligned} \operatorname{Min} C(Q_{ij}, R_{ij}) &= \sum_i \sum_j \frac{A_{ij} D_{ij}}{Q_{ij}} \\ &+ \sum_i \sum_j h_{ij} \left[\left(\frac{Q_{ij}}{2} \right) + \left(R_{ij} - \mu_{x_{ij}} \right) \right] + \sum_i \sum_j \frac{B_{ij} D_{ij}}{Q_{ij}} \left[\int_{R_{ij}}^{\infty} (x_{ij} - R_{ij}) f(x_{ij}) dx_{ij} \right] \\ &+ b_n + \sum_i \sum_j C_n \left(Q_{ij} + \left(R_{ij} - \mu_{x_{ij}} \right) \right) \end{aligned} \tag{1}$$

The first term of the objective function (1) refer to the annual ordering cost, which is the order cost multiplied by the number of cycles. The second term refer to the annual approximated holding cost. The third term represent the annual backorder cost, which is equal to the backorder cost multiplied by the expected number of shortages per cycle. The fourth term refer to the annualized fixed cost of acquiring S_n. The fifth term refer to variable storage cost per unit item en el size warehouse S_n, the annualized variable cost component related to the actual quantity stored and handled. Reorder point is determined by the following equation $R_{ij} = \mu_{x_{ij}} + z_{ij}\sigma_{ij}$ formulated as a function of the safety factor z_{ij} so $z_{ij}\sigma_{ij} = \left(R_{ij} - \mu_{x_{ij}}\right)$.

The space constraint

$$\sum_i \sum_j \gamma_{ij}\left(Q_{ij} + z_{ij}\sigma_{ij}\right) - S_n - \mu_Y - Z_{1-\alpha} * \sigma_Y \leq 0 \tag{2}$$

Space constraint (2) considers the space occupied by the sum of all the orders plus the security inventory minus the space that the average demand would occupy during the lead time Eq. (4). Because of uncertain demand, it is set the probability that the total simultaneous items inventory within the warehouse space when the order is received will not be smaller than α (Alawneh et al. 2017). If the DDLT for item i in stage j is in a situation where the demand and lead time are normally distributed and statistically independent, then the mean and standard deviation of the DDLT are:

$$\mu_{x_{ij}} = \mu_{L_{ij}} \times \mu_{d_{ij}} \sigma_{x_{ij}} = \sqrt{\mu_{L_{ij}} \times \sigma_{d_{ij}}^2 + \mu_{d_{ij}}^2 \times \sigma_{L_{ij}}^2} \tag{3}$$

$$\mu_Y = \sum_i \sum_j \gamma_{ij} \mu_{ij};\, \sigma_Y^2 = \sum_i \sum_j \gamma_{ij}^2 \sigma_{ij} \tag{4}$$

The expected shortage per cycle (Kundu and Chakrabarti 2012; Alawneh et al. 2017).

$$ESC(R_{ij}) = \int_{R_{ij}}^{\infty} (x_{ij} - R_{ij}) f(x_{ij}) dx_{ij} = \frac{\sigma_{x_{ij}}}{2}\left(\sqrt{1 + z_{ij}^2} - z_{ij}\right) \tag{5}$$

We consider a Lagrange function

$$\begin{aligned} L(Q_{ij}, z_{ij}, \theta) = {} & \sum_i \sum_j \frac{A_{ij} D_{ij}}{Q_{ij}} + \sum_i \sum_j h_{ij}\left[\left(\frac{Q_{ij}}{2}\right) + z_{ij}\sigma_{x_{ij}}\right] + \sum_i \sum_j \frac{B_{ij} D_{ij}}{Q_{ij}}\left[\frac{\sigma_{x_{ij}}}{2}\left(\sqrt{1 + z_{ij}^2} - z_{ij}\right)\right] \\ & + b_n + \sum_i \sum_j C_n (Q_{ij} + z_{ij}\sigma_{ij}) + \theta\left[\sum_i \sum_j \gamma_{ij}\left(Q_{ij} + z_{ij}\sigma_{x_{ij}}\right) - s_n - \mu_Y - z_1 - \alpha\right] \end{aligned} \tag{6}$$

Using the necessary Karush-Kuhn-Tucker (KKT) conditions for minimization problems, we obtain

$$\frac{\partial L}{\partial Q_{ij}} = 0,\ -\frac{A_{ij} D_{ij}}{Q_{ij}} + \frac{h_{ij}}{2} - \frac{B_{ij} D_{ij}\left(\frac{\sigma_{x_{ij}}}{2}\left(\sqrt{1 + z_{ij}^2} - z_{ij}\right)\right)}{Q_{ij}^2} + C_n + \theta_{\gamma_{ij}} = 0 \tag{7}$$

Getting:

$$Q_{ij} = \sqrt{\frac{2D_{ij}\left[A_{ij} + B_{ij}\left(\frac{\sigma_{x_{ij}}}{2}\left(\sqrt{1 + z_{ij}^2} - z_{ij}\right)\right)\right]}{h_{ij} + 2\gamma_{ij}\theta + 2C_n}} \tag{8}$$

$$\frac{\partial L}{\partial z_{ij}} = 0,\ h_{ij}\sigma_{ij} + \frac{B_{ij} D_{ij}}{2Q_{ij}}\left[\sigma_{x_{ij}}\left(\frac{z_{ij}}{\sqrt{1 + z_{ij}^2}}\right)\right] + C_n \sigma_{x_{ij}} + \theta \gamma_{ij} \sigma_{x_{ij}} = 0 \tag{9}$$

If we substitute (7) into (8), we have

$$\frac{B_{ij}D_{ij}}{\sqrt{\frac{2D_{ij}\left[A_{ij}+B_{ij}\left(\frac{\sigma x_{ij}}{2}\left(\sqrt{1+z_{ij}^2}-z_{ij}\right)\right)\right]}{h_{ij}+2\gamma_{ij}\theta+2C_n}}}\left[\sigma_{x_{ij}}\left(\frac{z_{ij}}{\sqrt{1+z_{ij}^2}}-1\right)\right]+h_{ij}\sigma_{ij}+C_n\sigma_{x_{ij}}+\theta\gamma_{ij}\sigma_{x_{ij}}=0 \tag{10}$$

The value of z_{ij} is determined using Matlab software given the complexity of solving z_{ij}

$$\frac{\partial L}{\partial \theta}=\sum_i\left(\gamma_{i2}(Q_{i2}+\sigma_{x_{i2}}z_{i2})+\gamma_{i1}(Q_{i1}+\sigma_{x_{i1}}z_{i1})\right)-S_n-\mu_Y-z_{1-\alpha}\leq 0 \tag{11}$$

5 Methodology

Data: This is a theoretical investigation which is based on secondary data from the numerical examples proposed by (Alawneh et al. 2017) inventory management data such as: order cost, inventory maintenance cost, cost of pending orders are also selected Specification of the product based on the demand and space requirements of each product in each storage area, for the warehouse sizing, the parameters of storage cost by parts of the work performed by (Fern 2007) are taken as reference: cost storage variable per item unit, annualized fixed cost to acquire a quantity of space (warehouse size), available warehouse sizes.

Inventory model. The inventory model used is the Probabilistic Economic Order Quantity (EOQ) model with a continuous review system for probabilistic demand conditions and warehouse sizing. The determination of the inventory model parameters is carried out with the following steps: (1) Determine the size of the order lot and the order points with pending order without space restriction; (2) Apply the bisection algorithm to find the solution of the Lagrange multiplier starting from the verification of the space restriction; (3) Calculate the area requirements for products in each storage area; (4) Calculate the quantity of the order and the point of order, with pending orders and restrictions for the warehouse area in which it is taken into account for the dimensioning of the warehouse is available in discrete sizes in which the total size of warehouse and storage areas; (5) Finally, determine the total expected inventory cost. (6) Sensitivity Analysis: Sensitivity analysis was conducted to determine the effect of changes in parameters on optimal ordering lot size and total inventory costs.

6 Modified Bisection Method

It consists of starting from an interval [x0, x1] such that f (x0) f (x1) < 0, so we know that there is at least one real root. From this point the interval is reduced successively until it becomes as small as the precision that we have decided correctly requires. $a_n \le k_n \le b_n$ where the initial values are given by: a_0: = a and b_0: = b (Fern 2007)

$$k_n = x_{n+1} = \frac{(a_n + b_n)}{2} \tag{12}$$

$$a_{n+1} = \{a_n \text{ if } f(a_n).f(k_n)\langle 0;\ k_n \text{ if } f(a_n).f(k_n)\rangle 0\} \tag{13}$$

$$b_{n+1} = \{b_n \text{ if } f(b_n).f(k_n)\langle 0;\ k_n \text{ if } f(b_n).f(k_n)\rangle 0\} \tag{14}$$

Based on the algorithm proposed by Alawneh et al. 2017) adapted bisection search to the sizing problem, calculating the best solution for each of the warehouse sizes to solve the Lagrangian dual problem. For a given value of θ, Qij and zij can be calculated using (7) or (9); then they can be substituted into Eq. (10). This reduces the problem to a solution for one equation with one unknown θ in which each of the warehouse sizes is evaluated:

$$g(\theta) = \sum_i \left(\gamma_{i2}(\widetilde{Q}_{i2} + \sigma_{x_{i2}}\widetilde{z}_{i2}) + \gamma_{i1}\left(\widetilde{Q}_{i1} + \sigma_{x_{i1}}\widetilde{z}_{i1}\right)\right) - S_n - \mu_Y - z_{1-\alpha} = 0 \tag{15}$$

1. Let $a_0 = \theta_1 = 0$ and let $b_0 = \theta_2$ must be a number, such that $g(\theta) < 0$.
2. Let $\widetilde{Q}_{a,}\widetilde{z}_a$ be the solution when $\theta = \theta_1$, $\widetilde{Q}_{b,}\widetilde{z}_b$ and let be the solution when $\theta = \theta_2$.
3. Let $k_n = \theta = \frac{\theta_1+\theta_2}{2}$ and solve for $\widetilde{Q}$ and $\widetilde{z}$; find $g(\theta)$.
4. If $g(\theta) > 0$, then $\theta_1 = \theta$, $\widetilde{Q}_1 = \widetilde{Q}$, and $\widetilde{z}_1 = \widetilde{z}$; if $g(\theta) < 0$, them $\theta_2 = \theta$, $\widetilde{Q}_2 = \widetilde{Q}$, and $\widetilde{z}_2 = \widetilde{z}$.
5. If $(g(\theta_1) - g(\theta_2)) < \varepsilon_g$, then stop. Otherwise, go to 3.
6. This process is repeated for each of the available sizes of warehouse S_n.

7 Numerical Results and Sensitivity Analysis

Numerical examples and results Table 1 were taken as reference for the warehouse sizing an example of four discrete choices of warehouse sizes for the inventory costs proposed by Yao and Huang (2017) and Table 2 shows an example proposed by Alawneh et al. (2017) for an item normal distribution demand and stochastic lead time.

Table 1 An example of four discrete choices of warehouse sizes

n	1	2	3	4
Sn	(0, 3817)	(3817, 11391)	(11391, 17840)	(17840, ∞)
bn	157.4	212.4	338.2	457.5
Cn	9.35	8.03	6.42	4.61

Table 2 Example 1 normal distribution demand and stochastic lead time

Parameter	Value	Parameter	Value	Parameter	Value
i	1	$\sigma12$	50	$\gamma11$	100
j	1,2	A11	40	$\gamma12$	50
D_{11}	120	A12	4000	z1-α	−1.3
D_{12}	1600	h11	20		
μ_{11}	30	h12	10		
μ_{12}	750	b11	50		
σ_{11}	10	b12	2000		

Table 3 Results for Example 1 with normal distribution demand and stochastic lead time

Sn	g(θ)	θ	R_{11}	R_{12}	Q_{11}	Q_{12}	Total cost
S_1	−0.001988	0.73795	36	1019	16	573	39457
S_2	−0.000104	0.45143	37	1045	19	690	35152
S_3	0.0097095	0.30443	38	1063	22	794	32051
S_4	−0.003945	0.19814	39	1079	25	913	28982

Table 3 shows the results for each of the warehouse sizes. being the warehouse size S4 the one with the lowest total cost for example 1. approximate total warehouse size 25000, stage j: 1 = 2500; stage j: 2 = 22500;

Table 4 shows the inventory and storage costs in which size S_4 has the highest Annual holding cost and Annualized fixed cost of acquiring Sn, but the costs of

Table 4 Inventory and storage costs results for Example 1

Sn	Annual ordering cost	Annual holding cost	Annual backorder cost	Annualized fixed cost of acquiring Sn	Variable storage cost per unit item
S_1	11460	5834.6	13922	157.4	8082.3
S_2	9525.9	6737.1	10554	212.4	8122.5
S_3	8282.2	7481.8	8649.4	338.2	7299.1
S_4	7198.4	8284.9	7157.9	457.5	5883.8

Table 5 An example two items normal distribution demand and stochastic lead time

Parameter	Value	Parameter	Value	Parameter	Value
i	1,2	σ11	0.5	B21	0.5
j	1,2	σ12	4	B22	8
D_{11}	240	σ21	0.3	h11	1
D_{12}	2400	σ22	2.9	h12	10
D_{21}	350	A11	50	h21	0.5
D_{22}	4500	A12	125	h22	8
μ_{11}	3	A21	40	γ11	25
μ_{12}	120	A22	100	γ12	12
μ_{21}	2.5	B11	10	γ21	15
μ_{22}	100	B12	60	γ22	7

Annual ordering cost, Annual backorder cost and Variable storage cost. per unit item are less than others.

Table 5 shows the data corresponding to an example with 2 products with normal demand and stochastic lead time.

Table 6 shows the results for each of the warehouse sizes. being the warehouse size S_1 the one with the lowest total cost for example 2. approximate total warehouse size 3817, stage j: 1 = 625; stage j: 2 = 3192.

Table 7 shows the inventory and storage costs in which size S_1 has the highest Annual ordering cost and Annual backorder cost, but the costs of Annual holding cost, Annualized fixed cost of acquiring Sn and Variable storage cost per unit item are less than others.

Table 6 Results for example 2 two items with normal distribution demand and stochastic lead

Sn	g(θ)	θ	R_{11}	R_{12}	R_{21}	R_{22}	Q_{11}	Q_{12}	Q_{21}	Q_{22}
S_1	−0.0003	0.840	3.242	136.7	2.65	106.0	22.01	235.7	29.06	286.2
S_2	0.0014	0.115	3.352	143.9	2.70	109.0	40	559.2	49.49	689.9
S_3	−0.0157	0.026	3.361	146.1	2.70	110.0	47.15	846.9	56.46	1072
S_4	0.0005	0.025	3.361	146.1	2.70	110.0	47.27	854.0	56.57	1082

Table 7 Inventory and storage costs results for Example 2

Sn	Annual ordering cost	Annual holding cost	Annual backorder cost	Annualized fixed cost of acquiring Sn	Variable storage cost per unit item
S_1	3871.5	439.21	160.89	157.4	5575.2
S_2	1771.5	883.64	50.585	212.4	11018
S_3	1276.2	1189.7	32.347	338.2	13225
S_4	1268.3	1196.8	32.088	457.5	9576.2

Table 8 Effect of model parameters on the optimal solutions

Scenario	R_{11}	R_{12}	R_{21}	R_{22}	Q_{11}	Q_{12}	Q_{21}	Q_{22}	Total cost
D	3.24	136.75	2.65	106.0	22.01	235.74	29.06	286.22	10204
D + 10%	3.25	136.78	2.65	106.09	22.14	235.42	29.30	285.73	10602
D-10%	3.23	136.72	2.65	106.08	21.86	236.13	28.78	286.81	9807.2
μ + 10%	3.55	149.07	2.90	116.21	22.73	244.93	29.96	297.46	10282
μ-10%	2.93	124.43	2.40	95.95	21.28	226.58	28.15	275.03	10138
B + 10%	3.27	137.45	2.65	106.37	21.76	237.61	28.70	283.10	10222
B-10%	3.20	136.01	2.65	105.77	22.27	233.77	29.45	289.56	10186
S1-10%	3.22	136.20	2.65	105.86	20.76	220.24	27.51	267.28	10101
S1-20%	3.20	135.63	2.65	105.63	19.49	204.81	25.90	248.46	10039

7.1 *Sensitivity Analysis*

Table 8 shows the results obtained, in terms of reorder point and order quantity for an increase and decrease in 10% each parameter: demand (total expected demand and the average of DDLT), the cost of pending order and the available warehouse space. Increasing and decreasing the expected annual demand by 10%, the total cost increases and decreases by an average of 3.90% and 3.88% respectively. Increasing and decreasing DDLT by 10%, the total cost increases and decreases by an average of 0.76% and 0.64% respectively. If the warehouse size is decreased by 20% compared to 10% the total cost decreases by 1.67%. Finally, If the warehouse size is decreased by 20% compared to 10%, the total cost decreases by 1.67%, the reorder point is not affected by the change.

8 Conclusion

In this work, a joint dimensioning and inventory management model for a dual-channel warehouse was developed, which sought to establish the indicated size of the warehouse and the inventory policy, minimizing the total expected cost, bearing in mind that the cost of storage behaved as a piecewise linear function that provides an approach to current storage costs. considering the uncertainty in the demand and in the lead time. it was solved using the bisection search method and validated using numerical examples.

Performing warehouse sizing and inventory management together provides flexibility in defining the appropriate warehouse size where the required inventory level can be met for each of the sales channels, be it for a single item in which it was determined that the size that provided the least cost was the largest, unlike the example of multiple items, the most appropriate warehouse size was the smallest, taking as a reference that the costs that most influence the decision are Annual holding cost,

Annual backorder cost and Variable storage cost per unit item. In the sensitivity analysis, it was shown that the parameter that has the greatest impact on total cost is the increase in demand and the average decrease between 3.90% and 3.88% respectively. For future research, consider joint replenishment for each of the stages to make an approach to the current electronic commerce environments. Additionally, consider a random storage using the warehouse design with mixed shelves for the online picking area as a reference. Also, future investigations formulate different methods of solution such as heuristics and metaheuristics in order to evaluate the problem with larger examples.

References

Agatz NAH, Fleischmann M, van Nunen JAEE (2008) E-fulfillment and multi-channel distribution—a review. Eur J Oper Res 187(2):339–356

Alawneh F, Zhang G (2018) Dual-channel warehouse and inventory management with stochastic demand. Transp Res Part E Logist Transp Rev 112:84–106

Allgor R, Graves S, Xu PJ (2004) Traditional inventory models in an e-retailing setting: atwo-stage serial system with space constraints. In: Proceedings on 2004 SMA Conference, pp 6–12

Batarfi R, Jaber MY, Zanoni S (2016) Dual-channel supply chain: a strategy to maximize profit. Appl Math Model 40(21–22):9454–9473

Bendoly E (2004) Integrated inventory pooling for firms servicing both on-line and store demand. Comput Oper Res 31(9):1465–1480

Cattani K, Gilland W, Heese HS, Swaminathan J (2006) Boiling frogs: pricing strategies for a manufacturer adding a direct channel that competes with the traditional channel. Prod Oper Manag 15(1):40–56

Chiang WK, Monahan GE (2005) Managing inventories in a two-echelon dual-channel supply chain. 162:325–341

Clemons E, Gu B, Row M (2000) eCommerce and eDistribution : understanding the role of power when selecting alternatives channel strategies. Melbourne institute of applied economic and social research working paper, pp 1–39

De Koster RBM, Johnson AL, Roy D (2017) Warehouse design and management. Int J Prod Res 55(21):6327–6330

Dumrongsiri A, Fan M, Jain A, Moinzadeh K (2008) A supply chain model with direct and retail channels. Eur J Oper Res 187(3):691–718

Fern C (2007) Convergencia Lineal y el algoritmo de bisección. Rev Educ Matemática 22(3):16–21

Geng Q, Mallik S (2007) Inventory competition and allocation in a multi-channel distribution system. Eur J Oper Res 182(2):704–729

Goh M, Jihong O, Chung-Piaw T (2001) Warehouse sizing to minimize inventory and storage costs. Nav Res Logist 48(4):299–312

Hübner A, Holzapfel A, Kuhn H (2015) Operations management in multi-channel retailing: an exploratory study. Oper Manage Res 8(3–4):84–100

Ishfaq R, Bajwa N (2019) Profitability of online order fulfillment in multi-channel retailing. Eur J Oper Res 272(3):1028–1040

Karim MT, Xu Q (2018) Review Dual-channel supply chain pricing and coordination. Int J Bus Appl Soc Sci 4(9):93–108

Kembro JH, Norrman A, Eriksson E (2018) Adapting warehouse operations and design to omnichannel logistics: a literature review and research agenda. Int J Phys Distrib Logist Manage 48(9):890–912

Kundu A, Chakrabarti T (2012) A multi-product continuous review inventory system in stochastic environment with budget constraint. Optim Lett 6(2):299–313
Pires M, Amorim P (2016) Design of retail backroom storage : a research opportunity ? Design of retail backroom storage: a research opportunity? No. September, 201
Seifert RW, Thonemann UW, Sieke MA (2006) Integrating direct and indirect sales channels under decentralized decision-making. Int J Prod Econ 103(1):209–229
Takahashi K, Aoi T, Hirotani D, Morikawa K (2011) Inventory control in a two-echelon dual-channel supply chain with setup of production and delivery. Int J Prod Econ 133(1):403–415
Teimoury E, Mirzahosseinian H, Kaboli A (2008) A mathematical method for managing inventories in a dual channel supply chain. Int J Ind Eng Prod Res 19(4):31–37
Statista (2019) Statista. Number of digital buyers worldwide from 2014 to 2021 (in billions)
Xu PJ (2005) Order Fulfillment in Online Retailing: What Goes Where. MIT dissertation
Yao MJ, Huang JY (2017) Optimal lot-sizing and joint replenishment strategy under a piecewise linear warehousing cost structure. J Intell Manuf 28(3):791–803
Yao DQ, Yue X, Mukhopadhyay SK, Wang Z (2009) Strategic inventory deployment for retail and e-tail stores. Omega 37(3):646–658

Lagrangian Dual Decomposition for Two-Echelon Reliable Facility Location Problems with Facility Disruptions

Mohammad Rohaninejad, Zdeněk Hanzálek, and Reza Tavakkoli-Moghaddam

Abstract This chapter considers the two-echelon supply chain network design with unreliable facilities when nodes related to facilities in both echelons fail under disruptions. A new mixed-integer programming (MIP) model is proposed for a reliable facility location with possible customer reassignment in different probabilistic scenarios. The maintaining of the materials flow between different echelons of the network is investigated under network disruptions. The performance of global optimization is investigated by comparing this approach with independent and non-integrated optimization. The objective function of the problem seeks to minimize expected costs, including fixed and service costs in the supply chain, such that maintaining the demand flow in both echelons of the network interconnects them. The medium- and large-sized problems are solved using a custom-designed Lagrangian dual decomposition algorithm. Our computational results show that the proposed algorithm is efficient for the given problems, efficiently overcomes the computational complexity of the problems, and provides good-quality solutions within an acceptable time.

Keywords Supply chain optimization · Reliable facility location · Dual decomposition · Location and allocation

1 Introduction

Facility location problems involve the study of determining locations of $|J|$ facilities with the finite or infinite capacity among $|I|$ demand points, and making assignment decisions to serve a set of clients. The objectives of location and allocation are to achieve a trade-off between first-stage setup costs and second-stage service

M. Rohaninejad (✉) · Z. Hanzálek
Czech Institute of Informatics Robotics and Cybernetics, Czech Technical University in Prague, Prague, Czech Republic
e-mail: rohaninejad.sm@gmail.com

R. Tavakkoli-Moghaddam
School of Industrial Engineering, College of Engineering, University of Tehran, Tehran, Iran

© The Author(s), under exclusive license to Springer Nature Switzerland AG 2020
P. Golinska-Dawson et al. (eds.), *Smart and Sustainable Supply Chain and Logistics – Trends, Challenges, Methods and Best Practices*, EcoProduction,
https://doi.org/10.1007/978-3-030-61947-3_25

and transportation costs. Facility location problems have extensively been studied; various types of facilities (e.g., factories, warehouses, stores, airports, hospitals, and emergency departments) have been examined (Rohaninejad et al. 2017). In classical facility location problems, it is assumed that clients always get service from the facilities and the facilities are always available. Considering realistic situations, these assumptions are not likely valid.

Every year, many companies are faced with unexpected events in their supply chain networks. From time to time, network performance can be disrupted for various reasons; for example, natural disasters, power outages, changes of ownership, operational risks, and strike actions. The strategic nature of supply chain decisions differentiates them from operational-level decisions, such as machine scheduling and short-term material requirement planning. Hence, uncertainties in a supply chain can impose heavy costs on the system by completely stopping the flow of the network for a lengthy period. In this paper, this default assumption is used to design a reliable network, whose facilities are unreliable. Continuous changes in the network structure (i.e., first-stage decisions) are impossible and without justification in facility location problems with facility disruptions. Designing optimal supply chains when facilities are subject to random disruptions is an appropriate response to dealing with this type of uncertainty, which is located in the class of provider-side uncertainty (randomness in facility capacity or reliability of facilities, etc.).

Considering reliability in facility location problems makes it possible that a system continues to function with the minimal loss when its components fail. Reliable design of a facility location problem considers a change in second-stage decisions by reassigning client demand to other facilities far from their regularly assigned facilities or arranging backup facilities for failed facilities. The reliable uncapacitated facility location problem (RUFLP), was first studied by Snyder and Daskin (2005). They formulated the problem as a linear MIP, called "implicit formulation" (IF), which employed Lagrangian relaxation for efficient solutions. A similar study was presented by Berman et al. (2007), who relaxed the assumption of identical failure probabilities, which was a basic assumption of the model presented by Snyder and Daskin (2005). Then, they developed several exact and heuristic approaches. Other studies, such as Cui et al. (2010) and Shen et al. (2011), have formulated the RUFLP with non-identical failure probabilities. In addition to a scenario-based model of the problem, Shen et al. (2011) also provided a mixed-integer nonlinear programming model and provided several approximation algorithms that can produce near-optimal solutions. Cui et al. (2010) proposed a compact mixed-integer programming (MIP) model and a continuum approximation (CA) model to study the problem and used Lagrangian relaxation to solve the models. In a reliable capacitated facility location problem (RCFLP), Aydin and Murat (2013) developed a model for the RCFLP and presented a novel hybrid method, namely swarm intelligence-based sample average approximation (SIBSAA), to solve their model. Peng et al. (2011) presented an RCFLP to minimize the nominal cost (when no disruptions occur) and at the same time reducing the disruption risk by applying the p-robustness criterion, in which bounds the cost in disruption scenarios. Li et al. (2013a) investigated reliable facility location problems, while the failure of the facilities was correlated. Fan et al. (2018) proposed a reliable location model for a nexus of interdependent infrastructure systems. Their model aims to locate the optimal facility locations in multiple heterogeneous systems to balance the trade-off between the facility investment and the expected nexus operation performance. Afify et al. (2019) are proposed an evolutionary learning technique to near-optimally solve Reliable p-Median Problem and Reliable Uncapacitated Facility Location Problem considering heterogeneous facility failure probabilities, one layer of backup and limited facility fortification budget.

Multi-echelon supply chain design has been extensively studied in classical facility location problem. To the best of the authors' knowledge, there have been a few studies on the multi-echelon in reliable facility location problems. These studies assumed that facility disruption occurs only at one echelon of the network. For example, Razmi et al. (2013) proposed a scenario base mixed-integer linear programming (MILP) model for redesigning a reliable single product, single period, and two-echelon logistics network. In this paper, we consider the two-echelon reliable facility location with unreliable facilities when nodes related to facilities in both echelons fail under disruptions. The analysis of problems that involve single-echelon supply chains and decisions made under that assumption may lead to many errors. In other words, the single-echelon perspective on a supply chain of goods and services lacks the required accuracy and efficiency in cases, in which the network itself is part of a larger network and interacts with other echelons and layers of the network. Decisions, actions, and reactions at each echelon of the supply chain can have significant effects on other echelons. For this reason, we intend to study the RFLP by taking into account all the echelons of the supply chain. Rohaninejad et al. (2018) first considered the RCFLP in a multi-echelon manner with the possibility of disruption in all echelons. They presented the new scenario-based formulation to maintain the materials flow between different echelons of the network under facilities disruptions. In continue of their study, this paper examines whether the design of reliable multi-echelon facility location problems with a global optimization approach (i.e., integrated optimization in all echelons) is more effective than independent and non-integrated optimization. The answer is to provide an insight for the owners of the supply chain, in which "how effective is the cooperation between the owners of each echelon of the network?"

A review of the literature on the RFLP regarding complexity shows that the RUFLP is NP-hard (Li et al. 2013b). The RCFLP, which involves the addition of capacity constraints, is NP-hard as well because the RCFLP reduces to the RUFLP when the capacities of the facilities tend towards infinity. In another hand, using the implicit formulation provided by Snyder and Daskin (2005) to provide low-complexity formulation has some drawbacks. This type of formulation has limitations when it is extended to problems with different failure probabilities for facilities, partial allocation of client demand to facilities, the possibility of partial disruption of facilities, facility capacity constraints and multi-echelon networks. Tracking demand flows across different network echelons is not possible or not easy in multi-echelon networks by implicit formulation. So using the scenario-based formulation is in priority. The scenario-based approach is flexible and can be used for our considered problem; however, there might be numerous scenarios, (in our case, if J facilities are assumed that each of them can fail independently, there are 2^J failure scenarios) and the complexity of the model increases exponentially by increasing the number of scenarios. Therefore, due to the complexity of the model, a custom-designed Lagrangian dual decomposition algorithm is proposed for this problem.

The literature on the RFLP presents various solution procedures: heuristic or approximation procedures (Shen et al. 2011; Rohaninejad et al. 2015; Aboolian et al. 2013); Lagrangian relaxation algorithms (Snyder and Daskin 2005; An et al. 2015); continuum approximation (Li and Ouyang 2010); exact methods (Rohaninejad et al. 2018; Hatefi and Jolai 2014) and metaheuristic algorithms (Aydin and Murat 2013; Peng et al. 2011).

The remaining sections are organized as follows. The problem definitions are defined in Sect. 2. Section 3 presents a new MILP model of the problem. Section 4 illustrates the proposed solution method. Section 5 provides the computational results. Finally, Sect. 6 outlines the conclusion and some suggestions for future research.

2 Problem Definition

In this paper, we consider a two-echelon reliable facility location problem (TE-RFLP). While considering a two-echelon instead of single-echelon RFLP, we assume that there is the possibility of a failure of facilities at both echelons of the network. In the TE-RFLP, we assume that facilities at the first echelon of the chain are clients at the second echelon of the chain and vice versa. In other words, the client demand at the second echelon of the supply chain is equal to the coefficient of the total client demand at the first echelon assigned to it as a facility at the first echelon. Demand in the first echelon is predetermined; however, the demand in the second echelons is uncertain and dependent on the first echelon. Therefore, it is necessary to examine two-echelon systems from an integrated perspective to enhance the reliability of the whole system. Figure 1 depicts a schematic representation of the relationship between the echelons in a supply chain. This figure shows that potential sites at each echelon are potential clients at a higher echelon. As well, actual sites (openings) at each echelon are actual clients at a higher echelon. In this figure, it is clear that the flow of demand is interdependent at all echelons. It is assumed, that the flow of demand may be discontinued from one echelon to a higher echelon, and demand is supplied from outside the network by paying a higher cost (i.e., penalty cost). In other words, all open facilities need to supply their client demand from inside the network (its higher echelons), or from the outside network by paying a penalty cost.

In the TE-RFLP, the facilities have two operational levels, namely active and inactive. In an active mode, the facilities are fully available and inactive facilities cannot provide any services to the clients and are out of reach. The aim of solving this problem is to design a reliable network for a two-echelon supply chain to minimize the expected total fixed costs and service costs (including the expected transportation

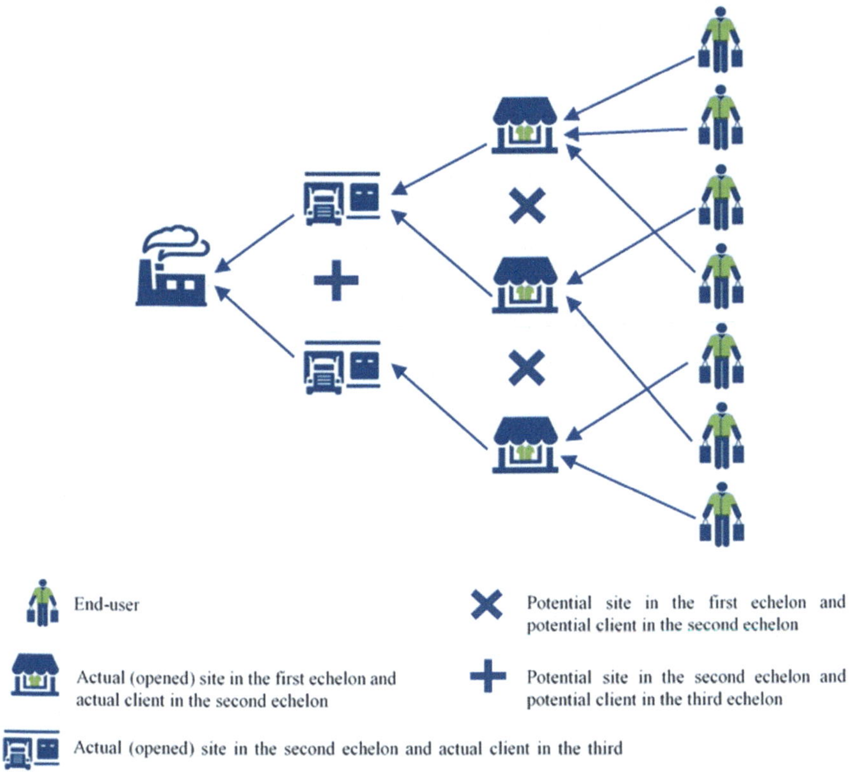

Fig.1 Relationship between echelons in a two-echelon supply chain design

costs and costs incurred for failing to meet client demand). Other assumptions of the problem are as follows:

- The potential sites for establishing facilities are predefined and discrete.
- There is no relationship between facilities at an echelon.
- The problem is a single-product model.
- The allocation of a demand node to more than one facility node is not allowed. In other words, partial or fractional allocation of demand to a facility is not allowed.
- There is a fixed cost to establish a facility.
- There is a transportation cost to allocate a demand node to a facility node.
- There is a penalty cost for supplying demand from outside the network.
- The maximum number of facilities that can be established in each location is one.
- The network is two-echelon and all facilities in each echelon are the same (e.g., all facilities are wholesalers or retailers).
- The problem parameters (e.g., demand in the first echelon, distance, and failure probabilities) are specified.

3 Notation and Formulation

The indices, parameters and variables of the proposed model are as follows:

Model indices:

i Client index.
j Potential site index.
l Echelon index; (l = 1, 2).
s Scenario index $s \in S$

Model parameters:

d_i Demand quantity of client i in the first echelon.
$s_{l,j}$ Fixed cost of opening facility j in echelon l.
$c_{l,i,j}$ Transportation cost per unit demand of client i in echelon l that is serviced by facility j.
$h_{l,i}$ Penalty cost per unit of client i demand in echelon l if its demand is not met.
β_j Conversion coefficient of demand volume from the first echelon to the second echelon for facility j in the first echelon (1 ≤ β_j ≤ 2).
I_l Total number of clients in echelon l.
J_l Total number of potential sites in echelon l.
q_s Probability of scenario s.
$b_{l,j,s}$ 1 if facility j in echelon l is active and functional under scenario s; 0, otherwise.
G A positive large number.

Model variables:

$y_{l,j}$ 1 if facility j at echelon l is opened; 0, otherwise
$X_{l,i,j,s}$ 1 if client i assigned to facility j at echelon l under scenario s; 0, otherwise
$Z_{l,i,s}$ 1 if demand of client i at echelon l is not met under scenario s; 0, otherwise
$u_{i,j,s}$ A positive variable specifies the demand of client i provided by facility j at second echelon under scenarios
$v_{i,s}$ A positive variable specifies the demand of client i at second echelon that is not met under scenarios

The mathematical model of the TE-RFLP based on the scenario-based formulation is as follows:

$$\text{Min} \sum_{l} \sum_{j \le J_l} s_{l,j} y_{l,j} + \sum_{s} \left(\sum_{i \le I_1} \sum_{j \le J_1} c_{1,i,j} d_i X_{1,i,j,s} + \sum_{i \le I_2} \sum_{j \le J_2} c_{2,i,j} u_{i,j,s} \right) q_s$$
$$+ \sum_{s} \left(\sum_{i \le I_1} h_{1,i} d_i Z_{1,i,s} + \sum_{i \le I_2} h_{2,i} v_{i,s} \right) q_s \qquad (1)$$

$$s.t \sum_{j \le J_1} X_{1,i,j,s} + Z_{1,i,s} = 1 \qquad \forall i \le I_1; s \qquad (2)$$

$$\sum_{j \le J_2} X_{2,i,j,s} + Z_{2,i,s} \le 1 \qquad \forall i \le I_2; s \tag{3}$$

$$\sum_{j \le J_2} u_{i,j,s} + v_{i,s} = \sum_{i' \le I_1} \left(d_{i'} \beta_{j'} X_{1,i',j',s} \right) \qquad \forall i \le I_2; j' = i; s \tag{4}$$

$$u_{i,j,s} \le G X_{2,i,j,s} \qquad \forall i \le I_2; j \le J_2; s \tag{5}$$

$$v_{i,s} \le G Z_{2,i,s} \qquad \forall i \le I_2; s \tag{6}$$

$$\sum_{i \le I_l} X_{l,i,j,s} \le G b_{l,j,s} y_{l,j} \qquad \forall l; j \le J_l; s \tag{7}$$

Objective function (1) consists of the total expected cost of facility fixed charge, transportation, and unmet demands penalty. Constraints (2) and (3), which refer to the first and second echelons respectively, ensure that all actual client demand is supplied from inside or outside the network. Since not all the second-echelon clients will necessarily become actual clients in Constraint (3), inequality is used instead of equality. For this reason, Constraint (3) is not enough to establish the demand flow in the second echelon; we need to add Constraint (4) to the model. Constraint (4), which refers to the second echelon ($l = 2$) of the supply chain, has conditions similar to Constraint (2). However, the right side of this constraint ensures when a facility in the first echelon becomes a client in the second echelon, its demand in the second echelon is a function of the number of orders it accepted in the first echelon. The combination of Constraints (5) and (6) with Constraint (4) prevents a partial assignment (i.e., assigning a part of the client's demand to a facility). Constraint (7) prevents assignment to a facility that has not been opened.

4 Lagrangian Dual Decomposition

Lagrangian dual decomposition (LDD) is a classical method for combinatorial optimization. This method is an important special case of the Lagrangian relaxation algorithm, in which the original problem decomposes to two or more combinatorial optimization problems. In this algorithm, we assume that there is a minimization problem as follows:

$$\min_{(x,y,z)} f(x,z) + g(y,z) \tag{8}$$

By decoding the problem, it can be found:

$$\min_{(x,y,z_1,z_2)} f(x,z_1) + g(y,z_2) \tag{9}$$

s.t.

$$z_1 = z_2 \tag{10}$$

We introduce a vector of Lagrange multiplier λ. The Lagrangian is now.

$$L(x, y, z_1, z_2, \lambda) = f(x, z_1) + g(y, z_2) + \lambda(z_1 - z_2) \tag{11}$$

And the dual objective is as follows:

$$\begin{aligned} L(\lambda) = \min_{(x,y,z_1,z_2)} L(x, y, z_1, z_2, \lambda) &= \min_{(x,y,z_1,z_2)} \{f(x, z_1) + g(y, z_2) + \lambda(z_1 - z_2)\} \\ &= \min_{(x,z_1)}\{f(x, z_1) + \lambda z_1\} + \min_{(y,z_2)}\{g(y, z_2) - \lambda z_2\} \\ &= \min_{(x,z)}\{f(x, z) + \lambda z\} + \min_{(y,z)}\{g(y, z) - \lambda z\} \end{aligned} \tag{12}$$

Equation (12) is a lower bound for the original problem and the tightest lower bound can be shown by:

$$\max_{(\lambda)} L(\lambda) = \max_{(\lambda)} \left\{ \min_{(x,z)}\{f(x, z) + \lambda z\} + \min_{(y,z)}\{g(y, z) - \lambda z\} \right\} \tag{13}$$

We solve each sub-problems of the relation (13) for the current λ separately and then update the λ for improve the sub-gradient level.

4.1 Custom Designed LDD

To solve the proposed model, we decompose the model based on the N subsets ($e_n; n = 1, \ldots, N$) of scenarios, so that $\bigcup_{n=1}^{N} e_n = S$. For creating each subset with the similar importance and weight, we first sort all scenarios based on their probability value (q_s) in a list, then for each subset, we select one scenario from the first of the list and then one another scenario from the end of the list. We repeat this procedure until the subset is filled (i.e., the number of members reaches a predefined value), and then the next subset will create until all possible scenarios assign to a subset. Then, to do the LDD method, we introduce the new first stage variables $y_{l,j}^{n}$ for each subsets of scenarios ($y_{l,j}^{1}, y_{l,j}^{2}, \ldots, y_{l,j}^{N} | l = 1, 2; j = 1, \ldots, J_l$). Finally, we add the non-anticipative constraints $y_{l,j}^{1} = y_{l,j}^{2} = \cdots = y_{l,j}^{N}$ so that establish these constraints by equations as follows:

$$\sum_{n=1}^{N} k^n y_{l,j}^{n} = 0 \quad \forall l; j \le J_l \tag{14}$$

where $k^1 = 1 - N$ and $k^n = 1$ for $n = 2, \ldots, N$.

Then, we relax the non-anticipative constraints and set the λ as a vector of Lagrangian multi-players that are related to these constraints. The objective function of Lagrangian relaxation for each set of scenarios are as follows:

$$L(\lambda) = \text{Min} \sum_{l} \sum_{j \le J_l} \left(\sum_{n=1}^{N} \frac{1}{N} s_{l,j} \, y_{l,j}^n \right) + \sum_{l} \sum_{j \le J_l} \left(\lambda_{l,j} \sum_{n=1}^{N} k^n y_{l,j}^n \right) + \sum_{n=1}^{N} \vartheta(y^n, e_n)$$

$$= \text{Min} \sum_{n=1}^{N} \left(\sum_{l} \sum_{j \le J_l} \left(\frac{1}{N} s_{l,j} \, y_{l,j}^n + \lambda_{l,j} k^n y_{l,j}^n \right) + \vartheta(y^n, e_n) \right) \tag{15}$$

where

$$\vartheta(y^n, e_n) = \sum_{s \in e_n} \left(\sum_{i \le I_1} \left(h_{1,i} d_i Z_{1,i,s} + \sum_{j \le J_1} c_{1,i,j} d_i X_{1,i,j,s} \right) + \sum_{i \le I_2} \left(h_{2,i} v_{i,s} + \sum_{j \le J_2} c_{2,i,j} u_{i,j,s} \right) \right) q_s \tag{16}$$

.

Now, we can decompose the problem to N subproblem with minimization of Relation (17) for each $n \in \{1, \ldots, N\}$. Also, the feasible solution space of the problem will be divided based on each subset of scenarios.

$$\underset{n \le N}{\text{Min}} \left\{ \sum_{l} \sum_{j \le J_l} \left(\frac{1}{N} s_{l,j} \, y_{l,j}^n + \lambda_{l,j} k^n y_{l,j}^n \right) + \vartheta(y^n, e_n) \right\} \tag{17}$$

4.2 Updating the Lagrangian Multiplayer

The Lagrangian multipliers $\lambda_{l,j}$ are updated at per iteration using standard sub-gradient optimization by Fisher (Fisher 2004). So, given the initial value of multi-player $(\lambda_{l,j}^{(0)})$, a sequence value of multiplayer in iteration $(t\lambda_{l,j}^{(t)})$ is generated by:

$$\lambda_{l,j}^{(t)} = \max\left\{0, \lambda_{l,j}^{(t-1)} + \beta^{(t-1)} \left(\sum_{n=1}^{N} k^n y_{l,j}^n \right) \right\} \tag{18}$$

$y_{l,j}^n$ is optimal to the previous iteration and $\beta^{(t)}$ is a positive scaler step size.

$$\beta^{(t)} = \Delta^{(t-1)} \frac{Z_{UB}^* - Z_{LB}}{\| \sum_{n=1}^{N} k^n y_{l,j}^n \|^2} \tag{19}$$

In Eq. (19), Z_{UB}^{*} is the best upper bound so far. Z_{LB} is the lower bound at the current iteration. Also, $\Delta^{(t-1)} \epsilon (0, 2]$ and we start with $\Delta^{(0)} = 2$ and $\Delta^{(t)}$ is halved if the improvement in the Lagrangean lower bound is not more than 0.2% in $L_{\max}$ consecutive iterations.

4.3 Generate the Upper Bound

Gradient method, a feasibility procedure is applied to generate an upper bound for the Lagrangian problem. So, after obtaining the lower bound, we propose a simple heuristic method to find a feasible solution at per iteration of the Lagrangian procedure, but possibly not an optimal solution. Since the non-anticipative Constraints (14) are relaxed, therefore, the $y_{l,j}^{n}$ are not equal in different subsets of scenarios. So, we generate a feasible solution in three steps as follows.

Step (1) Fixing the value of $y_{l,j}$ to 1 if they are 1 in minimum 50% of the optimal solution related to each subset of scenarios; 0, otherwise.
Step (2) Fixing the value of $X_{l,i,j,s}$ and their dependent variables ($Z_{l,i,s}$, $u_{i,j,s}$, $v_{i,s}$) for each $i \leq I_l$ and s to their optimal value if Constraint (7) is not violated after inserting the value of $y_{l,j}$.
Step (3) Solve the simplified original model for finding the optimum value of remained variables in the feasible state.

If the value of the new feasible solution is better than the incumbent upper bound, then the new value becomes the incumbent upper bound.

4.4 Pseudo Code of the Algorithm

The step-by-step pseudo code of the proposed algorithm is outlined in Fig. 2.

Step 0	Initialize the parameters (Δ_t, L_{max}, Maxiter, Maxtime, $L_{count} = 0$, $Z_{LB} \rightarrow -\infty$, $Z_{UB}^{*} \rightarrow +\infty$)
Step 1	Solve each subproblems seperately and find the $Z_n^{(t)*}$ $(n = 1, \ldots, N)$
Step 2	If $\sum_{n=1}^{N} Z_n^{(t)*} \leq Z_{LB}$ Then set $L_{count} = L_{count} + 1$ Otherwise $L_{count} = 0$
Step 3	$Z_{LB} = Max\left\{Z_{LB}, \sum_{n=1}^{N} Z_n^{(t)*}\right\}$
Step 4	Implementation of feasibility phase and find $\bar{Z}^{(t)}$
Step 5	$Z_{UB}^{*} = Min\left\{Z_{UB}^{*}, \bar{Z}^{(t)}\right\}$
Step 6	**If** $t \geq$ Maxiter Or CPU Time$\geq$ Maxtime Or $\frac{Z_{UB}^{*} - Z_{LB}}{Z_{UB}^{*}} \leq \varepsilon$ **Then** Stop and output Z_{UB}^{*} as a best objective function Otherwise go to 7
Step 7	Update the Lagrangian Multiplayer
Step 8	**If** $L_{count} > L_{max}$ **Then** $\Delta^{(t+1)} = \frac{\Delta^{(t)}}{2}$ and $L_{count} = 0$ Otherwise set $\Delta^{(t+1)} = \Delta^{(t)}$
Step 9	Set $t = t + 1$ and go to 1

Fig.2 Pseudo code of the proposed LDD algorithm

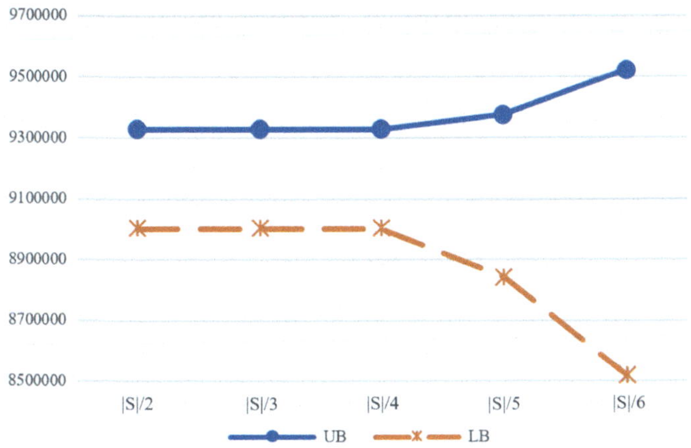

Fig. 3 Compare the results obtained sp100;8;2 by changing the size of the scenario's subsets

Fig. 4 Compare the CPU time (seconds) of sp100;8;2 by changing the size of the scenario's subsets

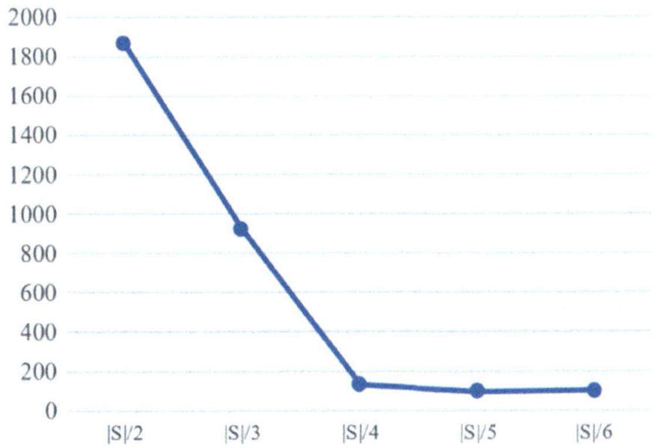

5 Computational Results

In this section, the credibility of the presented mathematical model and the performance and effectiveness of the presented heuristic algorithm are evaluated and compared with each other. The mathematical model and the LDD method are coded in the GAMS 24.1.2 software and solved by the CPLEX solver for the MIP models and run on a PC with 4 GHz processor and 8 GB of RAM. To compare the results of the proposed MIP formulation and LDD method two Relative Percentage Deviation criteria that called RPD is used. These performance measures are calculated based on the deviation of solutions to the best solution that achieved by MIP model and LDD. The RPD criterion for solution method A is calculated as follows (Note that the index A denotes a solution method.):

$$\text{RPD}_{\text{A}} = \frac{(\text{objectve function})_{\text{A}} - \text{Minimum}(\text{objectve functions})}{\text{Minimum}(\text{objectve functions})} \tag{20}$$

Also, to demonstrate the effectiveness and quality of LDD algorithm, we use three criteria (i.e., GAP1, GAP2, GAP3). The gap between the lower bound (LB) and the upper bound (UB) is calculated by:

$$\text{GAP1} = 2 * \frac{\text{UB} - \text{LB}}{\text{UB} + \text{LB}} \tag{21}$$

The distances between the upper bound (UB) and the optimal solution (OP) are denoted as GAP2, and between the lower bound (LB) and the optimal solution (OP) are denoted as GAP3. The distances are calculated by:

$$\text{GAP2} = 2 * \frac{\text{UB} - \text{OP}}{\text{UB} + \text{OP}} \tag{22}$$

$$\text{GAP3} = 2 * \frac{\text{OP} - \text{LB}}{\text{OP} + \text{LB}} \tag{23}$$

5.1 Generating Random Instances of the Problem

This section describes the testing of the proposed algorithms on 12 data sets that are generated randomly. The data set consists of 25–325 nodes for small to large instances. In each case, demands d_i in the first echelon are drawn from a normal distribution with ($\mu = 100; \delta = 30$) and rounded to the nearest integer. Fixed costs $s_{l,j}$ for $l = 1$ are drawn from U[30,000; 80,000] and rounded to the nearest integer. Fixed costs $s_{l,j}$ for $l = 2$ are drawn from U[120,000; 200,000] and rounded to the nearest integer. Penalty costs $h_{l,i}$ are drawn from U[1000; 3000] and rounded to the nearest integer.

Also, for each instance with $\sum_l I_l$ clients and $\sum_l J_l$ facilities, the locations of clients and facilities are determined randomly within a square, whose length and width are equal to $\partial = \left(\sum_l I_l + \sum_l J_l\right)\left(2.8 - 0.01\left(\sum_l I_l + \sum_l J_l\right)\right)$. Transportation costs $c_{l,i,j}$ are set equal to the Euclidean distance between i and j. The full failure probability of facilities in the first echelon is equal to 0.15 and in the second echelon is equal to 0.12. The scenarios of each instance are generated simply by taking into account all the possible combinations of active and inactive facilities. The probability of each scenario is computed by multiplying the probabilities of facilities according to their situation in an active or inactive scenario.

Ultimately, each instance is labelled with (a; b; c), where a indicates the number of clients in the first echelon; b indicates the number of potential sites in the first echelon (potential clients in the second echelon) and c indicates the number of potential sites in the second echelon. The size of these parameters influences both the solution quality and the efficiency of the proposed procedures. To find the best trade-off between the algorithm speed and solution quality, the runtime of the solution methods is limited to 3600 s.

5.2 Numerical Examples

In this subsection, we present the results of the numerical examples to show the validity of the proposed mathematical model and the efficiency of the presented solution method. Table 1 shows a comparison of the optimal results obtained by the scenario-based MIP model and the proposed LDD algorithm. This table shows that the proposed LDD algorithm can provide the optimal solution in all cases, which the scenario-based formulation is provided with a feasible solution in predefined time restriction (i.e., 3600 s).

When the size of the problem increases, the scenario-based formulation loses its effectiveness; solving problems of sp100;10;2 and larger is not possible with this formulation in predefined time restriction. Instead, the LDD can obtain a feasible solution for larger-sized problems. Therefore, considering the average difference of 0% between the best upper bound of the LDD algorithm and the optimal solution of the scenario-based formulation can be found by the LDD algorithm has the reasonable performance for larger-sized problems. This algorithm can also reduce the total solution time of sp18;5;2 to sp100;8;2 cases from 7910 to 294 s. Finally, concerning the inefficiency of the scenario-based formulation for large-sized problems, we can

Table 1 Comparison of the results obtained by the proposed mathematical model and LDD

Sample problems	Scenario-based formulation (CPLEX Solver)			Proposed LDD algorithm (CPLEX Solver)		
	Time (s)	Best upper bound	RPD (%)	Time (s)	Best upper bound	RPD (%)
sp18;5;2	45	2,241,719	0.0	12	2,241,719	0.0
sp30;5;2	753	3,712,944	0.0	26	3,712,944	0.0
sp50;8;2	2152	4,964,309	0.0	47	4,964,309	0.0
sp80;8;2	2607	7,319,741	0.0	77	7,319,741	0.0
sp100;8;2	2353	9,327,565	0.0	132	9,327,565	0.0
sp100;10;2	> 3600	-	-	440	7,892,730	-
sp120;12;3	> 3600	-	-	1361	8,622,641	-
sp150;15;4	> 3600	-	-	2714	8,718,256	-
sp200;15;4	> 3600	-	-	2615	12,350,919	-
sp200;20;5	> 3600	-	-	> 3600	14,748,366	-
sp250;20;5	> 3600	-	-	> 3600	17,362,641	-
sp300;20;5	> 3600	-	-	> 3600	22,477,543	-

Table 2 Criteria related to the quality of the LDD algorithm

Sample problems	Proposed LDD algorithm					Optimal objective function
	Best upper bound	Best lower bound	Gap 1 (%)	Gap 2 (%)	Gap 3 (%)	
sp18;5;2	2,241,719	2,241,719	0,00	0,0	0,00	2,241,719
sp30;5;2	3,712,944	3,703,125	0,26	0,0	0,26	3,712,944
sp50;8;2	4,964,309	4,941,372	0,46	0,0	0,46	4,964,309
sp80;8;2	7,319,741	7,285,340	0,47	0,0	0,47	7,319,741
sp100;8;2	9,327,565	9,002,588	3,55	0,0	3,5	9,327,565
sp100;10;2	7,892,730	7,621,409	3,50	-	-	-
sp120;12;3	8,622,641	8,370,264	2,97	-	-	-
sp150;15;4	8,718,256	8,405,831	3,65	-	-	-
sp200;15;4	12,350,919	11,518,202	6,98	-	-	-
sp200;20;5	14,748,366	13,564,572	8,36	-	-	-
sp250;20;5	17,362,641	14,355,617	18,9	-	-	-
sp300;20;5	22,477,543	16,539,741	30,4	-	-	-

propose the LDD algorithm as a suitable alternative to the scenario-based formulation and an efficient algorithm for two-echelon reliable uncapacitated facility location problems.

Table 2 shows the criteria related to the performance of the LDD algorithm. The table shows that the LDD algorithm allows using of the scenario-based model for larger-sized problems. On the other hand, according to the gap criterion, we are confident that the LDD algorithm can provide high-quality solutions (i.e., optimal or most closely optimal solution) in problems. The Gap 1 criterion is less than 5% in the 8 first samples and also less than 10% in the10 first samples {sp18,5;2 to sp200;20;5}. Therefore, we can recommend the use of the LDD algorithm to solve the scenario-based model of the TE-RFLP, especially for medium-sized problems.

Figures 3 and 4 show the comparative graphs of the results obtained from the LDD algorithm by changing the size of scenario's subsets, for the criteria of the best upper bound (UB) and lower bound (LB) values and also the CPU time. These graphs are shown for sp100;8;2. In this examination, we consider five levels for the size of the scenario's subsets, including $\left\{\frac{|S|}{2}, \frac{|S|}{3}, \frac{|S|}{4}, \frac{|S|}{5}, \frac{|S|}{6}\right\}$. These figures show the complexity of the problem is increased and the gap between the upper and lower bounds is decreased by increasing in the size of the subset. Based on these figures, creating a trade-off between the quality and complexity is so important by selecting the optimal size of subsets.

Table 3 shows the performance results of the LDD algorithm under the influence of changes in the failure probabilities of the facilities in the sp100,8;2. Based on this table results, the LDD algorithm shows sustainable performance under different scenarios of the failure probability and provided an optimal or near-optimal solution in many cases. In the LDD algorithm, except for $p_1 = p_2 = 0.9$, where we see a

Table 3 Performance of the LDD as a result of changes in the failure probabilities

Failure probability	Proposed LDD algorithm		MIP Model
	Best upper bound	RPD (%)	Optimal objective function
0.1	9,264,505	0.00	9,264,505
0.2	12,814,921	0.00	12,814,921
0.3	16,641,569	0.00	16,641,569
0.4	16,936,220	0.00	16,936,220
0.5	18,639,821	1.02	18,450,932
0.6	18,611,058	0.00	18,611,058
0.7	19,762,360	2.58	19,265,783
0.8	21,787,049	10.87	19,651,377
0.9	25,919,553	28.67	20,144,518
1.0	20,787,349	0.00	20,787,349

significant drop in quality (RPD = 28.67%), we do not notice any significant changes in the performance of this method in other failure probabilities. The average RPD, regardless of $p_1 = p_2 = 0.9$, is 1.61%.

Our models are based on the assumption of integrated optimization at all echelons. Establishing this assumption requires cooperation between the owners of all echelons of the network. Table 4 examines how effective is this cooperation. This table shows the results obtained from solving the model in the two approaches: the integrated optimization and hierarchical optimization. Each echelon is independently optimized from other echelons in the hierarchical approach. At first, the lowest echelon of the network (i.e., echelon 1) is optimized and its results are fixed. Next, the second

Table 4 The results obtained from two integrated and hierarchical optimization

Sample problems	Integrated optimization (LDD algorithm)		Hierarchical optimization (MIP formulation)	
	RPD (%)	Objective function	RPD (%)	Objective function
sp18;5;2	0.0	2,241,719	10,9	2,486,900
sp30;5;2	0.0	3,703,125	10,8	4,102,477
sp50;8;2	0.0	4,941,372	14,0	5,632,101
sp80;8;2	0.0	7,285,340	9,0	7,941,365
sp100;8;2	0.0	9,002,588	23,7	11,139,247
sp100;10;2	0.0	7,621,409	14,4	8,720,741
sp120;12;3	0.0	8,370,264	24,5	10,422,348
sp150;15;4	0.0	8,405,831	15,8	9,730,512
sp200;15;4	0.0	11,518,202	14,1	13,137,920

echelon is optimized and its results are fixed. According to the results of Table 4, the priority of the integrated approach to the hierarchical approach is evident. The average RPD criterion in this approach is 15.8% better than the hierarchical approach.

6 Conclusions

We developed a new mixed-integer programming (MIP) model based on the scenario-based formulation for two-echelon reliable facility location problems. However, the computational complexity of the scenario-based model made it less applicable for medium- and large-sized problems. Therefore, we were able to develop the use of this formulation for larger-sized problems by providing a Lagrangian dual decomposition algorithm. The computational results showed that the proposed algorithm was an efficient method for the scenario-based formulation in medium- and large-scale problems that provided a high quality of solutions with reasonable running time and resilience to change the failure probabilities of facilities. Also, the computational results showed that the adoption of an integrated approach to make a decision and simultaneous optimization at all echelons are much more effective than hierarchical optimization. Therefore, interaction and collaboration between owners of different echelons of the network are strongly recommended.

Future studies can focus on developing exact solution methods to solve the problem. Also, in multi-echelon systems, different types of relationships between echelons can be studied. For example, the failure probability of a facility or the cost of providing demand from a facility depends on the planned level of reliability in the other echelons of the network. For this reason, the lower reliability of the system in an echelon increases the costs of providing demand and the probability of failure in a lower echelon, and vice versa. It is also recommended to develop the implicit formulation for the problem with the aim of reduction in the complexity of the problem in future studies.

Acknowledgements The research has been supported by the Ministry of Education, Youth and Sports within the dedicated program ERC CZ under the project POSTMAN with reference LL1902.

References

Aboolian R, Cui T, Shen ZJM (2013) An efficient approach for solving reliable facility location models. INFORMS J Comput 25(4):720–729

Afify B, Ray S, Soeanu A, Awasthi A, Debbabi M, Allouche M (2019) Evolutionary learning algorithm for reliable facility location under disruption. Expert Syst Appl 115:223–244

An S, Cui N, Bai Y, Xie W, Chen M, Ouyang Y (2015) Reliable emergency service facility location under facility disruption, en-route congestion and in-facility queuing. Transp Res Part E Logist Transp Rev 82:199–216

Aydin N, Murat A (2013) A swarm intelligence based sample average approximation algorithm for the capacitated reliable facility location problem. Int J Prod Econ 145(1):173–183
Berman O, Krass D, Menezes MB (2007) Facility reliability issues in network p-median problems: strategic centralization and co-location effects. Operat Res 55(2):332–350
Cui T, Ouyang Y, Shen ZJM (2010) Reliable facility location design under the risk of disruptions. Operat Res 58(4-part-1):998–1011
Fan H, Ma J, Li X (2018) A reliable location model for heterogeneous systems under partial capacity losses. Transp Res Part C Emerg Technol 97:235–257
Fisher ML (2004) The Lagrangian relaxation method for solving integer programming problems. Manage Sci 50(12_supplement):1861–1871
Hatefi SM, Jolai F (2014) Robust and reliable forward–reverse logistics network design under demand uncertainty and facility disruptions. Appl Math Model 38(9–10):2630–2647
Li X, Ouyang Y (2010) A continuum approximation approach to reliable facility location design under correlated probabilistic disruptions. Transp Res Part B Methodol 44(4):535–548
Li X, Ouyang Y, Peng F (2013a) A supporting station model for reliable infrastructure location design under interdependent disruptions. Transp Res Part E Logist Transp Rev 60:80–93
Li Q, Zeng B, Savachkin A (2013b) Reliable facility location design under disruptions. Comput Oper Res 40(4):901–909
Peng P, Snyder LV, Lim A, Liu Z (2011) Reliable logistics networks design with facility disruptions. Transp Res Part B Methodol 45(8):1190–1211
Razmi J, Zahedi-Anaraki A, Zakerinia M (2013) A bi-objective stochastic optimization model for reliable warehouse network redesign. Math Comput Model 58(11–12):1804–1813
Rohaninejad M, Navidi H, Nouri BV, Kamranrad R (2017) A new approach to cooperative competition in facility location problems: mathematical formulations and an approximation algorithm. Comput Oper Res 83:45–53
Rohaninejad M, Amiri AH, Bashiri M (2015) Heuristic methods based on MINLP formulation for reliable capacitated facility location problems. Int J Indus Eng Prod Res 26(3):229–246
Rohaninejad M, Sahraeian R, Tavakkoli-Moghaddam R (2018a) Multi-echelon supply chain design considering unreliable facilities with facility hardening possibility. Appl Math Model 62:321–337
Rohaninejad M, Sahraeian R, Tavakkoli-Moghaddam R (2018b) An accelerated Benders decomposition algorithm for reliable facility location problems in multi-echelon networks. Comput Ind Eng 124:523–534
Shen ZJM, Zhan RL, Zhang J (2011) The reliable facility location problem: Formulations, heuristics, and approximation algorithms. INFORMS J Comput 23(3):470–482
Snyder LV, Daskin MS (2005) Reliability models for facility location: the expected failure cost case. Transp Sci 39(3):400–416

IoT Quality-Controlled Demand Pricing Model in Food Supply Chains

Zied Bahroun, Mohamed Ben-Daya, and Elkafi Hassini

Abstract With the increasing availability and use of internet of things (IoT) sensors and platforms in food supply chains, there is an interest in studying their efficiency and impact on supply chain operations. To illustrate how classical operations management models could be affected by the introduction of IoT, we present an inventory model with deteriorating quality that can be monitored by IoT-enabled time–temperature indicator (TTI) sensors with application in the food supply chain. We develop a novel demand function that incorporates quality and its deterioration through three supply chain parties: producer, distributor and retailer. We use the model to analyse the impact of IoT on the retailer and distributor in a Stackelber game context.

Keywords Internet of things · Quality-controlled demand · Pricing · Food supply chain · Stackelber game

1 Introduction

The Food and Agriculture Organization (FAO) of the United Nations and the Natural Resources Defense Council (NRDC) estimate that cumulative food loss and waste accounts from up to 40% of the food produced (Gustavsson et al. 2011; Gunders 2012). The food is lost throughout the supply chain, from production to storage, processing, distribution, retail and consumption (Muth et al. 2019). Managing food loss and waste does not only contribute to the efficiency of the food supply chain, it also enhances the sustainability of the food supply chain by reducing the use

Z. Bahroun · M. Ben-Daya
Industrial Engineering, American University of Sharjah, Sharjah, UAE
e-mail: zbahroun@aus.edu

M. Ben-Daya
e-mail: mbendaya@aus.edu

E. Hassini (✉)
DeGroote School of Business, McMaster University, Hamilton, ON L8S 4M4, Canada
e-mail: hassini@mcmaster.ca

© The Author(s), under exclusive license to Springer Nature Switzerland AG 2020
P. Golinska-Dawson et al. (eds.), *Smart and Sustainable Supply Chain and Logistics – Trends, Challenges, Methods and Best Practices*, EcoProduction,
https://doi.org/10.1007/978-3-030-61947-3_26

of resources such as land, water and energy. Internet of things (IoT) is gradually being adopted in different functions of a supply chain to enhance its visibility and traceability capabilities. In particular, IoT-enabled time–temperature indicator (TTI) sensors have been successfully adopted in the food supply chain (Ben-Daya et al. 2019). The use of TTIs allows for the possibility of quality-based control of supply chains (Bowman et al. 2009). This is giving rise to the area of quality-controlled supply chain management. Despite the importance of managing food waste, we find that there is a scarcity of studies that looks at these issues in the Operations Management (OM) area (Ben-Daya et al. 2020). There is also a call for studying the impact of IoT on supply chain management by looking at the efficiency of using IoT sensors and how they can be incorporated in classical OM models. As recent examples of studies, consider Leithner and Fikar (2019) who study the role of quality information in organic fresh food supply chains. The use simulation to study a supply chain of organic strawberries in Austria and study the impact of using quality data to decide about which distribution channel to use, e.g., direct sale, sale through a retailer or sale to a processing plant. They find that the quality threshold and customer behavior are important parameters to consider in quality-controlled supply chain management. Another example study is that of Heising et al. (2017) where they look at the role of smart packaging in operationalizing quality-controlled logistics. They suggest that using dynamic expiry dates and pricing based on data from intelligent package sensors can substantially reduce food loss.

As an effort to contribute to the growing field of quality-controlled supply chain management, the goal of this paper is to introduce a model for studying the efficiency of IoT sensors and their impact on supply chain operations. To illustrate how classical OM models could be affected by the introduction of IoT, we present an inventory model with deteriorating quality that can be monitored by IoT-enabled time–temperature indicator (TTI) sensors with application in the food supply chain. Our main contribution is the development of a novel demand function that incorporates quality and its deterioration through three supply chain parties: producer, distributor and retailer. We use the model to analyse the impact of IoT on the retailer and distributor.

2 Demand Model

Consider a retailer that orders a produce and sells it at a fixed market price. To highlight the impact of IoT we consider two scenarios: (1) the retailer does not use TTIs and (2) the retailer makes use of TTIs.

2.1 Notation and Assumptions

We use the following notation.

t	Time
i	Supply chain entity, $i \in 1$(producer), 2(distributor), 3(retailer)
q_c	Critical quality level below which a customer considers the product unsalable,
q_h	Highest quality level beyond which a customer is indifferent to quality, $0 < q_c < q_h$
q_i	Quality at supply chain echelon i, $q_h \leq q_i \leq 1$
t_i	Product arrival time at supply chain echelon i
μ_i	Instantaneous quality deterioration rate at supply chain echelon i
t_h	Time at which customer starts recognizing different quality levels
t_c	Time beyond which a customer finds the product unsalable due to bad quality
$d(p, q_3)$	Demand function for a given price p and retailer initial quality
$D(p, q_3)$	Total demand at as a function of price p and quality q_3
c_i	Unit production/purchase cost/price at supply chain echelon i
s_i	Ordering cost per cycle at supply chain echelon i
h_i	Inventory holding cost per unit per unit time at supply chain echelon i
a	Market size
b	Demand sensitivity to price
β	Demand sensitivity to quality, $\beta \leq b$
T	Replenishment period, $T = t_c - t_3 = (q_3 - q_c)/\mu_3$
p	Unit selling price
π_j	Profit for distributor ($j = 2$), retailer ($j = 3$), and supply chain ($j = SC$)

In our model we focus on replenishment from a backroom storage. It is therefore reasonable to make the following assumptions:

(1) Lead times are negligible, and the replenishment is instantaneous.
(2) At time zero all items are of perfect quality at echelon 1, the producer.
(3) Product quality remains at a premium level until it reaches the retailer premises at time t_3.
(4) Shortages are not allowed.
(5) We assume a linear quality loss function
(6) The retailer starts selling at time t_3 with quality level q_3
(7) Demand decreases in price and increases in quality

2.2 Demand Function

The quality deterioration process is illustrated in Fig. 1. At time t_1, the product quality level at the producer is q_1 and deteriorates at a rate μ_1. By time t_1, the product reaches the distributor premises at a quality level q_2 and start deteriorating at a rate μ_2. By time t_3, the product reaches the retailer's location at a quality level q_3 where it deteriorates at a rate μ_3. At the retailers shelves, where the demand is satisfied, the product goes through three stages: (1) *premium* quality stage from time

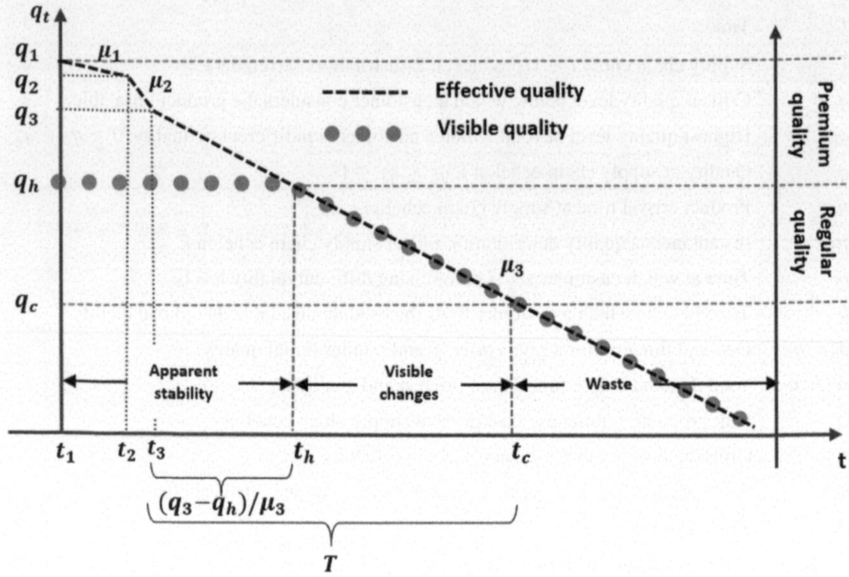

Fig. 1 Quality deterioration process. (Adapted from Ozbilge et al. 2019)

t_3 to t_h where the product quality is above q_h indicating that it is of premium quality. This period is referred to as *apparent stability* where the customer does not visibly see the changes in quality. (2) At time t_h the product quality becomes *regular* until time t_c. During this *visible changes* period the customer can discern different quality levels. For example, a customer may use colour, firmness or texture of a fruit to assess its quality level. (3) Finally, at time t_c the product quality deteriorates to a level q_c which makes it unacceptable to the customer and so the product goes to *waste*.

During this process of quality deterioration we observe two types of quality: *Effective* quality, which represents the actual measurable quality levels, and *visible* quality that may be assessed on an ad hoc basis by a customer.

To model the demand function at the retailer, we assume that demand will depend on price, using the well know downward sloping demand curve $a - bp$, and quality. As per Fig. 1, the quality component will be fixed up to t_h and then linearly decreasing at a rate of μ_3 with respect to time up to time t_c. Thus, starting at time t_3, the demand function at the retailer can be formulated as follows, for a given price p, quality q_3 and time point t:

$$d(p, q_3) = \begin{cases} a - bp + \beta q_h, t \le (q_3 - q_h)/\mu_3 \\ a - bp + \beta(q_3 - \mu_3 t), t > (q_3 - q_h)/\mu_3 \end{cases}$$

The first demand function corresponds to the time when the product has premium perceived quality, while the second function corresponds to the time when the product quality is perceived to be regular by the customers. Noting that $(q_3 - q_c)/\mu_3 = T$, the total demand is.

$$
\begin{aligned}
D_3(p, q_3) = \int_0^T d(p.q_3)dt = \int_0^{(q_3-q_h)/\mu_3} (a - bp + \beta q_h)dt + \int_{(q_3-q_h)/\mu_3}^{(q_3-q_c)/\mu_3} (a - bp + \beta(q_3 - \mu_3 t)) \\
dt = (a - bp + \beta q_h)((q_3 - q_h)/\mu_3) + (a - bp + \beta q_3)((q_3 - q_c)/\mu_3 - (q_3 - q_h)/\mu_3) \\
+ \frac{1}{2}\beta\mu_3\left(((q_3 - q_h)/\mu_3)^2 - ((q_3 - q_c)/\mu_3)^2\right) = (a - bp + \beta q_h)((q_3 - q_h)/\mu_3) + (a - bp + \beta q_3)((q_h - q_c)/\mu_3) \\
+ \frac{1}{2}\beta(2q_3 - (q_h + q_c))(q_c - q_h)/\mu_3 = A + Bp + \gamma q_3 + \delta p q_3
\end{aligned} \tag{1}
$$

Where

$$A = -(\beta(q_h^2 + q_c^2)/2 + a q_c)/\mu_3 \tag{2}$$

$$B = \frac{b q_c}{\mu_3} \tag{3}$$

$$\gamma = \frac{a + \beta q_h}{\mu_3} \tag{4}$$

$$\delta = -b/\mu_3 \tag{5}$$

2.3 Model Analysis

In the following propositions we investigate the role of pricing and quality in the proposed demand model.

Proposition 1 (Retailer Threshold Quality) *To ensure that there is demand at the retailer level, the retailer incoming quality threshold should be* $\frac{-Bp-A}{\gamma+\delta p}$

Proof To ensure that demand is nonnegative at the retailer we require that $D_3(p, , q_3) \geq 0$ and the results follows immediately. □

We note that the threshold quality depends on the market size, demand sensitivity to quality and price, premium and critical perceived quality levels. Interestingly, the threshold does not depend on the quality deterioration rate at the retailer.

Proposition 2 (Role of Price in Total Demand) *The rate of change of total demand to price is* $-\frac{b(q_3-q_c)}{\mu_3}$

Proof From (1) we get that $\frac{\partial D(p,q_3)}{\partial p} = B + \delta q_3 = -\frac{b(q_3-q_c)}{\mu_3}$. □

We note that the total demand sensitivity to price depends on the difference between the retailer's incoming quality level, q_3, and the customers' lowest acceptable perceived quality level, q_c, in addition to the quality deterioration level at the retailer, μ_3. In particular, demand sensitivity to prices increases when the customers are prepared to buy low quality levels. This is intuitive as a low value of q_c indicates that customers are not too sensitive to quality. We also see that a retailer that chooses to start with low quality levels will cater to customers with low price sensitivity.

Proposition 3 (Role of Quality in Total Demand) *The rate of change of total demand to quality is* $\frac{a-bp+\beta q_h}{\mu_3}$

Proof From (1) we have $\frac{\partial D(p,q_3)}{\partial q_3} = \gamma + \delta p = \frac{a-bp+\beta q_h}{\mu_3}$. □

Proposition 3 indicates that the total demand sensitivity to quality is proportional to the premium demand; as the demand from premium quality increases so does the sensitivity of demand to quality. In addition, the higher the level at which a customer discerns premium quality the more the demand is sensitive to quality. Finally, if the price-only sensitive demand is high then the demand sensitivity to quality will also tend to be high.

3 Stackelberg Pricing Game

In this section we assume that the retailer and distributor are independent supply chain entities. To analyse the retailer and supplier's pricing decisions, we adopt a Stackelberg (2010) game approach. The retailer is the leader who determines price p, given their proximity to the customer. The distributor is the follower who takes the price p and sets the wholesale price c_3 accordingly.

The retailer's profit is.

$$\pi_3(p) = (p - c_3)D(p, q_3) = (p - c_3)(a' + b'p) \tag{6}$$

where

$$a' = A + \gamma q_3 \tag{7}$$

$$b' = B + \delta q_3 \tag{8}$$

The distributor's profit is.

$$\pi_2(c_3) = (c_3 - c_2)D\big(p^*, q_3\big) = (c_3 - c_2)(a' + b'p^*) \tag{9}$$

where p^* is the optimal retailer's profit obtained by maximizing $\pi_3(p)$.

The following proposition establishes the optimal retail and wholesale prices.

Theorem 1 (Optimal Stackelberg Prices) *The retailer's optimal price is*

$$p^* = \frac{c_3}{2} + \frac{\beta\left(q_h^2 + q_c^2\right) + 2aq_c - 2q_3(a + \beta q_h)}{4b(q_c - q_3)}$$

and the distributor's optimal wholesale price is

$$c_3^* = \frac{c_2}{2} + \frac{\beta\left(q_h^2 + q_c^2\right) + 2aq_c - 2q_3(a + \beta q_h)}{4b(q_c - q_3)}.$$

Proof From (6) we have.

$$\frac{\partial \pi_3(p)}{\partial p} = a^{'} - b^{'} c_3 + 2b' p \tag{10}$$

and

$$\frac{\partial^2 \pi_3(p)}{\partial p^2} = 2b^{'} = B + \delta q_3 = -\frac{b(q_3 - q_c)}{\mu_3} < 0.$$

Therefore $\pi_3(p)$ is strictly concave in p and from (10) we deduce that.

$$p^* = \frac{c_3}{2} - \frac{a'}{2b'} \tag{11}$$

The result follows by substituting the values of a' and b' using Eqs. (7–8) and (2–5).

To find the optimal distributor's wholesale price we first substitute (11) in (9) to obtain:

$$\pi_2(c_3) = (c_3 - c_2)\left(\frac{a'}{2} + \frac{b'}{2} c_3\right).$$

It follows that.

$$\frac{\partial \pi_2(c_3)}{\partial c_3} = \frac{a^{'} - b^{'} c_2}{2} + b' c_3 \tag{12}$$

and

$$\frac{\partial^2 \pi_2(c_3)}{\partial {c_3}^2} = b^{'} = \frac{B + \delta q_3}{2} = -\frac{b(q_3 - q_c)}{2\mu_3} < 0.$$

Therefore $\pi_3(p)$ is strictly concave in c_3 and from (12) we deduce that.

$$c_3^* = \frac{c_2}{2} - \frac{a'}{2b'} \tag{13}$$

The result follows by substituting the values of a' and b' using Eqs. (7–8) and (2–5)□.

We note that the optimal distributor and retailer prices do not depend on the quality deterioration rates.

4 Numerical Study

In this section we conduct a numerical study to investigate the role of pricing, quality and IoT. We use the parameters in Table 1.

With the above values, we find that customers will start differentiating between the product quality levels at time $t_h = (q_3 - q_h)/\mu_3 = 90$ h. The selling horizon is $T = (q_3 - q_c)/\mu_3 = 130$ h. The distributor should set it wholesale price at \$10.35 and the retailer would sell at prices of \$12.03 resulting in a demand of 326.75 units. This results in profits of \$547.52 and \$1,095.03 for the distributor and retailer, respectively.

4.1 Demand Model: Role of Price and Quality Sensitivity Parameters

An interesting question is how price and quality parameters interact in the demand model. To see the joint impact of price and quality sensitivity parameters on profit, we plot, in Fig. 2, total profit versus the ration of price and quality sensitivity parameters. To obtain this graph we set an initial price and quality sensitivity parameters of 0.1 and 0.5 respectively. These are then increased by increments of 0.02 and 0.001, respectively.

We observe that when $\frac{b}{\beta} \geq 1$ (vertical dashed line, in Fig. 2) the profit decreases at a faster rate when that ratio is increasing. This suggests that demand quality-sensitivity is more impactful, in terms of changes in profits. when the price sensitivity is more prominent. This is an interesting observation, as it suggests that we should not ignore

Table 1 Parameters for numerical study

Parameter	Value
μ_3	0.05/hour
q_h	0.4
q_3	0.85
q_c	0.2
a	20/hour
b	1.5
β	1.5
c_2	\$7/unit

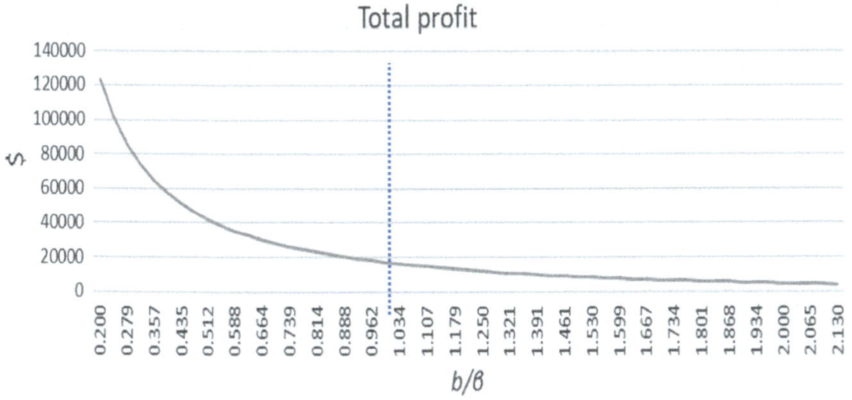

Fig. 2 Profit versus ratio of price and quality sensitivity parameters

the impact of quality in demand models even when the customers seem to be more cognizant of price changes.

4.2 *Demand Model: Role of Consumer Quality Expectation Range*

One important feature of our demand model is that it explicitly includes a quality range $[q_c, q_h]$ to represent the consumer quality expectations. In Fig. 3 we show how the total profits change with changes in the customer quality expectation range. To construct this graph we started with an initial indifference quality level q_h of 0.7 and waste quality value of 0.2. The indifference quality os then decrease by increments of 0.005.

We can make two observation based on Fig. 3. First, the larger the customer quality expectation range, the higher the profit. This is intuitive as the wider the range the

Fig. 3 Profit versus customer quality expectation range

more the retailer will sell to the customers. Second, as the quality expectation interval becomes smaller, the profits decrease almost linearly ($R^2 = 0.9853$).

4.3 *Role of IoT*

To understand the role of IoT sensors in improving the supply chain profits, we look at three scenarios where IoT sensors are deployed at the distributor, retailer or both of them.

4.3.1 IoT Deployment at the Retailer Only

The retailer investment in IoT sensors will impact the quality deterioration, μ_3, at its location. For example, the deployment of sensors will alert the retailer to quality deterioration at individual fruits items so they can be taken out before they can cause the rest of the lot to deteriorate. In Fig. 4 we show how improving the quality deterioration rate, through investment in IoT sensors by the retailer, impacts the total profits. Note that the x-axis is plotted in reverse order to show the impact of improving μ_3. We constructed this graph by taking an initial value of 0.001 for μ_3 and then increasing it by increments of 0.0005.

The total profits as a function of the deterioration rate changes according to a power function ($R^2 = 0.9955$). This suggests that the impact of investing in IoT sensors is not significant up to a certain threshold, beyond which the marginal benefit of investments becomes significant. In our numerical study, to improve total profits by more than 5% we need to invest in enough IoT sensors to bring the quality deterioration rate down to 0.0095 (vertical dashed line in the Fig. 4).

Figure 4 can be used by the retailer to assess the worth of investing in IoT sensors. For example, if current quality deterioration rate is 0.01 and we want to improve it

Fig. 4 Impact of IoT deployment by the retailer only

to 0.007, i.e., a 30% improvement, then we can invest up to \$1,173.25–\$821.27 = \$351.97 in sensors to break even. Note that the breakeven amount also represents the improvement in profits which account for 42.86% increase in profits. Such graphs can be very useful in practice as practitioners have been struggling with the question of assessing the return of investment on IoT (Ben Day et al. 2019).

4.3.2 IoT Deployment at the Distributor Only

The distributor investment in IoT sensors will impact the retailer's initial quality level, q_3. The deployment of sensors will allow the distributor to improve its quality control and remove any low-quality items to avoid further deterioration to the quantity that will be shipped to the retailer. In Fig. 5 we show how improving the initial retailer quality level, through investment in IoT sensors by the distributor, impacts the total profits. We constructed this graph by taking an initial value of 0.8 for q_3 and then increasing it by increments of 0.003.

We observe that the total profits increase linearly with the increase of the retailer's initial quality level ($R^2 \approx 1$). Thus, unlike the impact of IoT investment at the retailer, investments by the distributor have a monotonic increasing marginal benefit.

Figure 5 can be used by the distributor to assess the worth of investing in IoT sensors. For example, if current initial retailer's quality level 0.8 and we want to improve it to 0.962, i.e., a 20.0% improvement, then we can invest up to \$1928.17–\$1,515.04 = \$413.13 in sensors to break even. Note that the breakeven amount also represents the improvement in profits which account for 27.3% increase in profits.

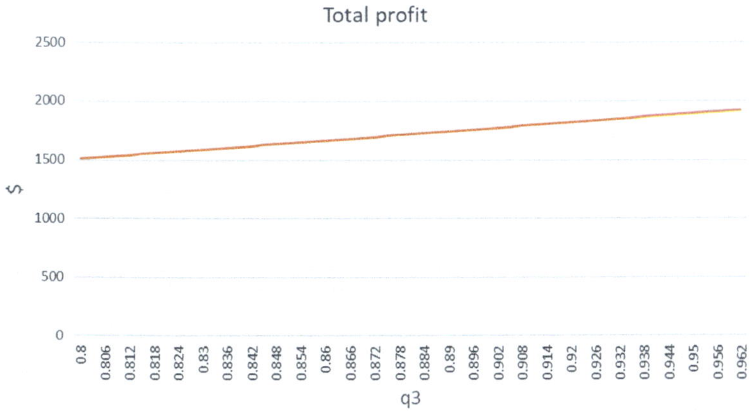

Fig. 5 Impact of IoT deployment by the distributor only

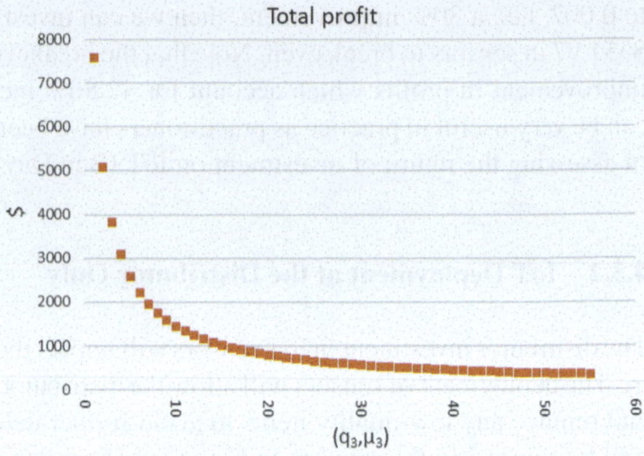

Fig. 6 Impact of IoT deployment throughout the supply chain (distributor and retailer)

4.3.3 IoT Deployment Throughout the Supply Chain

In this section we investigate the case when both the distributor and retailer invest in IoT sensors thus improving the retailer's initial quality and the retailer's quality deterioration rate at the same time. In Fig. 6 we show how the joint investment by the distributor and the retailer impacts the total profits. We constructed this graph by taking an initial value of 0.8 and 0.0001 for q_3 and μ_3, respectively, and then increasing them by increments of 0.001 and 0.003, respectively. Note that the x-axis has pairs of q_3 and μ_3 ordered in a decreasing order of the total profit.

We observe that when the total profits as a function of the deterioration rate changes according to a power function ($R^2 = 0.9981$). This suggests that the retailers return on investment in IoT sensors shadows that of the distributor. Similar to the investment curve for the retailer, Fig. 6 also suggest that there is a threshold beyond which the marginal return on investment diminishes.

4.4 *IoT Impact on Reducing Food Loss*

In Fig. 8, we show how demand changes depending on where IoT has been deployed.

We note that in all cases the demand increase and therefore the food loss decreases. The behaviour of how demand change are similar to those of the changes of profits described in Sects. 4.3.1–4.3.3. In particular, we also note that the benefit of IoT deployment has diminishing marginal benefits. i.e., there is a point beyond which additional investment in IoT do not lead to as much food waste reduction as the prior investment.

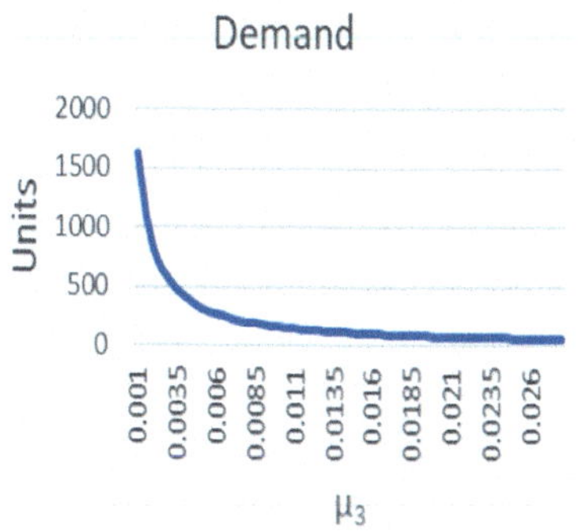

(a) Retailer only

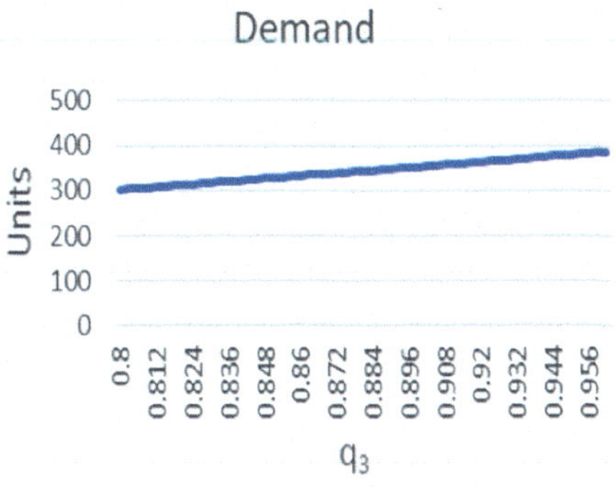

(b) Distributor only

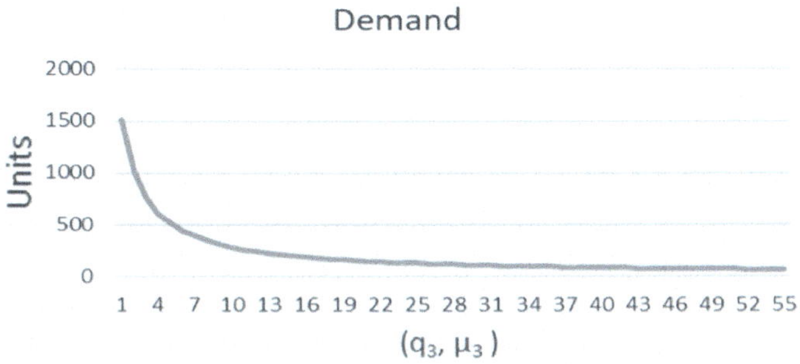

(c) Retailer and distributor.

Fig. 8 Impact of IoT deployment on food loss reduction

5 Conclusion

In this paper we propose an inventory model with deteriorating quality that can be monitored by IoT-enabled time–temperature indicator (TTI) sensors with application in the food supply chain. We develop a novel demand function that incorporates quality and its deterioration through three supply chain parties: producer, distributor and retailer. We use the model to analyse the impact of IoT on the retailer and distributor profits as well as food loss.

We find that the retailer's investment on IoT shadows that of the distributor. This suggest that in supply chains where the retailer and distributor may be considering investment in IoT as part of a contractual agreement then it would be better to delegate the investment to the retailer. We also find that there is s threshold beyond which the retailer's marginal return on investment in IoT diminishes. One practical implication of this finding is that it is not beneficial to invest in IoT sensors technology that may not significantly improve the quality deterioration at the retailer. Similar observations hold for IoT impact on food loss prevention.

Our paper can be extended in several ways, First, it is worth looking at how IoT investment may coordinate a decentralized supply chain, for example u incorporating a cost of ensuring quality at different supply chain entitities. Second, with IoT technology it would be possible for a retailer to have accurate quality levels at the *visible changes* period. In such circumstances it would be interesting to look at the impact of price discounts in avoiding food losses. These two lines of research are currently being investigated by the authors of this paper. A third line of research is to incorporate uncertainty in quality and demand.

Acknowledgements The authors acknowledge support from the American University of Sharjah Office of Research Enhanced Faculty Research Grant Award EFRG18-SCR-CEN-33 as well as the Natural Sciences and Engineering Research Council (NSERC) of Canada Discovery Grant RGPIN-2014-04827.

References

Ben-Daya M, Hassini E, Bahroun Z (2019) Internet of things and supply chain management: a literature review. Int J Prod Res 57(15–16):4719–4742

Ben-Daya M, Hassini E, Bahroun Z, Banimfreg B (2020) The role of internet of things in food supply chain quality management: a review. Working paper

Bowman P, Ng J, Harrison M, Lopez TS, Illic A (2009) Sensor based condition monitoring. In: Building radio frequency identification for the global environment (Bridge) Euro RFID project

Gunders D (2012) Wasted: how America is loosing up to 40 percent of its food from farm to fork to landfill. NRDC issue paper, August 2012 IP:12–06-B

Gustavsson J, Cederberg C, Sonesson U, Van Otterdijk R, Meybeck A (2011) Global food losses and food waste. In: International congress SAVE FOOD! Düsseldorf, Germany

Heising JK, Claassen GDH, Dekker M (2017) Options for reducing food waste by quality-controlled logistics using intelligent packaging along the supply chain. Food Addit Contamin Part A 34(10):1672–1680

Leithner M, Fikar C (2019) A simulation model to investigate impacts of facilitating quality data within organic fresh food supply chains. Annals Operat Res 1–22

Muth MK, Birney C, Cuéllar A, Finn SM, Freeman M, Galloway JN et al (2019) A systems approach to assessing environmental and economic effects of food loss and waste interventions in the United States. Sci Total Environ 685:1240–1254

Ozbilge A, Hassini E, Parlar M (2019) Donate more to earn more. Submitted for publication

Von Stackelberg H (2010) Market structure and equilibrium. Springer

Fuzzy Goal Programming Based on a Taylor Series for a Pharmaceutical Supply Chain with a Marketing Mix Strategy and Product Life Cycle

Farzaneh S. Shakouhi, Reza Tavakkoli-Moghaddam, Armand Baboli, and Ali Bozorgi-Amiri

Abstract Considering environmental protection and sustainable development, this study proposes a multi-objective mixed-integer nonlinear programming model for the pharmaceutical supply chain network design to maximize the profit, improve consumer health and minimize environmental pollution. In this model, different scenarios are considered based on different product life cycle (PLC) stages (i.e., introduction, growth and maturity) based on marketing mixes (i.e., price, quality, place and promotion). Also, a fuzzy goal programming approach is used to solve the multi-objective problem with parameters of interval type 2 fuzzy numbers (IT2 FNs). A Taylor series is also used to linearize the health objective function. The model presented for two real supply chains with the same product and alternative used in the pharmaceutical industry to demonstrate an application to support better decision making for sustainable supply chain management. The results of this research can be a reference for sustainable supply chain research and practical management.

Keywords Pharmaceutical chain · Network design · Sustainability · Product lifecycle · Marketing mix · Fuzzy programming

---

F. S. Shakouhi (✉)
Department of Industrial Engineering, Albourz Campus, University of Tehran, Tehran, Iran
e-mail: f.shakouhi@ut.ac.ir

R. Tavakkoli-Moghaddam · A. Bozorgi-Amiri
School of Industrial Engineering, College of Engineering, University of Tehran, Tehran, Iran
e-mail: tavakoli@ut.ac.ir

A. Bozorgi-Amiri
e-mail: alibozorgi@ut.ac.ir

A. Baboli
LIRIS Laboratory, UMR 5205 CNRS, INSA of Lyon, 69621 Villeurbanne cedex, France
e-mail: armand.baboli@insa-lyon.fr

© The Author(s), under exclusive license to Springer Nature Switzerland AG 2020

P. Golinska-Dawson et al. (eds.), *Smart and Sustainable Supply Chain and Logistics – Trends, Challenges, Methods and Best Practices*, EcoProduction,
https://doi.org/10.1007/978-3-030-61947-3_27

1 Introduction

Investigations into health services, especially the supply chain, are growing rapidly (Nematollahi et al. 2018). In September 2015, the United Nation General Assembly (Nematollahi et al. 2018) approved a new agendum: sustainable development for all. Health is at the center of the 2030 Program as a sustainable development goal. Carter and Rogers (Carter and Rogers 2008) defined sustainability as "achieving the social, environmental, and economic goals of the organization by systematically coordinating essential inter-organizational processes to improve the long-term economic performance of the company and its supply chain". Considering the following goals at the same time, sustainability is an impartible part of the supply chain: Maximizing profits of the whole chain like other supply chains; many environmental laws are approved by the government to reduce environmental impacts and pollution; most governments, put pressure on stakeholders to increase the level of health and consumer satisfaction in the chain by applying appropriate laws.

Today, ensuring the sustaining design of any network of supply chain activities is dependent on maintaining and optimizing the use of the limited and non-replaceable resources (e.g., energy and environmental resources). This will lead to the adoption of strategies related to investment in improving the environmental performance of the supply chain, reducing pollutions, managing waste, recycling and reusing waste (Fahimi et al. 2017). A sustainable supply chain is one of the aspects of designing and supplying the pharmaceutical supply chain network. To measure supply chain sustainability, some studies have examined all three concepts including 1) economic (maximizing the profit (Moradinasab et al. 2018; Mota et al. 2018) or minimizing the cost (Orjuela-Castro et al. 2019; Narayana et al. 2019; Sazvar et al. 2018; Allaoui et al. 2018; Ghaderi et al. 2018; Tsao et al. 2018; Eskandari-Khanghahi et al. 2018; Baboli et al. 2011)), 2) environmental (Moradinasab et al. 2018; Mota et al. 2018; Orjuela-Castro et al. 2019; Narayana et al. 2019; Sazvar et al. 2018; Allaoui et al. 2018; Ghaderi et al. 2018; Tsao et al. 2018; Eskandari-Khanghahi et al. 2018), and 3) social (maximizing the number of jobs created (Moradinasab et al. 2018; Mota et al. 2018; Allaoui et al. 2018; Eskandari-Khanghahi et al. 2018), food security and consumer health (Orjuela-Castro et al. 2019; Narayana et al. 2019), efficiency of the supply chain (Allaoui et al. 2018), social benefits earned from establishing facilities (Ghaderi et al. 2018; Tsao et al. 2018), satisfaction of social responsibility (Roshan et al. 2019), and unmet demand (Roshan et al. 2019; Zahiri et al. 2018)) factors. Insufficient knowledge of market parameters and competitor reaction causes a kind of uncertainty in decision making that is seen in models. Khalilpourazari et al. (2019) and Babaee Tirkolaee et al. (2019) used triangular fuzzy numbers for demand parameters.

To fill the gap, for the first time, all the three aspects of sustainability including economic, social and environmental have been considered in PSC as a multi-objective model and fuzzy theory (fuzzy goal programming based on Taylor series) used to solve it. In this research, two pharmaceutical supply chains are considered for their different stages of the product life cycle (PLC). Innovations of this research are: using marketing strategies considering the PLC stages (product introduction, growth and maturity periods); sustainable supply chain network design (SCND) depending on marketing mixes (price, quality, place and promotion); insufficient knowledge of market parameters and competitor reaction causes a kind of uncertainty in decision making that is seen in the objective function with parameters of trapezoidal interval type-2 fuzzy numbers (IT2 FNs); using Taylor series for linearizing, the nonlinear objective functions of consumer health level; considering trapezoidal IT2 FNs disturbances in a supply chain (i.e., disturbances in producer, distributor and retailer). As for solving the uncertain problem, Dalman (Dalman 2018) presented an interactive fuzzy goal programming approach based on Taylor series to achieve the highest degree of membership function for multi-objective nonlinear programming problems with trapezoidal IT2 FNs. Because of these advantages, the fuzzy goal programming based on Taylor series is utilized for solving the multi-objective nonlinear programming problem with IT2 FNs in this study.

2 Model Formulation

2.1 Problem Definition

This research involves two pharmaceutical supply chains and three actors. Supply chains are assumed to each has one manufacturer, distributors and retailers. The pharmaceutical supply chain is considered at all stages of product such as introduction, growth and maturity. The SCND is based on three objective functions, maximizing the profit, minimizing pollution emission, and maximizing consumer health increase (number of product complaints according to product quality). The decision variables at this level are the amount of production, choosing distribution centers, and the amount of sent products from a manufacturer to a distributor and from a distributor to a retailer. It should be noted that at this stage, the issues of a producer, a distributor and retailer disruption are considered in the network design. So, the supply chain presented in issue is multi-objective, multi-echelon, which are depended to the objective functions marketing mix factors (i.e., price, quality, place and promotion). The assumptions assumed in the problem are as follows:

- Different stages of the PLC is considered for the supply chain.
- Supply chain and demand depend on the marketing mix.
- There are disruptions and capacity limitation of manufacturing centers, distribution centers, and retailor centers.
- Candidate locations for distribution centers are identified.

2.2 Model Description

Indices:		Tcc_{jr}	Transportation cost per unit of distance from a distribution *j* to a retailer *r*
i	Product index ($i = 1, \ldots, I$)	e_c	Cost per unit of CO_2 emissions
x	Set of manufacturers ($x = 1, \ldots, X$)	p'_i	Price of no-name product *i*
j	Set of distributors ($j = 1, \ldots, J$)	η_{qi}	Amount of CO_2 produced in the production process depends on the quality level of the product
r	Set of retailers ($r = 1, \ldots, R$)	ψ_i	Amount of CO_2 emissions in the product *i* production process
Parameters:		t_{ixj}	Amount of CO_2 emissions in the transship process from a manufacture *x* to a distribution *j*
p_i	Product price *i*	t'_{ijr}	Amount of CO_2 emissions in the transship process from a distribution *j* to a retailer *r*
D_i	Product demand *i*	σ_{xi}	Cost of eliminating or reducing the negative side effects of product *I* production by a manufacture *x*
m_{qi}	Product quality level *I*	δ_j	Efficiency coefficient for distributor *j*
m_{si}	Product access level *i*	$\tilde{\omega}_{ix}$	Percentage of product *I* disturbance at the manufacturing center *x*
m_{ai}	Product promotion level *i*	$\tilde{\omega}'_{ij}$	Percentage of product *i* disturbance at the distribution center *j*
f_j	Fixed costs for opening distribution *j*	$\tilde{\omega}''_{ir}$	Percentage of product *i* disturbance at the retailer center *r*
ε_{qi}	Variable cost of product quality *i*	cap_{ix}	Maximum production capacity of product *i* at the manufacturing center *x*
ε_{si}	Variable cost of product access *i*	cap'_{ij}	Maximum storage capacity of product *i* at the distribution center *j*
ε_{ai}	Variable cost of product promotion *i*	cap''_{ir}	Maximum storage capacity of product *i* at the retailer center *r*
ϕ_{qi}	Fixed cost of product quality *i*	*B*	Supply chain management budget
ϕ_{ai}	Fixed cost of product promotion *i*	**Decision variables:**	
Dis_{xj}	Distance from manufacture *x* to distribution *j*	z_j	1 if the distributor *j* is selected; 0, otherwise
$Diss_{jr}$	Distance from distribution *j* to retailer *r*	q_{ixj}	Quantity of product *i* shipped from manufacture *x* to distribution *j*

(continued)

(continued)

Tc_{xj}	Transportation cost per unit of distance from manufacture x to distribution j	q_{ijr}	Quantity of product i shipped from distribution j to retailer r

2.3 Objective Functions

$$\text{Max } Z_1 = \sum_{i\in I, x\in X, j\in J} p_i q_{ixj} - \sum_{j\in J} f_j z_j - \sum_{i\in I, x\in X, j\in J} (\tilde{\varepsilon}_{qi} m_{qi} + \tilde{\varepsilon}_{si} m_{si} + \tilde{\varepsilon}_{ai} m_{ai}) q_{ixj} -$$
$$\sum_{i\in I, x\in X, j\in J} (\tilde{\phi}_{qi} m_{qi}^2 + \tilde{\phi}_{si} m_{si}^2 + \tilde{\phi}_{ai} m_{ai}^2) q_{ixj} - \sum_{i\in I, x\in X, j\in J} Dis_{xj} Tc_{xj} q_{ixj} -$$
$$\sum_{i\in I, r\in R, j\in J} Diss_{jr} Tcc_{jr} q_{irj} \quad (1)$$

$$\text{Min } Z_2 = e_c \left(\sum_{i\in I, x\in X, j\in J} (\tilde{\eta}_{qi} m_{qi} + \tilde{\psi}_i + \tilde{t}_{ixj}) q_{ixj} + \sum_{i\in I, r\in R, j\in J} \tilde{t}'_{ijr} q_{ijr} \right) +$$
$$+ \sum_{i\in I, x\in X, j\in J} \tilde{\sigma}_{xi} q_{ixj} \quad (2)$$

$$\text{Min } Z_3 = \sum_{i\in I, x\in X, j\in J} \frac{(1 - \tilde{\theta}_{xi}) m_{qi}}{q_{ixj}} \quad (3)$$

The first objective represents the obtained profit using the difference in income and expense The first part shows the revenue from product manufacturing, Part 2 shows the cost of distribution opening, Part 3 and 4, respectively show variable and fixed cost of product based on quality, access and promotion done by supply chain and part five shows the transportation cost from a manufacturer to a distributor and part six shows the transportation cost from a distributor to retailer. In this section, quality, access and promotion cost parameters are IT2 FNs. The second objective represents the created pollution throughout the supply chain. The first part shows the total amount of pollutants released in the production process commensurate with the quality level plus the amount of pollutants released in the process of transportation from a manufacturer to a distributor, the second part represents the amount of pollutants released in the process of transportation from a distributor to a retailer and the second part is negative side effects decreasing cost of the production (e.g., the cost of removing industrial wastewater in production). In this objective function, the amount of CO_2 emission parameters are IT2 FNs. The third objective function represents the level of consumer health. The level of consumer health depends on a sensitive coefficient to the level of complaint of product quality and provided quality, in which manufacturer x and coefficient sensitive to the rate of complaint of product quality.$\tilde{\theta}_x$ is considered as an

IT2 FN according to the number of complaints made in the previous periods from the manufacturer x.

2.4 Constraints

$$\sum_{r \in R} q_{ijr} \leq \sum_{x \in X} q_{ixj} \quad \forall i \in I,\ j \in J \tag{4}$$

$$\sum_{j \in J} q_{ixj} \leq \tilde{\omega}_{ix} cap_{ix} \qquad \forall i \in I,\ x \in X \tag{5}$$

$$\sum_{j \in J, r \in R} q_{ijr} \leq D_i \qquad \forall i \in I \tag{6}$$

$$\sum_{x \in X} q_{ixj} \leq \tilde{\omega'}_{ij} cap'_{ij} z_j \qquad \forall i \in I,\ j \in J \tag{7}$$

$$\sum_{j \in J} q_{irj} \leq \tilde{\omega}''_{ir} cap''_{ir} \quad \forall i \in I,\ r \in R \tag{8}$$

$$\sum_{j \in J} f_j z_j \leq B \tag{9}$$

$$q_{ixj} \geq 0 \qquad \forall i \in I,\ x \in X,\ j \in J \tag{10}$$

$$q_{ijr} \geq 0 \qquad \forall i \in I,\ j \in J,\ r \in R \tag{11}$$

$$z_j \in \{0, 1\} \qquad \forall j \in J \tag{12}$$

Equation (4) guarantees the product current balance from a manufacture to a retailer. Equation (5) considers the disturbances in the production process. That means, with the numerating disturbances, the quantity of produced and transported products to the distributor must not exceed the maximum production capacity. Equation (6) guarantees that the product transported to the retailer is less than or equal to customer demand so that the retailer does not face unsold volume due to lack of space. Equation (7) considers the disturbances in the distributor. That means, with numerating the disturbances, the quantity of produced and transported products to the distributor is not more than the maximum storage capacity of the distributor. Equation (8) considers disturbances in the retailer. That means, with numerating the disturbances, the quantity of transported products to the retailer should not exceed the maximum storage capacity of the retailer. Equation (9) states that the select and opening cost of distribution locations should be less than the available budget. The considered disturbances are considered as IT2 FNs. Constraints (10)–(12) define the ranges for all the variables.

3 Methodology

According to Hu et al. (2013), a trapezoidal IT2 FN is defined that the upper and the lower membership functions are both trapezoidal fuzzy numbers, i.e.,

$$A = (\bar{A}, \underset{-}{A}) = (\bar{a}_1, \bar{a}_2, \bar{a}_3, \bar{a}_4;\ H_1(\bar{A});\ H_2(\bar{A})), (\underset{-1}{a}, \underset{-2}{a}, \underset{-3}{a}, \underset{-4}{a};\ H_1(\underset{-}{A});\ H_2(\underset{-}{A})) \tag{13}$$

where $H_i(\underset{-}{A})$ and $H_i(\bar{A})$ denote membership values of the corresponding elements $\underset{-i+1}{a}$ and $\bar{a}_{i+1}$, respectively. The expected value of A is determined as follows:

$$E(A) = \frac{1}{2}\left(\frac{1}{4}\sum_{i=1}^{4}(\bar{a}_i + \underset{-i}{a})\right) \times \frac{1}{4}\left(\sum_{i=1}^{2}(H_i(\bar{A}_i) + H_i(\underset{-i}{A}))\right) \tag{14}$$

According to the linearization approach proposed by Dalman (2018), transform nonlinear membership functions by using the Taylor series approach around the solution $\widetilde{x}^* = (\widetilde{x}_1^*, \widetilde{x}_2^*, \ldots, \widetilde{x}_n^*)$, which is the solution that is employed to maximize the k-th nonlinear membership function $\mu_k(f_k(x))$ associated with kth nonlinear objective $f_k(x)$:

$$\mu_k(f_k(x))_{k=1,2,\ldots,1} \cong \begin{bmatrix} \dfrac{\mu_k(f_k(\widetilde{x}_1^*))}{\partial x_1}\Bigg|_{\widetilde{x}_1^*}(x_1 - \widetilde{x}_1^*) + \dfrac{\mu_k(f_k(\widetilde{x}_1^*))}{\partial x_2}\Bigg|_{\widetilde{x}_1^*} \times (x_2 - \widetilde{x}_2^*) + \ldots \\ + \dfrac{\mu_k(f_k(\widetilde{x}_1^*))}{\partial k_n}\Bigg|_{\widetilde{x}_1^*}(x_n - \widetilde{x}_n^*) \end{bmatrix}$$

In multi-objective programming, if an imprecise aspiration level is injected to each of the objectives, then these fuzzy objectives are expressed as fuzzy goals. Now, consider the k-th fuzzy goal $f_k(x) \gtrsim s_k$. Its membership function can be defined by:

$$\mu_k(f_k(x)) \cong \begin{cases} 1 \; f_k(x) \geq s_k \\ \dfrac{f_k(x) - l_k}{s_k - l_k} \; l_k \leq f_k(x) \leq s_k \\ 0 \, f_k(x) \leq l_k \end{cases} \tag{15}$$

Then, an equivalent linear fuzzy goal programming model for problem (15) can be developed as follows:

$$\text{Min}\, \beta \begin{cases} \mu_k(f_k(x)) + \overline{d_k} = 1, & k = 1, 2, \ldots, l \\ \beta \geq \overline{d_k} & k = 1, 2, \ldots l \\ \underline{x_l} \leq x_l \leq \overline{x_l} & l = 1, 2, \ldots, n \\ \overline{d_k} \geq 0 & k = 1, 2, \ldots, l \\ & 0 \leq \beta \leq 1 \end{cases} \tag{16}$$

where $\underset{-l}{x} \leq x_l \leq \bar{x}_l$ denotes that the limits of decision variables derived from the individual optimal solutions of each objective and $d_k^- \geq 0$ represents the negative deviations from the aspired levels.

4 Illustrative Example

4.1 Example Description

A real case of two pharmaceutical supply chains is presented to illustrate the application of the proposed model to support better decision making for sustainable supply chain management. The study assumes that each supply chain has one manufacture, three distributors, and three retailers. Also, the supply chains products are similar to each other in the market.

4.2 Parameters Settings

Company information has been modified for business confidentiality. The parameter data are shown in Table 1. This table provides data on two pharmaceutical supply chains with the same product over different periods of their PLC stages. Naturally, the marketing mix data is sensitive to the competitor's strategy and changes according to the competitor's PLC stages. The fixed cost for establishing 400, 500 and 300

Table 1 Distribution centers and retailers data

Distributors	First distributor	Second distributor	Third distributor	Retailer	First retailer	Second retailer	Third retailer
Dis_{1j}/Tc_{1j}	5/1	5/1	4/0.5	$Diss_{1r}/Tcc_{1r}$	1/0.1	2/0.15	3/0.5
$E(\tilde{t}_{11j})$	0.5	0.1	0.1	$Diss_{2r}/Tcc_{2r}$	1.5/0.6	2.5/0.55	3.5/0.5
$E(\tilde{\omega}'_{1j})$	0.03	0.02	0.01	$Diss_{3r}/Tcc_{3r}$	0.5/0.5	1.5/0.45	2/0.5
cap'_{1j}	9800	10000	11000	$E(\tilde{t}'_{11r})$	5	5	5
				$E(\tilde{t}'_{12r})$	10	10	10
				$E(\tilde{t}'_{13r})$	1	1	1
				$E(\tilde{\omega}''_{1r})$	0.03	0.02	0.01
				cap''_{1r}	400	450	500

unit distribution centers and the variable costs dependent on quality, access and promotion are respectively equal to 1, 2 and 4. The fixed costs dependent on the quality, place, and the promotion are 2, 3 and 1, respectively. The amount of CO_2 released in the production process is 6 and the amount of CO_2 released based on the quality level is 1. The producer side costs are 1 and manufacture disturbances are 0.05. The manufacturer capacity is 1000.

5 Results

GAMS 22.2 software was employed to solve the proposed model using the test case data. The results of the calculation of the objective functions in two supply chains at different PLC stages are shown in Table 2. As observed in Table 2, the profit of the second supply chain diminishes with the increased lifecycle of the first supply chain product. This is natural because, with an increase in the lifecycle of the competitor supply chain product (second), the power of the competitor increases in the market, and the share of the market grows for the competitor.

In this way, the profit of the first supply chain diminishes. The notable point in Table 2, is that the profit of the introduction period of the first supply chain product is less than that of the product growth period, and the growth period profit is less than maturity period profit of the product. The reason is that based on the lifecycle of the product, with an increase in the PLC (from introduction to maturity), the profit acquired from product sales grows.

Table 2 PLC and the data of the objective functions for first supply chain

PLC (SC1)	Introduction	Introduction	Introduction	Growth	Growth	Growth	Maturity	Maturity	Maturity
PLC (SC2)	Introduction	Growth	Maturity	Introduction	Growth	Maturity	Introduction	Growth	Maturity
Price (SC1)/(SC2)	40/67	40/64	40/59	41/67	41/64	41/59	42/67	42/64	42/59
Quality (SC1)/(SC2)	22/36	22/36	22/38	22/36	22/36	22/38	22/36	22/36	22/38
Access (SC1)/(SC2)	13/11	13/11	13/10	13/11	13/11	13/10	13/11	13/11	13/10
Promotions (SC1)/(SC2)	23/39	22/43	22/44	23/39	23/43	23/44	23/39	23/43	23/44
Demand (SC1)/(SC2)	458/250	453/254	448/264	455/250	451/255	446/264	453/251	449/255	443/264
F1 (SC1)	1053	1088	1116	1084	1121	1144	1113	1154	1179
F2 (SC1)	357304	289996	196911	358020	355872	293576	358020	358020	356946
F3 (SC1)	0.0467	0.0465	0.0462	0.0464	0.0462	0.046	0.0461	0.0458	0.0456
F1 (SC2)	5407	4479	3515	5419	4549	3515	5489	4549	3527
F2 (SC2)	588338	596211	649892	589054	600506	635577	591917	599074	650608
F3 (SC2)	0.1356	0.1329	0.1355	0.1357	0.1325	0.1356	0.1353	0.1326	0.1357

SC1: First supply chain, SC2: Second supply chain

6 Conclusions

This study presented a multi-objective model for two sustainable pharmaceutical supply chains, which simultaneously considered economic, environmental, social, and uncertain parameters to maximize profit and consumer health and minimize pollution emissions. A fuzzy goal programming approach was used to solve the multi-objective model with parameters of IT2 FNs. The consumer health objective function was also linearized using the Taylor series. In this method, the objective functions became one objective function applied to two pharmaceutical supply chains with the same product in the market. Different scenarios for the supply chains were considered based on their PLC stages. The marketing mix was also examined proportional with supply chains marketing strategies at different periods of their PLC stages. For the future study, two avenues are considered: (1) replacing the Taylor series approach by other techniques for eliminating the nonlinear terms in the model, (2) replacing the IT2 FNs by type-1 fuzzy parameters and can be compared with the results of this paper.

References

Allaoui H, Guo Y, Choudhary A, Bloemhof J (2018) Sustainable agro-food supply chain design using two-stage hybrid multi-objective decision-making approach. Comput Oper Res 89:369–384

Babaee Tirkolaee E, Goli A, Weber G (2019) Multi-objective aggregate production planning model considering overtime and outsourcing options under fuzzy seasonal demand. In: 2019 Manufacturing, Poznan, Poland, 19–22 May 2019

Baboli A, Fondrevelle J, Tavakkoli-Moghaddam R, Mehrabi A (2011) A replenishment policy based on joint optimization in a downstream pharmaceutical supply chain: centralized vs. decentralized replenishment. Int J Adv Manuf Technol 57(1–4):367–378

Carter CR, Rogers DS (2008) Sustainable supply chain management: toward new theory in logistics management. Int J Phys Distrib Logist Manage 38(5):360–387

Dalman H (2018) Interactive fuzzy goal programming based on Taylor series to solve multiobjective nonlinear programming problems with interval type 2 fuzzy numbers. IEEE Trans Fuzzy Syst 26:2434–2449

Eskandari-Khanghahi M, Tavakkoli-Moghaddam R, Taleizadeh AA, Hassanzadeh Amin S (2018) Designing and optimizing a sustainable supply chain network for a blood platelet bank under uncertainty. Eng Appl Artif Intell 71:236–250

Fahimi K, Seyedhosseini SM, Makui A (2017) Simultaneous competitive supply chain network design with continuous attractiveness variables. Comput Ind Eng 107:235–250

Ghaderi H, Moini A, Pishvaee MS (2018) A multi-objective robust possibilistic programming approach to sustainable switchgrass-based bioethanol supply chain network design. J Clean Prod 179:368–406

Hu J, Zhang Y, Chen X, Liu Y (2013) Multi-criteria decision making method based on possibility degree of interval type-2 fuzzy number. Knowl-Based Syst 43:21–29

Khalilpourazari S, Mirzazadeh A, Weber G-W, Pasandideh SHR (2019) A robust fuzzy approach for constrained multi-product economic production quantity with imperfect items and rework process. Optimization, 1–28

Moradinasab N, Amin-Naseri MR, Jafari Behbahani T, Jafarzadeh H (2018) Competition and cooperation between supply chains in multiobjective petroleum green supply chain: A game theoretic approach. J Clean Prod 170:818–841

Mota B, Gomes MI, Carvalho A, Barbosa-Povoa AP (2018) Sustainable supply chains: An integrated modeling approach under uncertainty. Omega 77:32–57

Narayana SA, Pati RK, Padhi SS (2019) Market dynamics and reverse logistics for sustainability in the Indian pharmaceuticals industry. J Clean Prod 208:968–987

Nematollahi MR, Hosseini-Motlagh SM, Ignatius J, Goh M, Saghafi Nia M (2018) Coordinating a socially responsible pharmaceutical supply chain under periodic review replenishment policies. J Clean Prod 172:2876–2891

Orjuela-Castro JA, Aranda-Pinilla JA, Moreno-Mantillac CE (2019) Identifying trade-offs between sustainability dimensions in the supply chain of biodiesel in Colombia. Comput Electron Agric 161:162–169

Roshan M, Tavakkoli-Moghaddam R, Rahimi Y (2019) A two-stage approach to agile pharmaceutical supply chain management with product substitutability in crises. Comput Chem Eng 127:200–217

Sazvar Z, Rahmani M, Govindan K (2018) A sustainable supply chain for organic, conventional agro-food products: The role of demand substitution, climate change and public health. J Clean Prod 194:564–583

Tsao Y-C, Thanh V-V, Lu J-C, Yu V (2018) Designing sustainable supply chain networks under uncertain environments: Fuzzy multi-objective programming. J Clean Prod 174:1550–1565

Zahiri B, Jula P, Tavakkoli-Moghaddam R (2018) Design of a pharmaceutical supply chain network under uncertainty considering perishability and substitutability of products. Inf Sci 423:257–283

Resilience Analysis for Integrated Planning of Selective Harvesting and Post-harvest Operations

Thanaphorn Sornprom, Voratas Kachitvichyanukul, and Kanokporn Kungwalsong

Abstract This article analyses resource flexibilities and consequences of one disruption on an agricultural supply chain through a mixed-integer programming model. The disruption considered is inflation. This situation leads to high living costs, high demand, an increase in employment level, and a rise in production level. The decisions on harvest planning, harvesting equipment selection, hauling truck assignment, and product and processor assignments are simultaneously determined with the objectives to maximize profit and minimize lost sales cost. A set of non-dominated solutions is obtained for decision-makers to evaluate possible solutions for trade-offs between profit and lost sales. Evaluations of these trade-offs may have policy implications on resource allocation decisions. The purpose of this chapter is to assess the consequences of inflation on the key decisions in the integrated planning of an agricultural supply chain with selective harvesting. This chapter identifies the causes and effects of inflation at an operational level. The findings underscore the model's recommended decisions in an agricultural supply chain disruption. In this study, the authors analyse one disruption related to demand, available capacities, costs, and yields through a mixed-integer programming model. Numerical data sets for this disruption are derived based on real operations. CPLEX Optimization software package is used to find the possible solutions for both single and multi-objective problems.

Keywords Resilience · Integrated planning · Risk analysis · Supply chain disruptions

---

T. Sornprom (✉)
Faculty of Logistics, Burapha University, Saen Suk, Thailand
e-mail: thanaphorn.prae@gmail.com

V. Kachitvichyanukul
Asian Institute of Technology, Khlong Nueng, Thailand
e-mail: voratas@ait.ac.th

K. Kungwalsong
Graduate School of Management and Innovation, King Mongkut's University of Technology Thonburi, Thonburi, Thailand
e-mail: kanokporn.rie@gmail.com

© The Author(s), under exclusive license to Springer Nature Switzerland AG 2020
P. Golinska-Dawson et al. (eds.), *Smart and Sustainable Supply Chain and Logistics – Trends, Challenges, Methods and Best Practices*, EcoProduction,
https://doi.org/10.1007/978-3-030-61947-3_28

1 Introduction

Agricultural production is one of the principal revenue sources for many countries. The supply chain of agricultural production may take months or years and may cover a broad geographical area. The occurrences of unplanned events such as natural disasters, equipment failure, and unpredictable customers' demand may adversely affect the performance of a supply chain. In recent years, fierce competition in agricultural business leads to the awareness of the necessity of flexibility and adaptability in a supply chain to minimize risk. To reduce the impacts and severities of the supply chain risks, risk management and resilience assessment should be applied. Additionally, it is desirable to consider flexibility and adaptability in the design of a supply chain management system.

Supply chain resilience is often seen as the capacity to maintain a normal operation when disruptions occur. It directly relates to disaster recovery, risk management, and reduction of the supply chain vulnerability. Many researchers (Elleuch et al. 2016a; Hosseini et al. 2019; Kamalahmadi and Parast 2016; Ponomarov and Holcomb 2009; Singh et al. 2019) reviewed the literature related to the supply chain resilience and vulnerability. They (Kamalahmadi and Parast 2016; Ali et al. 2017) claimed that the supply chain resilience should cover three phases. Firstly, emergency plans should be prepared for urgencies. Secondly, the supply chain should be robust to deal with the disruptions (Tang 2006). Finally, effective responses need to minimize the negative impacts of supply chain disruptions.

Besides, resilience analysis has been applied in multiple kinds of supply chains such as energy (Jabbarzadeh et al. 2016; Urciuoli et al. 2014), dairy products (Bourlakis et al. 2014), fruits (Behzadi et al. 2017), food (Behzadi et al. 2017), meat (Leat and Revoredo-Giha 2013), and automobiles (Carvalho et al. 2012; Rezapour et al. 2017).

Decision-making problems under uncertainty have been implemented with stochastic programming, robust programming, and simulation (Borodin et al. 2016). Several solution techniques have been applied to improve resilience such as optimization models (Dixit et al. 2016; Septiani et al. 2016), mixed-integer linear programming model (Yavari and Zaker 2019; Goli et al. 2019), simulation (Carvalho et al. 2012), quality function deployment (QFD) (Elleuch et al. 2016b), questionnaire survey (Bourlakis et al. 2014), and interview (Leat and Revoredo-Giha 2013).

Agricultural supply chain risk management has been a topic of interest in the last two decades. The food supply chain with strategic resilience was built by Manning and Soon (Manning and Soon 2016). They proposed the strategic resilience indicator framework to develop organizations by considering in 3Rs: ready, respond, and recovery. Septiani and et al. (2016) found that most of the research works related to resilient supply chain were done in three perspectives: risk identification, risk assessment, and risk mitigation. First, risk identification can be conducted by qualitative method, semi-qualitative method, and quantitative method. Second, risk assessment can be done through various models such as simulation model, intelligent technique, optimization model, statistical analysis, and others. Last, Interpretative Structural Modelling (ISM) and life cycle assessment can be applied for risk mitigation.

Some studies applied optimization techniques to manage the agricultural supply chain under risk. In 2000, the complexity of selecting effective crops pattern and maximizing farmers' utility led to the notion of a conceptual framework (Romero 2000) which was implemented as a multi-objective linear programming model. A weighting method was applied to find the solutions for the multi-objective problems which attempted to maximize an expected gross margin, and to minimize variability. Through experiments, the authors found that the proposed framework can be an alternative method to deal with risk instead of the mathematical model based on portfolio theory (Freund 1956).

For the influence of weather in agricultural management, a crop planning model with stochastic values was proposed as a linear programming problem (Itoh et al. 2003). The model focused on the limitation of land, and labor. The purpose was to maximize the minimum total gain due to random profit coefficients. The experiments used the planning data of various crops; carrot, radish, cabbage, and Chinese cabbage within ten acres. The results showed that an optimal solution can be applied in real situations.

The uncertainties of supply and demand in the supply chain of perishable products have challenged decision-makers to develop optimization models. A mixed-integer linear programming model and Monte Carlo simulation model were applied to incorporate demand, yield, and harvest failure of tomatoes plantation (Merrill 2007). The objectives of this study were to find the appropriate time, quantity, and location to plant tomatoes to minimize customer service requirements and to maximize profit. Numerical experiments showed that the model can achieve a 90% customer service level with 20% less planted acreages with almost three times profit growth.

Another work that was related to tomato planting was done in 2012. Tan and Comden (2012) applied a non-linear equation to determine farm areas and seeding times to maximize the total profit over the planning period. The detailed model considered maturation time, harvest time, and yield uncertainty with random demand. The numerical experiments showed that the objective function increased exponentially with the number of farms in a single-period. For the multi-period problem, the approach with farm areas and the seeding times had approximately 16% higher profit than the mean-value approach.

In 2017, a two-stage stochastic model was developed to maximize the expected profit of the Kiwifruit supply chain in New Zealand (Behzadi et al. 2017). A fractional formulation was used to integrate perishability and a quality barrier into the model. This study investigated the effectiveness of robust, and resilient strategies to manage harvest time and yield disruptions. After the model was applied, the results showed the improvement of the company's benefits; however, the authors considered only two types of suppliers' risks: harvest time disruption and yield disruption.

In the same period, some scholars designed an integrated supply chain network for perishable products (Shrivastava et al. 2017). The researchers considered disruptions in the transportation links between manufacturers and retailers. They formulated the problem of locating and allocating facilities as a mixed-integer quadratic model.

The model was solved by the default settings of the CPLEX optimization software. Consequently, the total cost of the supply chain was higher in the resilient model. In the future, multi-products (Goli et al. 2019), multi-routes, and multi-periods should be considered.

As mentioned earlier, there was no attempted to address resilience through the assessments of capabilities, costs, and impacts and considering multi-objective optimizations (Elleuch et al. 2016a; Hosseini et al. 2019; Kamalahmadi and Parast 2016). Therefore, this study focuses on a modeling framework to examine the consequences of inflation on suppliers, operations, capacities, time availabilities, costs, resources, and demands. Two objective functions, profit maximization, and lost sales cost minimization and five decisions of inflation will be compared with those results obtained in a normal condition. Lastly, two scenarios will be evaluated while simultaneously considering both objective functions. The set of non-dominated solution values or the Pareto frontier will be obtained for decision-makers.

2 Methodology

Ahumada and Villalobos (2009) reviewed the literature on agricultural supply chain and found that many research works have been concentrated on developing tactical plans for the supply chain of non-perishable products. Only several researchers have worked on developing operational plans for agricultural supply chain for fresh produces and most of the studies have separately considered four functional areas: production, harvest, storage, and distribution, and the plans for each functional area were made sequentially. The modeling approaches used in agricultural planning included stochastic programming (SP), linear programming (LP), dynamic programming (DP), and mixed-integer programming (MIP). Common objective functions were cost minimization and profit maximization. Nevertheless, most of the prior researchers did not pay much attention to risk modeling, uncertain information, logistic integration, quality, and security of products. Besides, planning models have been developed for each functional area with very little attempt made on integration due to the complexity of the model.

The focus of this study is to assess the impacts of inflation on the key decisions in the integrated planning of an agricultural supply chain with selective harvesting. The integrated planning model that includes harvesting, hauling, production, transport, and distribution for selective harvesting by Sornprom et al. (2019) is used for risk analysis of the agricultural supply chain. It is a mixed-integer model that considers resource allocation such as time availabilities, resource capacities, and the number of haulers, vehicles, and harvesting tools. The key decisions include the quantities of harvested crops, hauled crops, finished products, delivered products, products sold, and product shortage. A mathematical model and detailed description can be found in the reference (Sornprom et al. 2019). The development of a framework for a resilience analysis along with a solution technique is given in the next subsection.

2.1 Framework Development

The chronological sequence of decisions in an agricultural supply chain is shown in Fig. 1. The arrow indicates the dependency of decisions from previous operations. For instance, a drought may result in the decrease of mature crop quantities to be harvested. Consequently, this may lead to less requirement for hauling equipment and fewer raw material available for production and eventually a reduction in products delivered to the distribution center.

The disruption considered in this study is inflation. The disruptions caused by this risk factor on various decisions in an integrated supply chain are illustrated in Fig. 2.

According to Fig. 2, it is apparent that inflation affects the value of money, leading to an increasing amount of wages to maintain living costs. In this economic variation, banks reduce interest so that many companies offer more jobs. As a consequence, individuals in society have money for basic needs, resulting in the rising demand for agriculture products. When the demand is higher than supply, companies raise the prices of products. A resilience analysis of this situation is carried out using the mixed-integer programming model. The data is a subset of a real system with the size reduced for suitable computational complexity. A baseline case and the two scenarios described above are evaluated and compared.

The supply chain system considered contains 20 fields from 4 suppliers. The system is operated with three harvesting machines, five haulers, four processors,

Fig. 1 Dependency of decisions in an agricultural supply chain

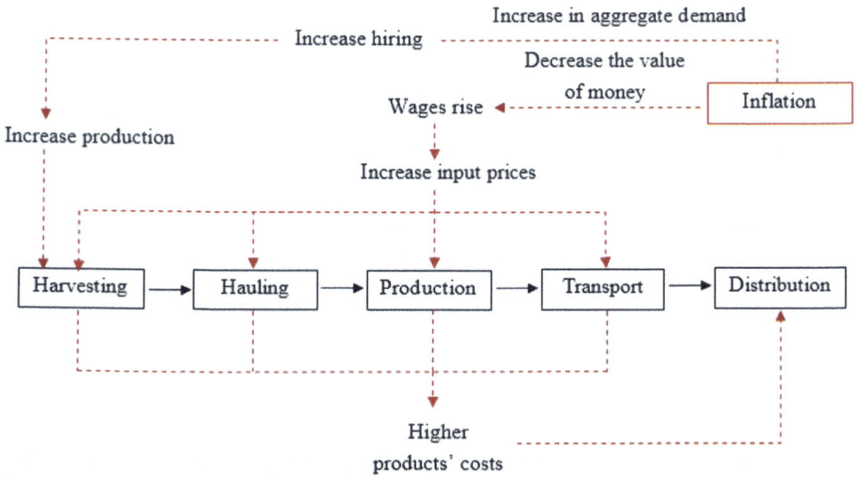

Fig. 2 Cause and effect framework

three products, and three vehicles. All the fields are near and can use the assigned machines without much transportation delay.

2.2 Solution Technique

CPLEX optimization software package is a well-established solver for single-objective mixed integer programming problems. This software uses a branch and bound method in seeking optimal solutions. To solve the multi-objective problem directly with CPLEX, one objective must be converted into a constraint as shown in Eqs. (2) and (4). From these equations, it clearly shows that the objective values from profit maximization and lost sales cost minimization acts as the bounds of possible multi-objective values and Pareto frontier. The "deviation" variables Δ and ξ are treated as parameters and the model is solved repeatedly with varying values of Δ and ξ to yield a set of non-dominated solutions.

$$\text{Objective function}: Max\ Profit = Revenue - Total\ direct\ cost \quad (1)$$

$$\text{Constraint}: Lost\ sales\ cost \leq Minimal\ lost\ sales\ cost + \Delta \quad (2)$$

$$\text{Objective function: Minimize Lost sales cost} = \sum_p \left[\sum_d J_{pd} \right] * \sigma_p$$

J_{pd} : Shortage quantity of product p on day d (box)

σ_p : Lost sales cost of product p (baht per box per day) (3)

$$\text{Constraint}: Profit \geq Maximal\ profit - \zeta \quad (4)$$

3 Results and Analysis

As mentioned earlier, this study concentrates on a possible risk called inflation. This situation will be compared with the normal one to study the consequential impacts on decisions, profit, and lost sales cost.

The math model proposed by Sornprom et al. (2019) was solved via the CPLEX Optimization software package. The results of a single objective problem are given in Table 1 for profit maximization (totally ignoring lost sales cost) and for lost sales cost minimization in Table 2. For brevity, the decisions on harvesting, harvester assignment, hauler assignment, processor assignments, etc. are not shown here. Note

Table 1 The solutions for profit maximization

Scenario	Maximal Profit (baht)	Lost sales cost (baht)
Normal situation	445,128.35	59,240.00
Inflation	462,351.85	67,152.00

Table 2 The solutions for lost sales cost minimization

Scenario	Profit (baht)	Minimal Lost sales cost (baht)
Normal situation	388,363.70	57,104.00
Inflation	417,699.00	64,784.00

that for each scenario, these decisions could be different when the model is solved with different objectives.

The mixed-integer programming model provides decisions for daily operations of the agricultural supply chain such as harvesting plots to be harvested, harvester assignments, hauler assignment, processor assignments, etc. The total direct cost is the summation of five direct costs: harvesting cost, hauling cost, processing cost, holding cost, and transportation cost while lost sales cost is focused only on the indirect cost which directly links to shortage amount.

When the two objectives are considered simultaneously, as mentioned earlier, one of the objective functions must be converted into a constraint to use CPLEX to solve the problem. For instance, if the objective is to maximize profit and the lost sales cost is turned into a constraint as shown in Eq. (2), the value of Δ is treated as a parameter and is varied from 0 to 4000 for scenario 0 as shown in Table 3. According to Eq. (2), the generated lost sales cost cannot be smaller than the minimal lost sales cost. By varying the value of Δ, solutions can be obtained with different profit and lost sales cost for each value of Δ as shown in Table 3.

The Δ value can be interpreted as the additional lost sales cost allowed above the minimal lost sales cost. As the lost sales cost constraint is relaxed by increasing the value of Δ, CPLEX finds different solutions with higher profits until it reaches the

Table 3 Solutions of multi-objective problem with varying values of Δ

Δ	Minimal Lost sales cost + Δ (baht)	Minimal Profit (baht)	Lost sales cost (baht)
0	57,104.00	442,144.45	57,104.00
85	57,189.00	442,481.03	57,152.00
540	57,644.00	443,063.99	57,616.00
865	57,969.00	443,516.71	57,964.00
978	58,082.00	443,639.95	58,080.00
1,256	58,360.00	443,946.43	58,312.00
1,500	58,604.00	444,245.91	58,544.00
2,500	59,604.00	445,128.35	59,240.00
4,000	61,104.00	445,128.35	59,240.00

absolute maximal profit when Δ is at 4,000. The effects of the Δ value on various cost elements are given in Table 3. Similarly, the value of ξ can be interpreted as the missed profit amount allowed from the absolute maximal profit. A similar result can be obtained by setting the lost sales cost minimization as the objective with the constraint on profit as shown in Eq. (4) earlier.

To obtain the set of non-dominated solutions or the Pareto front, each scenario must be solved multiple times with varying values of Δ or ξ as shown in Table 4 and Table 5, where appropriate. The sets of non-dominated solutions for all scenarios are shown in an X–Y plot of profit versus lost sales cost in Fig. 3.

For the objective values plotted in Fig. 3, these points are associated with five main decisions for harvesting, hauling, production, delivery, and distribution. The comparisons of these decisions will be made by using results from scenarios with normal condition and inflation.

Table 6 is the results when Δ value is equal to 0. This table compares cost components for the scenarios of normal condition and inflation. In the inflation situation, customers may buy more in anticipating rising price and this could lead to higher demand and if resources are not appropriately adjusted, production capacity may be limited and the lack of supplies may lead to higher lost sales. Consequently, the company may lose the opportunity to receive more benefits as shown in Table 6.

The operation decisions of both scenarios in Table 6 are displayed in Figs. 4, 5, 6, 7, 8 and 9 to illustrate the decisions for harvesting, hauling, processing, delivery, and distribution.

As can be seen in Fig. 4, the model recommends different daily harvesting decisions for the two scenarios. If one is operating under the normal condition when inflation occurs, the demand outstrips supply. As a result, a shortage may occur.

Figure 5 shows the daily hauling decisions for the two scenarios. The hauling decisions are dependent on the harvesting decisions made in the previous stage of operation. The number of trips for daily hauling is calculated from the planned harvesting quantities. Again, the model indicates that too many haulers are assigned for inflation and not enough haulers if the hauler assignment is made to follow the normal operating condition.

The processing decisions are shown in Fig. 6. In the test data, the product priority is product P1, P2, and P3 in that order. The production cost for processors 3 and 4 are lower than the others. The model logically allocates the supply to be processed into product P1 by process 3 on day 1. On day 2, the quantity is higher than the capacity of processor 3 so the remaining quantity is assigned to processor 4. After day 2, the demand for product P2 and P3 are assigned after all demand for product P1 are assigned.

Similar coordinated decisions are recommended by the model for delivery and distribution. Figure 7 presents the vehicle trips of vehicle type 2, which has the cheapest cost.

Daily product distribution decisions are shown in Fig. 8 along with the resulting shortages in Fig. 9.

Overall, Figs. 4, 5, 6, 7, 8 and 9 show the model's recommended decisions under inflation condition and solution variations of resources along with the corresponding

Table 4 The effect of changing Δ values to costs for normal situation

Δ	Harvesting cost (baht)	Hauling cost (baht)	Processing cost (baht)	Transportation cost (baht)	Revenue (baht)	Total cost (baht)	Profit (baht)	Lost sales cost (baht)
0	179,700	60,698.67	33,974	83,942.88	800,460	358,315.55	442,144.45	57,104
85	179,850	60,749.33	33,897	84,085.64	801,063	358,581.97	442,481.03	57,152
540	179,850	60,749.33	33,499	84,656.68	801,819	358,755.01	443,063.99	57,616
865	179,850	60,749.33	33,185	85,084.96	802,386	358,869.29	443,516.71	57,964
978	179,850	60,749.33	33,108	85,227.72	802,575	358,935.05	443,639.95	58,080
1,256	179,850	60,749.33	32,894	85,513.24	802,953	359,006.57	443,946.43	58,312
1,500	179,850	60,749.33	32,687	85,798.76	803,331	359,085.09	444,245.91	58,544
2,500	179,850	60,749.33	32,082	86,655.32	804,465	359,336.65	445,128.35	59,240
4,000	179,850	60,749.33	32,082	86,655.32	804,465	359,336.65	445,128.35	59,240

Table 5 The effect of changing Δ values to costs for inflation

Δ	Harvesting cost (baht)	Hauling cost (baht)	Processing cost (baht)	Transportation cost (baht)	Revenue (baht)	Total cost (baht)	Profit (baht)	Lost sales cost (baht)
0	201,300	67,994.67	35,923	88,082.92	862,413	393,300.59	469,112.41	73,096
56	201,300	67,994.67	35,923	88,082.92	862,413	393,300.59	469,112.41	73,096
89	201,300	67,994.67	35,923	88,082.92	862,413	393,300.59	469,112.41	73,096
179	201,300	67,994.67	35,816	88,225.68	862,602	393,336.35	469,265.65	73,212
284	201,300	67,994.67	35,746	88,368.44	862,791	393,409.11	469,381.89	73,328
315	201,300	67,994.67	35,709	88,368.44	862,791	393,372.11	469,418.89	73,328
415	201,300	67,994.67	35,632	88,511.2	862,980	393,437.87	469,542.13	73,444
585	201,300	67,994.67	35,532	88,653.96	863,169	393,480.63	469,688.37	73,560
698	201,300	67,994.67	35,425	88,796.72	863,358	393,516.39	469,841.61	73,676
767	201,300	67,994.67	35,318	88,939.48	863,547	393,552.15	469,994.85	73,792
912	201,300	67,994.67	35,211	89,082.24	863,736	393,587.91	470,148.09	73,908
1,212	201,300	67,994.67	34,927	89,510.52	864,303	393,732.19	470,570.81	74,256
1,557	201,300	67,994.67	34,646	89,938.8	864,870	393,879.47	470,990.53	74,604
1,857	201,300	67,994.67	34,429	90,224.32	865,248	393,947.99	471,300.01	74,836
2,053	201,300	67,994.67	34,208	90,509.84	865,626	394,012.51	471,613.49	75,068

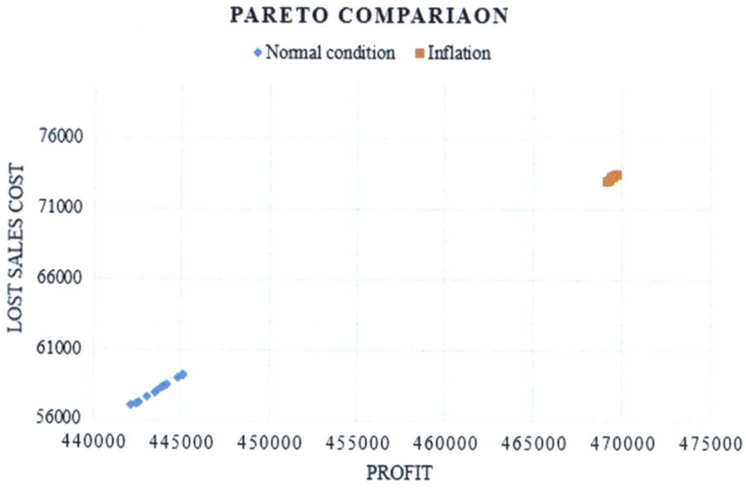

Fig. 3 Pareto frontier comparison

Table 6 An example of scenario comparison

Cost components (baht)	Normal situation	Inflation
Harvesting cost	179,700.00	201,300.00
Hauling cost	60,698.67	67,994.67
Processing cost	33,974.00	35,923.00
Holding cost	0	0
Transportation cost	83,942.88	88,082.92
Revenue	800,460.00	862,413.00
Total cost	358,315.55	393,300.59
Profit	442,144.45	469,112.41
Lost Sales Cost	57,104.00	73,096.00

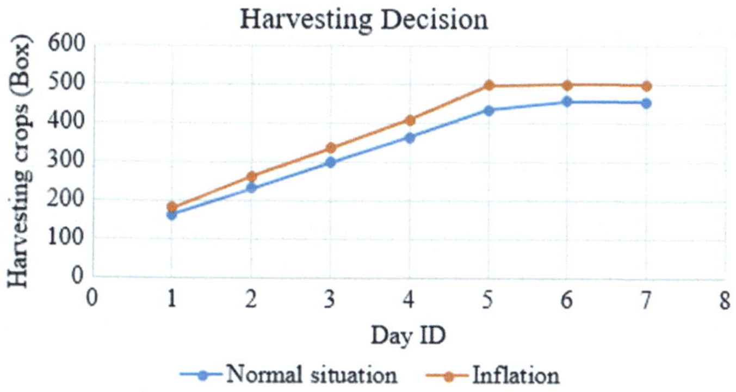

Fig. 4 The comparison of harvesting decision

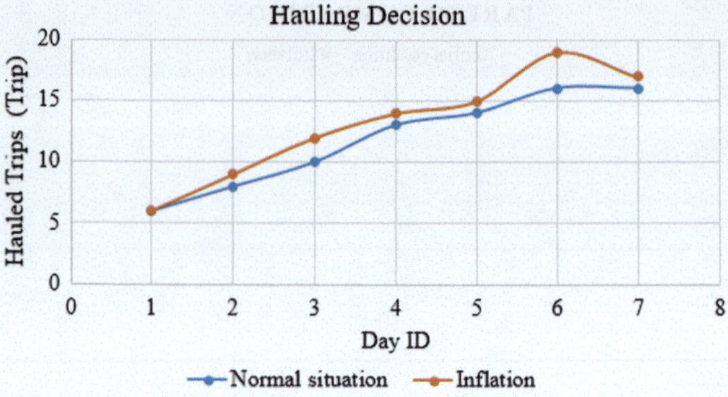

Fig. 5 The comparison of hauling decision

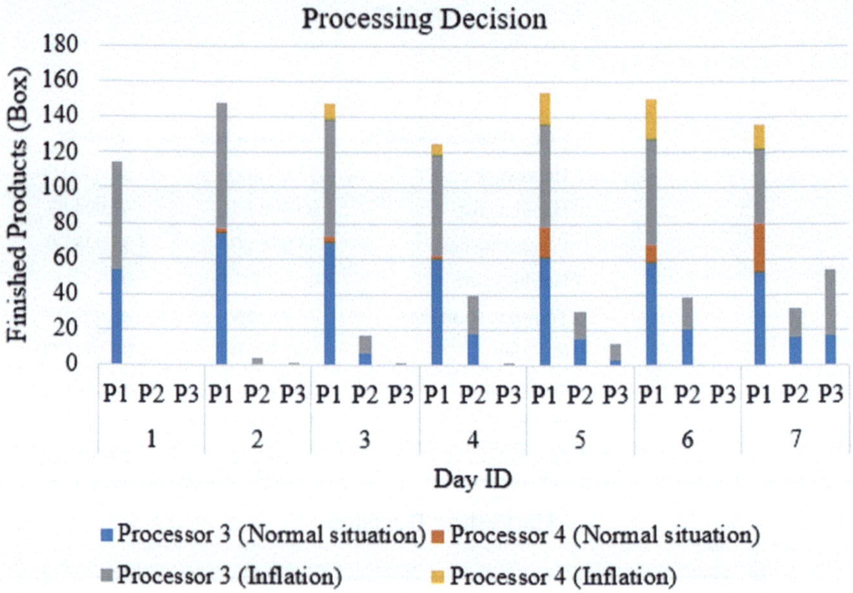

Fig. 6 The comparison of processing decision

decisions under the normal scenario. The model can be conveniently exercised to assess supply chain risks under various conditions before making final decisions.

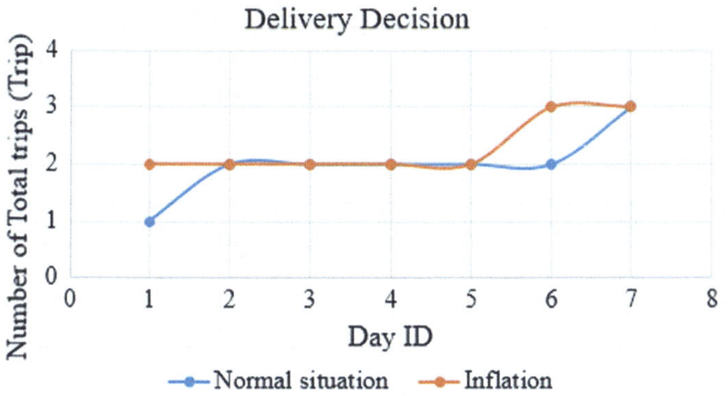

Fig. 7 The comparison of delivery decision

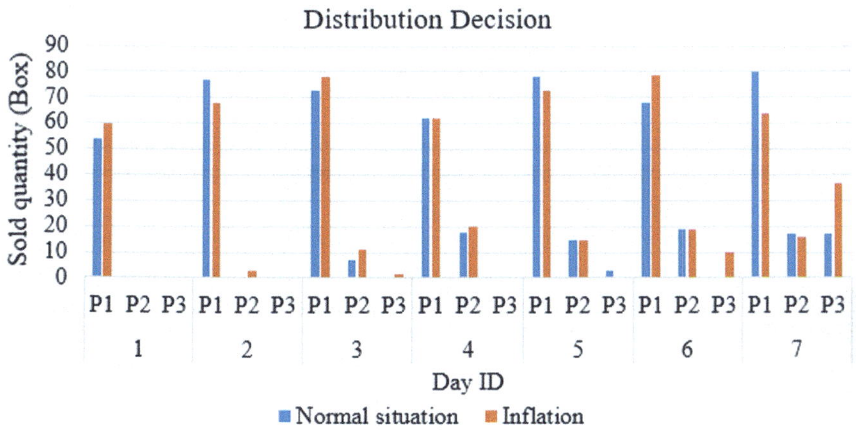

Fig. 8 The comparison of distribution decision

4 Conclusions

This study applies a mixed-integer programming model for integrated planning of selective harvesting and post-harvest operations (Sornprom et al. 2019) for resilience analysis of an agricultural supply chain. A possible disruption, inflation, is included to study the variations of decisions, profit, and lost sales cost. CPLEX Optimization software package is used to find solutions for both single and multi-objective problems.

For a single objective problem, CPLEX tries to find an optimal solution for each case without considering another objective. This program can generate optimal solutions and decisions for both objectives.

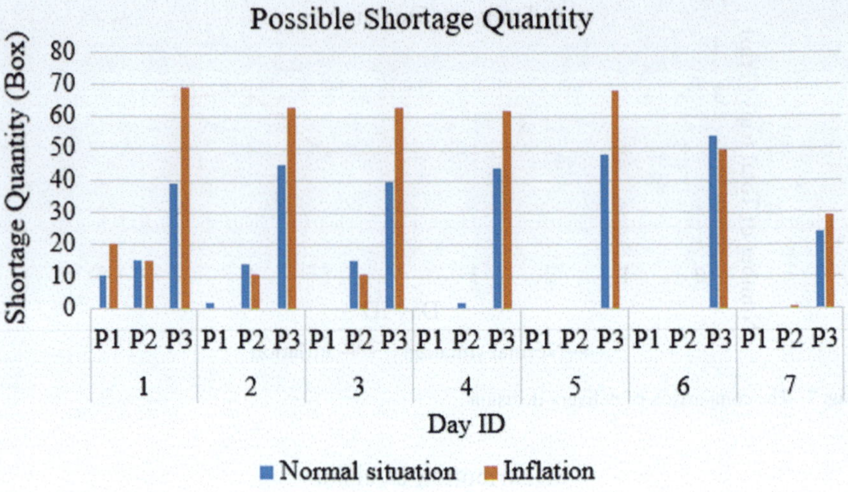

Fig. 9 The comparison of possible shortage quantity

For multi-objective cases, the Pareto frontiers provide a convenient way to tradeoff profit versus lost sales. The model also provides coordinated operating decisions for the corresponding profit and/or lost sales target.

For a larger size model, CPLEX solver may not be suitable as the solution tool because of the extremely long computational time. Metaheuristic methods might be a feasible solution method for larger industrial-scale problem. A possible direction of further studies would be to study other disruptions such as lack of workers, drought, equipment failure, and so on.

References

Ahumada O, Villalobos JR (2009) Application of planning models in the agri-food supply chain: a review. Eur J Oper Res 195:1–20

Ali A, Mahfouz A, Arisha A (2017) Analysing supply chain resilience: integrating the constructs in a concept mapping framework via a systematic literature review. Int J Supply Chain Manage 16–39

Behzadi G, O'Sullivan MJ, Olsen TL, Scrimgeour F, Zhang A (2017) Robust and resilient strategies for managing supply disruptions in an agribusiness supply chain. Int J Prod Econ 191:207–220

Borodin V, Bourtembourg J, Hnaien F, Labadie N (2016) Handling uncertainty in agricultural supply chain management: a state of the art. Eur J Oper Res 254:348–359

Bourlakis M, Maglaras G, Gallear D, Fotopoulos C (2014) Examining sustainability performance in the supply chain: the case of the Greek dairy sector. Ind Mark Manage 43:56–66

Carvalho H, Barroso AP, Machado VH, Azevedo S, Cruz-Machado V (2012) Supply chain redesign for resilience using simulation. Comput Indus Eng 62:329–341

Dixit V, Seshadrinath N, Tiwari M (2016) Performance measures based optimization of supply chain network resilience: a NSGA-II + Co-Kriging approach. Comput Ind Eng 93:205–214

Elleuch H, Dafaoui E, Elmhamedi A, Chabchoub H (2016a) Resilience and vulnerability in supply chain: literature review. IFAC-PapersOnLine 49(12):1448–1453
Elleuch H, Dafaoui E, Mhamedi AE, Chabchoub H (2016b) A quality function deployment approach for production resilience improvement in supply chain: case of agrifood industry. IFAC-PapersOnLine 49(31):125–130
Freund RJ (1956) The introduction of risk into a programming model. Econometrica 24(3):253–263
Goli A, Tirkolaee EB, Malmir B, Bian GB, Sangaiah AK (2019) A multi-objective invasive weed optimization algorithm for robust aggregate production planning under uncertain seasonal demand. Computing 101(6):499–529
Hosseini S, Ivanov D, Dolgui A (2019) Review of quantitative methods for supply chain resilience analysis. Transp Res Part E Logist Transp Rev 285–307
Itoh T, Ishii H, Nanseki T (2003) A model of crop planning under uncertainty in agricultural management. Int J Product Econ 555–558
Jabbarzadeh A, Fahimnia B, Sheu J-B, Moghadam HS (2016) Designing a supply chain resilient to major disruptions and supply/demand interruptions. Transp Res Part B 94:121–149
Kamalahmadi M, Parast MM (2016) A review of the literature on the principles of enterprise and supply chain resilience: major findings and directions for future research. Int J Product Econ 171:116–133
Leat P, Revoredo-Giha C (2013) Risk and resilience in agri-food supply chains: the case of the ASDA PorkLink supply chain in Scotland. Supply Chain Manage Int J 18(2):219–231
Manning L, Soon JM (2016) Building strategic resilience in the food supply chain. Br Food J 118(6):1477–1493
Merrill JM (2007) Managing risk in premium fruit and vegetable supply chains. Master thesis, Massachusetts Institute of Technology, Cambridge
Ponomarov SY, Holcomb MC (2009) Understanding the concept of supply chain resilience. Int J Logist Manage 20(1):124–143
Rezapour S, Farahani RZ, Pourakbar M (2017) Resilient supply chain network design under competition: a case study. Eur J Oper Res 259:1017–1035
Romero C (2000) Risk programming for agricultural resource allocation: a multidimensional risk approach. Ann Oper Res 94:57–68
Septiani W, Marimin Herdiyeni Y, Haditjaroko L (2016) Method and approach mapping for agri-food supply chain risk management: a literature review. Int J Supply Chain Manage 5(2):51-64
Shrivastava H, Dutta P, Krishnamoorthy M, Suryawanshi P (2017) Designing a resilient supply chain network for perishable products with random disruptions. In: International multiconference of engineers and computer scientists. Hong Kong.
Singh CS, Soni G, Badhotiya GK (2019) Performance indicators for supply chain resilience: review. J Indus Eng Int 1–13
Sornprom T, Kachitvichyanukul V, Luong HT (2019) A multi-objective model for integrated planning of selective harvesting and post-harvest operations. In: Environmental sustainability in Asian logistics and supply chains. Springer, Singapore, pp 245–260
Tang CS (2006) Robust strategies for mitigating supply chain disruptions. Int J Logist Res Appl 9(1):33–45
Tan B, Comden N (2012) Agricultural planning of annual plants under demand, maturation, harvest, and yield risk. Eur J Oper Res 220:539–549
Urciuoli L, Mohanty S, Hintsa J, Boekesteijn EG (2014) The resilience of energy supply chains: a multiple case study approach on oil and gas supply chains to Europe. Supply Chain Manage Int J 19(1):46–63
Yavari M, Zaker H (2019) Designing a resilient-green closed loop supply chain network for perishable products by considering disruption in both supply chain and power networks. Comput Chem Eng 134:106680

Roadmap for a Successful Implementation of a Predictive Maintenance Strategy

Marcel André Hoffmann and Rainer Lasch

Abstract Machine failure can have significant impacts on increasingly global orientated supply chains in the producing industry. Predictive maintenance (PdM) is a powerful method to avoid economic damage that can occur as a consequence of critical system breakdowns. The latest research shows that only a minority of industrial companies use an approach of residual lifetime prognosis in maintenance. Especially smaller and mid-sized enterprises have a lack of resources and knowledge to focus on a PdM strategy. The purpose of this article is to provide a structured approach on how to implement a PdM strategy in industrial companies in order to reduce maintenance costs and resources. It contains practical orientated recommendations for analyzing, decision making, and implementation of a smart data-based maintenance strategy. Most of the relevant literature in this field focuses on operational decision making in maintenance and residual lifetime prognosis. This chapter provides a structured integrative managerial approach of PdM with a focus on the implementation process of this strategy in an industrial context.

Keywords Predictive maintenance · Maintenance management · Maintenance framework · Smart factory · Implementation process

1 Introduction

Reliable production processes are crucial for industrial companies since machine downtime can lead to delivery failure and, consequently, to losses in sales. Therefore, predictive maintenance is considered as an adequate strategy to avoid unplanned downtimes, as a forecast of the residual useful life (RUL) allows to conduct necessary maintenance activities before a breakdown occurs. The development of data-driven approaches was particularly extended in the past years with the latest innovations in the analysis of high amounts of data (Feng and Shanthikumar 2018).

M. A. Hoffmann (✉) · R. Lasch
Chair of Business Management, esp. Logistics, Technische Universität Dresden, Dresden 01062, Germany
e-mail: marcel.hoffmann@tu-dresden.de

© The Author(s), under exclusive license to Springer Nature Switzerland AG 2020

P. Golinska-Dawson et al. (eds.), *Smart and Sustainable Supply Chain and Logistics – Trends, Challenges, Methods and Best Practices*, EcoProduction,
https://doi.org/10.1007/978-3-030-61947-3_29

Even though PdM provides the possibility to reduce total maintenance costs, only a minority of companies performed the implementation of this strategy successfully. According to the study from Haarman et al. (2018), where 268 companies in The Netherlands, Germany, and Belgium were asked about their maturity of maintenance, only 11% implemented a PdM strategy in their maintenance management. A structured and application-oriented procedure is seen as necessary to support companies in a prosperous realization of PdM. Therefore, the following research questions arise in this context:

1. Which management-oriented frameworks can be found in the literature considering the implementation of PdM?
2. How should a structured and application-oriented roadmap be developed to support the successful implementation of a predictive maintenance strategy in industrial companies?

The remainder of this chapter is organized as follows: Sect. 2 explains the general aspects of maintenance management and maintenance strategies. In Sect. 3, a systematic literature review (SLR) is conducted to give an overview of existing frameworks to implement a PdM strategy. Section 4 shows a management-oriented roadmap to implement PdM, and conclusions are discussed in Sect. 5.

2 Maintenance Management

A fundamental distinction is made between strategic and operational maintenance management. The necessary processes are divided into core, support, and management processes (Wald 2003). Within the core processes, those activities are grouped, which have a direct influence on the availability of the plants. Support processes have no direct impact on operational maintenance activities but ensure smooth operations and availability. Management processes serve to plan and control maintenance targets and communication. Strategic maintenance management carries out medium- and long-term planning, control, and monitoring of the company's maintenance objectives.

On the other hand, operational maintenance management is responsible for the short- and medium-term implementation of the goals. For this purpose, planning of maintenance program, capacities, and scheduling of maintenance activities are executed. The achievement of objectives is monitored through progress control (Wald 2003).

The main aim of maintenance management is to minimize maintenance costs. These are divided into direct and indirect costs. Direct maintenance costs include all expenses for carrying out maintenance activities. These mainly include costs for materials, personnel, and operating resources or external services. With increasing maintenance intensity (MI), the direct maintenance costs increase proportionally. The indirect costs result from the failure of plants and their economic consequences for the company. In the event of a breakdown within interlinked processes, capacity

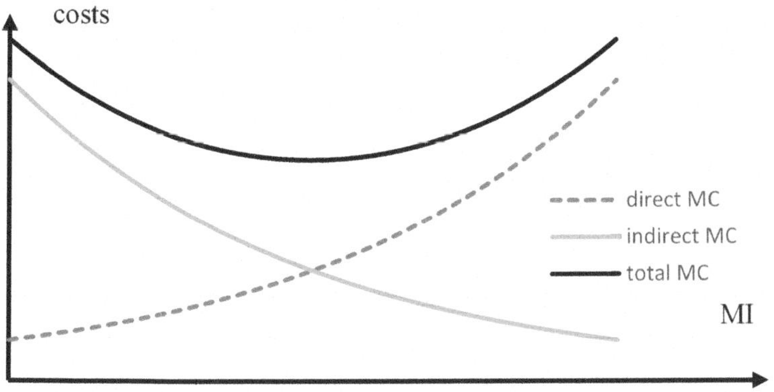

Fig. 1 Maintenance Costs (MC), based on Lasch (2018)

costs are incurred due to downtimes in subsequent plants. Besides, costs are incurred due to lost contribution margins (Matyas 2016; Pawellek 2016).

As the intensity of maintenance increases, indirect maintenance costs are negatively proportional. Compared to direct costs, indirect maintenance costs can reach a value that is three to five times as high, which underlines their high relevance (Kuhn et al. 2006). Figure 1 shows an exemplary course of the total maintenance costs as a function of the direct and indirect costs, according to Lasch (2018). Maintenance management aims to find the right balance between the expenses for maintenance activities and the resulting consequential costs for inadequate maintenance by choosing appropriate maintenance strategies.

2.1 *Maintenance Strategies*

According to the European standard DIN EN 13,306, there are four different maintenance strategies. These are reactive, preventive time-based, preventive condition-based, and predictive maintenance (CEN 2017). Within reactive maintenance, components are only replaced when damage has occurred, or the wear reserve has been exhausted. This strategy leads to high downtimes and unscheduled maintenance activities. Within time-based maintenance, fixed intervals are defined in which measures are carried out to prevent machine failure and breakdown. This strategy leads to great planning opportunities and low downtimes. However, components are replaced prematurely, which results in higher material and personnel costs. Time-based maintenance can also be used where it is not possible to observe the condition using sensors or inspection by employees (Pawellek 2016).

Condition-based maintenance (CBM) pursues the goal of using up the wear inventory as completely as possible and replacing it shortly before it breaks down. The condition is either recorded by sensors via condition monitoring systems or determined by trained employees. The PdM strategy goes beyond the characteristics of

CBM by forecasting the residual useful life (RUL) and thus predicting the time to failure (Pawellek 2016).

2.2 *Predictive Maintenance*

With the PdM strategy, trend analyses or RUL forecasts are used to determine at which point the wear reserve is optimally exhausted and a maintenance measure is necessary. This maintenance strategy makes it possible to carry out long-term planning for activities to make sufficient resources available at the right time. Also, the wear reserve of the components is used to the best possible extent. This leads to low material costs, which is particularly relevant for expensive spare parts (Pawellek 2016).

Another intention of PdM is to detect hidden faults through permanent plant monitoring and thus prevent unplanned downtimes. This strategy is used, where plants of high value are operated, and machine failure would have fatal consequences for the production process. PdM strategies are also used in areas in which the technical condition of a system is a decisive factor for human health and safety. A precise prediction of the objects' RUL is of central importance for this strategy, which requires a high degree of experience and efficient data processing (Pawellek 2016).

The approaches of the RUL prediction can be divided into three main methods. These are data-driven, model-driven, and hybrid procedures. Data-driven methods use sensor measurements as input data, and prognoses are conducted through statistical methods. On the other hand, there are data-driven approaches that use artificial intelligence or machine learning algorithms like artificial neuronal networks (ANN). Those methods require training data sets to set up the algorithms. Once the system is trained, it provides precise prognosis results without significant modelling expenditure (Si et al. 2017).

Modell driven approaches can come into use when sensor measurements are not possible. Therefore, complex physical and mathematical models of the deterioration process are necessary to predict the RUL of the investigated object (Sun et al. 2019). Hybrid approaches are combinations of data and model-based methods to compensate for specific disadvantages. Therefore, data-driven procedures can, e.g., be used to validate model-driven approaches or increase their precision (Di Maio et al. 2012).

Even though the scientific literature provides a wide range of PdM approaches, the study from Haarman et al. (2018) shows that only a minority of companies implemented this maintenance strategy successfully. This fact shows a disparity between research findings and application activities into industrial maintenance management.

3 Systematic Literature Review

3.1 Methodology

The SLR approach according to Cooper (1982), contains the five stages of problem formulation, data collection, data evaluation, analysis, and interpretation, as well as a public presentation. Considering research question 1, the problem formulation consists of the search for application-oriented managerial approaches to implement a PdM strategy in an industrial context. To determine the most relevant literature for this problem, the databases *Academic Search Complete, Business Source Complete,* and *EconLit with Full Text* are used because of their high amount of economic and management relevant literature with the following search string: (framework OR roadmap OR approach) AND (predictive maintenance OR condition based maintenance OR prognosis and health management OR preventive maintenance OR prescriptive maintenance OR PdM) AND (implement* OR applicat* OR apply OR execut*) NOT (ship* OR aviation OR aircraft OR mariti* OR medic* OR airport OR biolog* OR agricultur*).

Either the title of the paper, subject terms, or the abstract must contain a synonym of roadmap together with one phrase of the topic predictive maintenance and a synonym of implementation. Besides this, some terms are excluded from the search that are not relevant to solve the described task. The topic of this research is primarily practical orientated. Therefore, no restriction is made concerning the article ranking to not exclude any unranked application-oriented literature, e.g., from consulting companies.

To find any relevant articles for answering the research question 1, the paper underwent a title screening to exclude any literature without a reference to the topic of implementing a predictive maintenance strategy. An abstract screening of the remaining articles revealed 93 potentially relevant articles that underwent a full-text screening. General requirements for process-models such as accuracy, relevance, economy, clarity, comparability, and a systematic structure were further criteria of the analyzed frameworks (Becker et al. 2000; König 2009).

Most of the excluded articles focused on residual useful life (RUL) prediction strategies and had a lack of implementational or managerial approaches in the context of PdM. As a result, the authors selected nine relevant articles. A forward and backward search revealed six further papers that are relevant to answer the research question 1. In total, 15 articles were chosen for this purpose, that where published between 2007 and 2019.

3.2 Findings

Since this paper aims to provide a generic roadmap that faces managerial aspects of implementing a PdM strategy, the following analysis criteria were derived from the examined frameworks and considered relevant for an application-oriented implementation process of PdM. The framework should be management-oriented, and therefore contain a *cost–benefit* analysis of PdM implementation to make sure, the implementation creates a value addition in maintenance. The *implementation process* should be described with recommendations about how a PdM strategy can be applied to the maintenance management system, so a potential user of the proposed framework can execute the implementation step by step. Furthermore, the framework should contain approaches of RUL prognosis to be considered as *PdM related*, and include a *feedback loop* to enable improvements of the maintenance system from a practical point of view. To avoid stand-alone system solutions, and use information of the existing ERP-system, the framework also should take a *connection to existing IT* systems into account.

A comparison of the analyzed articles considering the criteria explained above is made in Table 1.

Every framework has strengths and weaknesses according to the proposed aspects. A cost–benefit analysis is not treated at all in two articles. General efficiency factors of implementing PdM are mentioned within the remaining 13 articles, whereas only Vogl et al. (2014) included a cost–benefit analysis as a full process step into their framework. The implementation process is described in the majority of the analyzed articles as a structured and iterative procedure. Only three of the screened paper did not provide practical descriptions of the application since they focused on PdM architecture building. A PdM relation is given by 13 of the proposed articles since these provide RUL prognosis approaches for the maintenance objects. Two of the analyzed articles are not considered as PdM related as they focus on CBM without implementing prediction methods. A make or buy decision is only partially considered by the article from Schmidt and Wang (2018). Their proposed cloud manufacturing approach enables the use of a Maintenance-as-a-Service model with the inclusion of external service providers. However, a detailed make or buy analysis is not included in the framework. The installation of a feedback loop can be found by four of the analyzed articles, but none of these focused on feedback as a tool to optimize the maintenance from a management-related view. Instead, the feedback is used to improve the precision of further prognosis results in the proposed approaches. To ensure consistent data and information flow, it is necessary to aim for an embedded RUL prediction and maintenance decision solution. This task is faced by eight of the evaluated articles, whereas especially the articles of Ansari et al. (2019) and Groba et al. (2007) contain detailed recommendations about the connection between the RUL prediction and the production and resource planning of the company.

The majority of the articles focus on RUL prediction approaches and the implementation of the framework. In contrast, management relevant tasks like cost–benefit-ratios, make or buy decisions, and feedback loop are rarely treated by the investigated literature. This paper aims to provide such a generic roadmap for industrial purposes. Therefore, not only technical criteria are seen as relevant tasks, but also management related aspects. Furthermore, the analysed articles did not contain explanations about the requirements of the individually proposed frameworks.

Table 1 Evaluation of the literature analysis

	Cost–benefit analysis	Implementation process	PdM relation	Make or buy decision	Feedback loop	Connection to existing IT
Ansari et al. (2019)	–	+	+ +	–	+ +	+ +
Márquez et al. (2019)	–	+	+ +	–	–	–
Nguyen and Medjaher (2019)	–	–	+ +	–	–	–
Qu et al. (2019)	+	+	+	–	–	–
Haarman et al. (2018)	+	+ +	+ +	–	–	+
Katona and Panfilov (2018)	+	+ +	+	–	–	+
Schmidt and Wang (2018)	+	+	+ +	+	+	+
Selcuk (2017)	+	+	+ +	–	–	+
Park et al. (2016)	–	+	–	–	–	–
Bousdekis et al. (2015)	+	+	+ +	–	+	–
Mehta et al. (2015)	+	–	+	–	+	+
Vogl et al. (2014)	+ +	+	+	–	–	–
Lewandowski and Scholz-Reiter (2013)	+	+	+	–	–	+
Sharma (2013)	–	+	–	–	–	–
Groba et al. (2007)	–	–	+	–	–	+ +

4 Implementation Roadmap

The SLR shows a lack of overall management and efficiency based frameworks to implement a PdM strategy that faces the explained requirements. According to research question 2, this paper aims to provide such a generic roadmap with a focus on management relevant tasks of the PdM application.

The structured implementation process is related to the framework of Haarman et al. (2018). The division of the roadmap in analysis, decision making, and implementation compares to the system engineering model proposed by Lewandowski and Scholz-Reiter (2013). The integration of the strategical and tactical aspects, as well as overall managerial and organizational criteria, are related to Qu et al. (2019). According to Vogl et al. (2014), a cost–benefit analysis is fused into the roadmap as a process step within the decision phase. A make or buy decision following the approach of Schmidt and Wang (2018), is also integrated into this part of the roadmap, and extended by decision support methods. The presented approach of Ansari et al. (2019) contains recommendations about implementing software and data processing for PdM related solutions and installing a feedback-loop. Therefore, a data warehouse system, as well as a RUL prediction toolbox, are integrated into the software side of this roadmap. The implementation of a feedback-policy aims to continuously improve maintenance management by using findings that are gained during the operation. Furthermore, this framework clarifies that implementing a PdM strategy is recommended to proceed in a project-oriented way (Pawellek 2016). The structure of the proposed roadmap is shown in Fig. 2, and the individual process steps are explained in the following.

4.1 Analysis

4.1.1 Analysis of Existing Maintenance Management

The maintenance management has to choose an optimal combination of the available maintenance strategies to minimize the total maintenance costs. A thorough analysis of the current maintenance management should determine the actual condition and show potential benefits as well as an effort for the introduction of PdM.

In the first step, the current maintenance management will be examined. Maintenance costs are used as an essential criterion for this. The direct costs are usually recorded comprehensively by the accounting department. The focus within the analysis is, therefore, more on indirect costs. As mentioned before, this is particularly important because indirect maintenance costs often have a much higher value than direct costs, and the choice of the maintenance strategy has a significant influence on the composition of the costs. The absolute amount of indirect maintenance costs is complicated to determine (Pawellek 2016; Rötzel and Rötzel-Schwunk 2017). The maintenance performance is rarely directly measurable and quantifiable. The use of meaningful KPI's can at least partially compensate for this disadvantage and compare the performance of different departments or companies. KPI's are divided into absolute values (basic figures) and ratios, which represent quotients of observation values and reference values. Maintenance relevant indicators can be personnel cost ratio, maintenance material cost ratio, failure rate, maintenance cost ratio, or external service ratio (Rötzel and Rötzel-Schwunk 2017). Further analysis criteria include structural and process organization, failure causes, and available information systems.

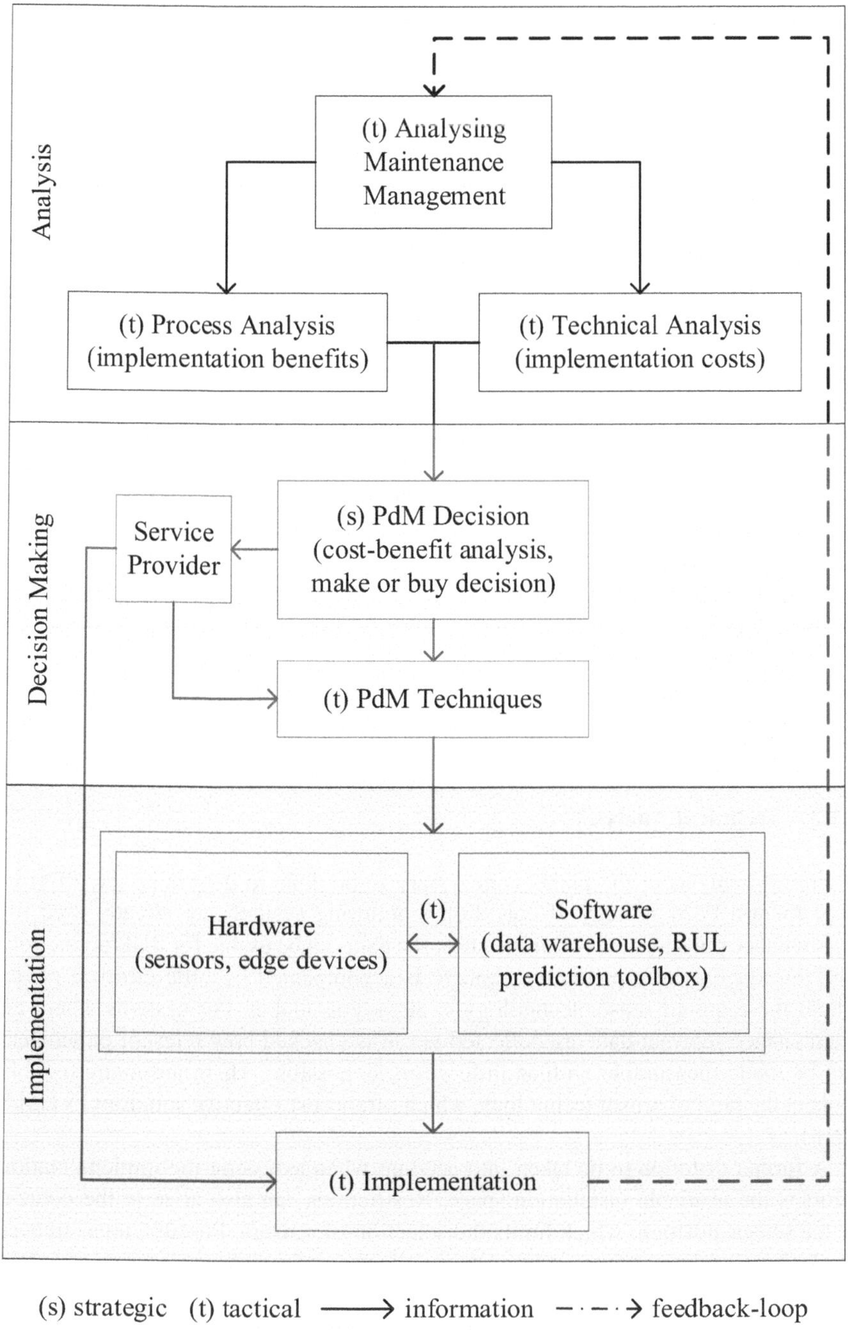

Fig. 2 Implementation Roadmap, own study

4.1.2 Process Analysis

Process analysis is a crucial factor to investigate the potential benefits or necessity of a PdM strategy. The degree of concatenation is a vital factor in industrial processes. In lean production processes, buffers are avoided wherever possible, and a minimum of safety stock is required. Hence, in the event of an unplanned failure, a plant shutdown can occur, which affects a large part of the production. With such critical processes, especially, it is important to obtain a reliable forecast of the expected RUL to avoid a production shutdown.

According to Pawellek (2016), the degree of automation of the production process also plays a decisive role in the classification of plants. The position of a plant under consideration within the production process has a further influence on maintenance. It is an advantage to consider the needs of subsequent maintenance right from the factory planning stage. Parallel systems, for instance, those in which there are multiple machines with the same performance, have a slight risk of total failure. Serial configurations may be cheaper to purchase, but their lack of redundancy leads to a much higher risk of causing high indirect maintenance costs due to unplanned plant downtimes (Strunz 2012).

Further criteria that are included in the process analysis are failure frequency, product requirements, as well as legal framework conditions that can require the use of a PdM strategy. A predictive strategy also allows a reduction of storage costs, since spare parts can be procured as required due to a minimal risk of unplanned failure (Pawellek 2016).

4.1.3 Technical Analysis

Technical analysis of the plants is necessary to be done to determine the effort to implement a PdM strategy. If condition monitoring systems are already used for maintenance purposes, it is examined how the data can be usable for RUL prediction. The introduction of a predictive strategy is accompanied by little effort at plants where the required sensor technology is already available. For systems where no maintenance-relevant data are collected yet, it is checked how relevant parameters can be made measurable with as little effort as possible. There are many useable tools in the field of sensor technology, which offers cost-effective solutions (VDMA 2018).

A further criterion to be taken into account when assessing the implementation effort is the available installation space. Restrictions can also arise in the context of the sensor position, which limits the selection of sensors. Possible interferences must also be taken into account in the application of sensors to exclude falsified measured values. These can be, for example, vibrations that affect the component from the outside. The available power supply and communication interfaces must also be included in the analysis (Pawellek 2016; Feng and Shanthikumar 2018).

It is necessary to classify the findings of the process analysis as well as the technical analysis to enable decision-making about the implementation of a predictive strategy.

Therefore, it is possible to use a Pareto-Analysis where the A category includes, for example, the 10% of plants that have the highest maintenance costs, downtimes, failure frequencies, etc. and thus have the highest benefit potential for predictive maintenance. Category B contains the systems that have descending values of the criteria between 10 and 30%. Category C includes the remaining 70%.

A similar classification is done for the aforementioned technical analysis with a XYZ categorization, whereas category X represents low, Y medium, and Z a high implementation effort. Other methods to prioritize the maintenance objects can be the Balanced Score Card or the FMEA (Strunz 2012; Pawellek 2016).

4.2 *Decision Making*

4.2.1 Decision Support Matrix

To create an overview of the existing portfolio of maintenance objects, the method of ABC-XYZ decision matrix is used. This is necessary to gain information about the potential and effort of implementing a PdM strategy to the considered plants. Therefore, the results of the ABC-classification of the process analysis are plotted on the ordinate. The prioritized objects of the technical analysis (XYZ) are deducted on the abscissa. The proposed decision matrix is shown in Fig. 3.

According to the classification of each maintenance object, the following recommendations can be made about the implementation of a predictive strategy:

AX–highly recommended implementation,
BX, AY–recommended implementation,

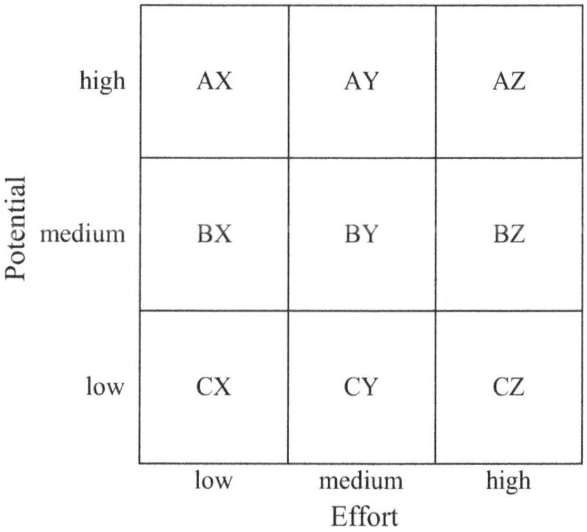

Fig. 3 Decision support matrix, own study

CX, BY, AZ–selective realization,
CY, BZ–implementation in individual cases,
CZ–currently no implementation.

The decision to implement a PdM strategy is followed by a make or buy analysis. Small and medium-sized companies often do not have the opportunity to set up a maintenance department that has the know-how for all necessary maintenance activities. Therefore, it is not only crucial for smaller companies to carefully consider which tasks belong to the core competences of their maintenance department and which should be performed by external service providers. The primary focus here is on cost-effectiveness (Strunz 2012).

4.2.2 Maintenance Service Provider

Due to their specialization, many service providers have an extensive range of offers in the field of maintenance services. The make or buy decision of maintenance activities can be supported by several analysis methods. These can be based on cost accounting, transaction costs, market orientation, resource orientation, and principal-agent approach (Lasch 2019).

Services that are not suitable for external contracting are those that have to be carried out by the company as a legal entity. These include, for example, occupational safety and environmental safety tasks (Strunz 2012).

Furthermore, it is not advisable to outsource activities that have a significant influence on the company's objectives. These include planning and quality assurance within plant maintenance. On the other hand, there are activities that, for legal reasons, must be performed by external service providers or tasks that are not part of the core areas of the company. The decision to outsource must be considered on a case-by-case basis. The maintenance activities can be carried out both by manufacturers of the equipment and by external maintenance service providers. Furthermore, it is crucial to fuse the maintenance services into the companies' maintenance management system (Pawellek 2016).

4.2.3 Choice of PdM Techniques

The choice of appropriate RUL prediction approaches is a crucial factor within the application of PdM. According to Lei et al. (2018), PdM approaches consist of the steps data acquisition, construction of the health indicator, division of the health stage and RUL prediction. Data acquisition is achieved by sensor measurements, data transmission and data storage. Within the construction of the health indicator, signal processing or AI algorithms are applied to the data set for measurement noise filtering, and feature extraction to achieve a representation of the health condition. The division of the health state, also known as fault detection, aims to divide the health indicator into a healthy and an unhealthy stage. A classification in more than two stages is also possible and depends on the individual wear pattern of the maintenance object (Lei et al. 2018).

The subsequent RUL prediction step focuses on the prognosis of the period between the current condition and the objects' end of life, that is set as a failure threshold. This threshold should be described in the form of a probability distribution as several criteria influence a breakdown. The choice of the particular RUL prognosis method depends on the deterioration process of the monitored part. An overview of the commonly used algorithms and their advantages and disadvantages can be found in the paper of Lee et al. (2014).

4.3 Implementation Process

4.3.1 Hardware

To successfully implement the methods of PdM, it is necessary to coordinate the required software and hardware, so complications at interfaces and expensive standalone solutions can be avoided. Furthermore, it is crucial to pay attention to the hard- and software components that are already implemented and to analyze, how they can be embedded into the required solutions for PdM. The global goal is to develop an integrated solution instead of numerous isolated applications (Schmidt and Wang 2018).

Relevant data have to be collected to gain information about the condition of a component. Therefore, a large number of different sensors are available and have to be chosen for the individual use case of condition monitoring. The most commonly deployed method is vibration analysis, which is performed by measuring accelerations and forces to diagnose imbalances, damage to the gear, or bearings (Selcuk 2017). The selection of the appropriate sensors must be made depending on the component to be monitored and is also influenced by the available space and environmental conditions. Further methods are thermography, electrophysical methods, flow measurement, or geometric quantities such as distance, angle, or position (Selcuk 2017; Sun et al. 2019).

Furthermore, mobile devices or terminals are necessary to inform the responsible worker about the condition of the maintenance objects or upcoming maintenance activities. Applications of augmented reality (AR) are also used in the context of maintenance and repair. Considering Palmarini et al. (2018), AR is used primarily in this field to support assembly or disassembly processes and repairs.

4.3.2 Software

The high amount of process, product, and machine data require a robust data warehouse system. Furthermore, a toolbox for data analytics is necessary for the prediction of RUL. This toolbox can be linked to machine learning algorithms to enable artificial intelligence applications (Ansari et al. 2019). Cloud computing solutions, in which external service providers deliver the required software and applications, make it possible to find inexpensive entry-level solutions for PdM. They convince through flexible and scalable usage models of the providers. The only requirement is sufficient internet access. By using IoT enabled sensors, collected data can be stored directly in a cloud. The prognosis algorithms can then either be applied directly on mobile devices or be carried out by another service provider connected to the cloud system (Schmidt and Wang 2018).

4.3.3 Implementation in Maintenance Management

The implementation of a PdM strategy should be seen as an integral part of the management processes. This is especially the case since the analysis and application of PdM is usually determined by the management. Further management criteria that are affected by this strategy are the objectives and targets of maintenance. The fundamental goal of maximum system availability at the lowest possible cost remains the same, but some specific tasks within plant maintenance change. For example, organizational and process structures are undergoing changes that must be adapted to new maintenance challenges by the management. This is because maintenance activities within a predictive strategy require fewer ad hoc measures, and better planning is predominant. Furthermore, sufficiently trained employees must be available who can meet the challenges, mainly due to the complexity of the information technology used (Wald 2003).

The use of PdM means that an optimum time for the replacement of a spare part is determined shortly before its failure. However, this conflicts with production planning. The interests of both areas must be taken into account to achieve the most efficient results within maintenance. This can be done either through a holistic approach within the ERP system used or through the use of algorithms that respect both the concerns of the data-based status information of plants and information about the production schedule to reach smart maintenance management (Feng and Shanthikumar 2018).

4.3.4 Feedback

Feedback from operational maintenance is essential to incorporate the findings from the practical implementation and operational phase of PdM into the strategic orientation and objectives of maintenance management (Ansari et al. 2019). Various quality management methods are available for this purpose, e.g., the Continual Improvement

Process (CIP) that is defined in the ISO 9001 and is also a relevant part of the Total Productive Maintenance concept. The aim is to achieve a permanent improvement of processes through regular suggestions for change made by employees. According to Sharma (2013), benchmarking is another efficient way to improve maintenance processes. Therefore, processes of cooperative best-practice companies are analyzed according to predefined benchmarking goals to enhance the maintenance processes of the analyzing company.

5 Conclusion

Predictive maintenance is a powerful method to avoid unplanned machine failure and losses in sales due to process downtimes. The research of the existing literature shows a lack of management and application orientated frameworks to support the implementation of PdM. This paper provides a structured recommendation of analyzing, decision making, and implementation of a predictive strategy. Nevertheless, it is not reasonable to apply this strategy to every maintenance object in the company. The proposed roadmap should be seen as a contribution to find an optimal mix of maintenance strategies, and therefore to minimize maintenance costs.

Further research has to be done to validate and implement the proposed roadmap in a practical environment, as well as to gain information about the reasons for the low percentage of successful implementations of PdM. Besides, the latest development on prescriptive maintenance focus on a holistic approach of maintenance and production management to receive smart and IoT supported solutions that also require further research in that field.

References

Ansari F, Glawar R, Nemeth T (2019) PriMa: a prescriptive maintenance model for cyber-physical production systems. Int J Comput Integr Manuf 32(4–5):482–503

Becker J, Kugeler M, Rosemann M (2000) Prozessmanagement, 2nd edn. Springer, Heidelberg

Bousdekis A, Magoutas B, Apostolou D, Mentzas G (2015) A proactive decision making framework for condition-based maintenance. Indus Manage Data Syst 115(7):1225–1250

CEN (2017) DIN EN 13306: maintenance—maintenance terminology. Beuth Verlag, Berlin

Cooper HM (1982) Scientific guidelines for conducting integrative research reviews. Rev Educ Res 52(2):291–302

Di Maio F, Tsui KL, Zio E (2012) Combining relevance vector machines and exponential regression for bearing residual life estimation. Mech Syst Signal Process 31:405–427

Feng Q, Shanthikumar JG (2018) How research in production and operations management may evolve in the era of Big Data. Product Operat Manage 27(9):1670–1684

Groba C, Cech S, Rosenthal F, Gössling A (2007) Architecture of a predictive maintenance framework. In: Proceedings—6th international conference on computer information systems and industrial management applications, CISIM 2007, pp 59–64

Haarman M, Klerk P, de Decaigny P, Mulders M, Vassiliadis C, Sijtsema H, Gallo I (2018) Beyond the hype: PdM 4.0 delivers results. Predictive Maintenance 4.0. Retrieved from https://www.pwc.de/de/industrielle-produktion/pwc-predictive-maintenance-4-0.pdf

Katona A, Panfilov P (2018) Building predictive maintenance framework for smart factory environment application systems. In: 29th DAAAM international symposium on intelligent manufacturing and automation, pp 460–471

König S (2009) Ein Wiki-basiertes Vorgehensmodell für Business Intelligence Projekte. In: Perspektiven Der Betrieblichen Management- Und Entscheidungsunterstützung, pp 33–52

Kuhn A, Schuh G, Stahl B (2006) Nachhaltige Instandhaltung: Trends, Potenziale und Handlungsfelder. In: Ergebnisbericht Nachhaltige Instandhaltung. Available from https://www.ifm.com/obj/VDMA_Nachhaltige_Instandhaltung.pdf

Lasch R (2018) Strategisches und operatives Logistikmanagement: Prozesse, 2nd edn. Springer, Wiesbaden

Lasch R (2019) Strategisches und operatives Logitikmanagement: Beschaffung, 2nd edn. Springer, Wiesbaden

Lee J, Wu F, Wenyu Z, Ghaffari M, Liao L, Siegel D (2014) Prognosis and health management design for rotary machinery systems—reviews, methodology and applications. Mech Syst Signal Process 42:314–334

Lei Y, Li N, Guo L, Li N, Yan T, Lin J (2018) Machinery health prognostics: a systematic review from data acquisition to RUL prediction. Mech Syst Signal Process 104:799–834

Lewandowski M, Scholz-Reiter B (2013) Condition-based maintenance systems: application at a German Sea Port. Int J Indus Eng 20(1–2):2–11

Márquez AC, de la Fuente Carmona A, Antomarioni S (2019) A process to implement an artificial neural network and association rules techniques to improve asset performance and energy efficiency. Energies 12(18)

Matyas K (2016) Instandhaltungslogistik, 6th edn. Hanser Verlag, München

Mehta P, Werner A, Mears L (2015) Condition based maintenance-systems integration and intelligence using Bayesian classification and sensor fusion. J Intell Manuf 26(2):331–346

Nguyen KTP, Medjaher K (2019) A new dynamic predictive maintenance framework using deep learning for failure prognostics. Reliab Eng Syst Safety 188:251–262

Palmarini R, Erkoyuncu JA, Roy R, Torabmostaedi H (2018) A systematic review of augmented reality applications in maintenance. Robot Comput Integrated Manuf 49:215–228

Park C, Moon D, Do N, Bae SM (2016) A predictive maintenance approach based on real-time internal parameter monitoring. Int J Adv Manuf Technol 85(1–4):623–632

Pawellek G (2016) Integrierte Instandhaltung und Ersatzteillogistik, 2nd edn. Springer, Berlin

Qu Y, Ming X, Qiu S, Zheng M, Hou Z (2019) An integrative framework for online prognostic and health management using internet of things and convolutional neural network. Sensors 19(10)

Rötzel A, Rötzel-Schwunk I (2017) Instandhaltung - eine betriebliche Herausforderung, 5th edn. VDE Verlag GmbH, Berlin

Schmidt B, Wang L (2018) Cloud-enhanced predictive maintenance. Int J Adv Manuf Technol 99(1–4):5–13

Selcuk S (2017) Predictive maintenance, its implementation and latest trends. Proc Instit Mech Eng Part B J Eng Manuf 231(9):1670–1679

Sharma SK (2013) Maintenance reengineering framework: a case study. J Quality Maint Eng 19(2):96–113

Si X, Zhang Z-X, Hu C-H (2017) Data-driven remaining useful life prognosis techniques: stochastic models, methods and applications. Springer, Berlin

Strunz M (2012) Instandhaltung. Springer, Heidelberg

Sun B, Li Y, Wang Z, Ren Y, Feng Q, Yang D, Chen X (2019) Remaining useful life prediction of aviation circular electrical connectors using vibration-induced physical model and particle filtering method. Microelectron Reliab 92:114–122

VDMA (2018) Guideline sensors for industrie 4.0—options for cost-efficient sensor systems, 0–29

Vogl GW, Weiss BA, Donmez MA (2014) Standards for prognostics and health management (PHM) techniques within manufacturing operations. In: PHM 2014—proceedings of the annual conference of the prognostics and health management society 2014, 2013, pp 576–588
Wald G (2003) Prozessorientiertes Instandhaltungsmanagement. Shaker Verlag, Aachen

The manufacturer's authorised representative in the EU is Springer Nature Customer Service Centre GmbH, Europaplatz 3, 69115 Heidelberg, Germany. If you have any concerns regarding our products, please contact ProductSafety@springernature.com

Printed and bound by CPI Group (UK) Ltd, Croydon, CR0 4YY
30/09/2026
02225695-0013